NanoScience and Technology

NanoScience and Technology

Series Editors:
P. Avouris B. Bhushan D. Bimberg K. von Klitzing H. Sakaki R. Wiesendanger

The series NanoScience and Technology is focused on the fascinating nano-world, mesoscopic physics, analysis with atomic resolution, nano and quantum-effect devices, nanomechanics and atomic-scale processes. All the basic aspects and technology-oriented developments in this emerging discipline are covered by comprehensive and timely books. The series constitutes a survey of the relevant special topics, which are presented by leading experts in the field. These books will appeal to researchers, engineers, and advanced students.

Nanoelectrodynamics
Electrons and Electromagnetic Fields in Nanometer-Scale Structures
Editor: H. Nejo

Epitaxy of Nanostructures
By V.A. Shchukin, N.N. Ledentsov and D. Bimberg

Applied Scanning Probe Methods I
Editors: B. Bhushan, H. Fuchs, and S. Hosaka

Nanostructures
Theory and Modeling
By C. Delerue and M. Lannoo

Nanoscale Characterisation of Ferroelectric Materials
Scanning Probe Microscopy Approach
Editors: M. Alexe and A. Gruverman

Magnetic Microscopy of Nanostructures
Editors: H. Hopster and H.P. Oepen

Silicon Quantum Integrated Circuits
Silicon-Germanium Heterostructure Devices: Basics and Realisations
By E. Kasper, D.J. Paul

The Physics of Nanotubes
Fundamentals of Theory, Optics and Transport Devices
Editors: S.V. Rotkin and S. Subramoney

Single Molecule Chemistry and Physics
An Introduction
By C. Wang, C. Bai

Atomic Force Microscopy, Scanning Nearfield Optical Microscopy and Nanoscratching
Application to Rough and Natural Surfaces
By G. Kaupp

Applied Scanning Probe Methods II
Scanning Probe Microscopy Techniques
Editors: B. Bhushan, H. Fuchs

Applied Scanning Probe Methods III
Characterization
Editors: B. Bhushan, H. Fuchs

Applied Scanning Probe Methods IV
Industrial Application
Editors: B. Bhushan, H. Fuchs

Nanocatalysis
Editors: U. Heiz, U. Landman

Roadmap 2005 of Scanning Probe Microscopy
Editors: S. Morita

Nanostructures - Fabrication and Analysis
Editor: H. Nejo

H. Nejo (Ed.)

Nanostructures - Fabrication and Analysis

With 178 Figures and 3 in Color

Professor Hitoshi Nejo
National Institute for Materials Science
Tsukuba 305-0047, Japan
E-mail: nejo.hitoshi@nims.go.jp

ISSN 1434-4904
ISBN-10 3-540-37577-5 Springer Berlin Heidelberg New York
ISBN-13 978-3-540-37577-7 Springer Berlin Heidelberg New York

Library of Congress Control Number: 2006933054

Springer is a part of Springer Science+Business Media.
springer.com

Cover background image: Fig. 1.12a

Typesetting by the authors and SPi using a Springer LaTeX macro package.
Cover design: WMX Design GmbH, Heidelberg

Printed on acid-free paper SPIN: 11768548 57/3100/SPi 5 4 3 2 1 0

Preface

The many wonders of nanophysics were first anticipated by Richard Feynman in a speech he gave in late 1959, subsequently published under the title "There's Plenty of Room at the Bottom". After almost half a century has passed since his provocation, we have reached some of the expectations which he envisioned. One of the reasons for not fully attempting the objectives he gave is that there are still a lot of unsolved problems to connect the nanometer–scale world and macroscopic–scale world. The main problem is that there are different effects in the nanoworld and the macroworld. We hope that this book contributes to a better understanding.

A part of this book is based on the results accumulated by the Center–of–Excellence (COE) Project under the Science and Technology Agency Japan, at the National Institute for Materials Science Tsukuba. The authors would like to express their gratitude to all the co–workers of this COE project. The editor also would like to thank those contributors who did not actually join the COE project but contributed to this book.

We would also like to thank Dr. Claus Ascheron, the Springer editor of this book series, as well as Ms. Alice Blanck and Ms. Adelheid Duhm, Springer, for their assistance in editing this book. The editor of this book would also like to thank Professor Danilo Pescia, ETH Zuerich, for giving him the time to edit this book during his stay at ETH.

Tsukuba, Japan
September, 2005

Hitoshi Nejo

Contents

List of Contributors

Rodion V. Belosludov
Institute for Materials Research
Tohoku University, Sendai 980-8577
Japan
rodion@imr.edu

Zhen-Chao Dong
University of Science and
Technology of China 230026
China
zcdong@ustc.edu.cn

Amir A. Farajian
Institute for Materials Research
Tohoku University, Sendai 980-8577
Japan
amir@imr.edu

Andrea Goldoni
Sincrotrone Trieste S.C.p.A.
s.s.14 Km. 163.5, 34012 Trieste
Italy
goldonia@elettra.trieste.it

Yoshiyuki Kawazoe
Institute for Materials Research
Tohoku University, Sendai 980-8577
Japan
kawazoe@imr.edu

X.H. Kong
Institute of Chemistry
Chinese Academy of Sciences,
Beijing 100080
China
kongxh@iccas.ac.cn

S.B. Lei
Institute of Chemistry
Chinese Academy of Sciences
Beijing 100080, China
leisb@nanoctr.cn

Hiroshi Mizuseki
Institute for Materials Research
Tohoku University, Sendai 980-8577
Japan
mizuseki@imr.edu

Hitoshi Nejo
National Institute for Materials
Science, 1-2-1 Sengen
Tsukuba 305-0047
Japan
nejo.hitoshi@nims.go.jp

Maurizio Prato
INSTM, Unit of Trieste, Università
degli Studi di Trieste, Dipartimento
di Scienze Farmaceutiche, Piazzale
Europa 1, 34127 Trieste, Italy
prato@univ.trieste.it

Olga V. Pupysheva
Institute for Materials Research
Tohoku University, Sendai 980-8577
Japan
olga@imr.edu

Duncan Rogers
Texas Instruments Incorporated
13560 North Central Expressway
MS 3737, Dallas, 75243
USA
duncan-rogers@ti.com

Dimitrios Tasis
Università degli Studi
di Trieste, Dipartimento
di Scienze Farmaceutiche,
Piazzale Europa 1 34127
Trieste, Italy
dtassis@yahoo.it

Lisa Vaccari
Università degli Studi
di Trieste, Dipartimento
di Scienze Farmaceutiche
Piazzale Europa 1 34127
Trieste, Italy

and

Sincrotrone Trieste S.C.p.A.,
s.s.14 Km. 163.5, 34012
Trieste, Italy
vaccari@tasc.infm.it

Bing Wang
Hefei National Laboratory
for Physical Sciences at Microscale
University of Science
and Technology of China, Hefei
Anhui 230026, P.R. China
bwang@ustc.edu.cn

C. Wang
National Center for Nanoscience
and Technology, Beijing 100080
China
wangch@nanoctr.cn

D.X. Wu
Institute of Chemistry
Chinese Academy of Sciences
Beijing 100080
China
dxwu@iccas.ac.cn

Y.L. Yang
National Center for Nanoscience
and Technology, Beijing 100080
China
yangyl@nanoctr.cn

Q.D. Zeng
Institute of Chemistry
Chinese Academy of Sciences
Beijing 100080
China
stmzqd@iccas.ac.cn

Yunshen Zhou
Hefei National Laboratory for
Physical Sciences at Microscale
University of Science and
Technology of China, Hefei
Anhui 230026, P.R. China
bwang@ustc.edu.cn

1 Atomic-Scale Chains: Fabrication and Evaluation Technologies

Zhen-Chao Dong and Hitoshi Nejo

1.1 Introduction

There are several motivations to study nanometer-scale structures. First, the silicon device industry has almost reached the fabrication limit of line widths using lithography and hence needs to search for alternative techniques of wire fabrication for device purposes. Second, the structures at the atomic-scale or nanometer-scale show quantum-mechanical effects such as quantized steps of conductance [1], Peierls distortions and charge density waves [2], resonant tunneling, single electron tunneling and quantum interference. Future-generation devices are likely to operate according to new principles; one of the candidates is quantum devices. Third, nanometer-scale structures themselves demonstrate various nanosystem functionalities, as revealed elegantly by Nature. The study of the underlying law of various nanosystem behavior will lead to deeper understanding of Nature and may bring fruitful applications.

If the structures connected to outer electric circuits for evaluation are not small enough, the system has to be cooled down to low temperatures to eliminate electron scattering effects to study their transport characteristics, etc. The limitation of fabrication of nanostructures is due to beam spreading in the resist for making patterns. Computer simulation of electron-beam resist profiles was done in the early 1980s [3]. However, recent progress on fabrication has suggested that nanometer-scale structures are promising for practical applications [4–10]. After the invention of the novel technology, scanning tunneling microscopy (STM), many books and papers were published from the viewpoint of physics [11–13] and instrumentation. This chapter concentrates on the methods of fabrication of atomic-scale chains and the techniques to connect them to macroscopic electric pads as well as the instrumentation for macroscopic electric measurements.

When the size of the structure of interest becomes smaller, quantum effects appears such as quantum conductance [14–19], quantum oscillations in a confined electron gas [20], unexpected periodicity in an electronic double-slit interference experiment [21], quantum reflection and transmission of ballistic 2D electrons by a potential barrier [22, 23]. A single-electron transistor is an important structure using nanostructures [24–41]. It is essential to define the capacitance of tunnel junctions when the tunnel junction is downsized so that atomic capacitance can be defined from the viewpoint of quantum mechanics [42]. From the viewpoint of the near-field effect, interaction of charged particles with surface plasmons in cylindrical channels in solids has been shown [43]. The scanning near-field optical microscope is very useful technology to evaluate nanostructures [44–51].

The fabrication of nanostructures may make use of the anisotropic feature of the substrate surface itself. For fabricating atomic chains, a well-defined substrate surface should be prepared. A Si(111)-2×1 surface is a good candidate to fabricate atomic-scale structures on it and hence a lot of work has been done on this surface [52] and the electronic structure has been clarified [53]. Anomalous surface reconstruction has been observed on sputtered and annealed Si(111) surfaces [54]. Furthermore, atomic-scale conversion of clean Si(111):H-1×1 to Si(111)-2×1 has been demonstrated by electro-stimulated desorption [55]. Adsorption and diffusion of Si atoms on the H-terminated Si(001) surface has been studied from the viewpoint of Si migration assisted by H mobility [56]. Further, π-bonded chains and surface disorder on Si(111)-2×1 have been shown [57].

Another strategy to fabricate nanostructures is to add another species on the Si substrate. Growth mode and surface structures of the Pb/Si(001) system have been observed [58]. Surface diffusion of Au on Si(111) has been observed and study of Pb diffusion on Si(111)-7×7 has been done [59–61]. Patterning of a substrate surface was one of the biggest motivations to study the surface. Since the early days, NH_3 dissociation on Si(001) has been well studied [62]. Recently, localized atomic reactions imprinting molecular structures have been shown [63]. Also, self-directed growth of molecular nanostructures on hydrogen-terminated Si(100) has been shown [64].

There exist practical problems to fabricate nanostructures. One drawback of manipulating individual atoms using scanning probe techniques is that it is time-consuming. To overcome this drawback, the use of metallic components via self-assembly can be a good candidate to fabricate 1D nanostructures [65–67]. Further, usage of self-assembly of metallic particles has been considered. Metallic particle self-organization is one of the candidates for fabricating nanostructures on substrates, and self-organization of large gold nanoparticle arrays has been shown [68–70].

Including measurement and fabrication technology, one advantage of using electron beam lithography is that it is possible to fabricate nanostructures and also the necessary electric circuits to outer measuring equipment. When nanostructures are fabricated, they have to be connected to outer macroscopic electrodes for measurements, for example, of electric conductance [71]. For this purpose, an ion source is one of the effective tools to fabricate intermediate structures [72–74].

1.1.1 Areas Covered in This Chapter

From the viewpoint of the electron transport along a metallic wire, the surface electron transport on Si(111)-$\sqrt{3}\times\sqrt{3}$-Ag has been measured by using probes with 10-μm distance [75]. It shows a fairy good surface conductivity. Also, instability and the charge density wave of indium linear chains on a Si(111) surface have been measured [2]. All these are 2D structures or one-monolayer (ML) films without isolated single chains. One of the purposes in this work is to fabricate a single chain between the macroscopic pads and then measure the electron transport along this chain. Such a structure will make it possible to realize the lateral atomic-scale 1D electron transport system and even atomic-scale single electron transistors. Investigation of such structures will also help to clarify the fundamental question of electron transport through a metallic wire where large electron interaction is expected.

1.1.2 Fabrication Strategy

To achieve the objectives of fabricating the lateral nanowires or single-electron transistors described in Sects. 1.7 and 1.8, we try to fabricate nanowires between macroscopic pads in three different ways:

1. Lead wire on a Si(111) substrate
2. Wire made of a series of gold dots on a Si(111) substrate
3. Gold wire both on a Si(111) and on a sapphire substrate

All the experiments use either STM or atomic force microscopy (AFM) in ultrahigh vacuum (UHV). Four-point probe chambers are attached to each STM or AFM chamber so that the sample can be transferred to four-point probe chambers without exposing it to air for the electric conductance measurement. All the scanning tunneling microscopes were obtained commercially, whereas the atomic force microscope and all the four-point probes used in these experiments were homebuilt.

Each detailed strategy is as follows:

1. Both the wire and the macroscopic pads of lead on the depassivated Si(111) surface are fabricated sequentially using the same scanning tunneling microscope tip so that the scan size of the scanning tunneling microscope extends over microns. By fabricating both the wire and the macroscopic pads sequentially, we can eliminate the difficulty of finding the relative position between the tip and the macroscopic pads, which is difficult when a wire and macroscopic pads are fabricated separately using different tools.
2. The second method is extracting gold clusters from a scanning tunneling microscope tip. By tuning the extraction conditions, we may able to control the size of the clusters which are deposited on a surface. Of course, the scanning tunneling microscope tip itself can be used to confirm the shape of the fabricated structure. Since the tip apex keeps a sharp shape, we can get a high-resolution image.
3. The third way is drawing a wire on a substrate by making contact with the gold-coated tip of an atomic force microscope cantilever. The advantage of using AFM is, of course, that we can fabricate a wire even

on an insulating substrate. Also, our atomic force microscope works also in noncontact mode, so the shape of the fabricated wire can be confirmed without destroying the structure. This method of using the same cantilever for both fabrication and imaging, again, eliminates the difficulty of finding the position of the wire on a substrate, since it is almost impossible to find such a small structure if another cantilever is used for the structure confirmation.

1.2 Adsorption and Tunneling of Atomic-Scale Lines of In and Pb on Si(100)

1.2.1 Introduction

Atomic-scale low-dimensional systems have been attracting enormous attention in the ongoing drive to downsize devices [76]. An atomic line, composed of a single row of atoms, is the ultimate limit in the lateral miniaturization of a wire. Owing to the atomic-scale dimension and boundary conditions imposed, the transport behavior of atomic chains is different from that in the bulk and could reveal a new intriguing aspect of physics. Characterization and understanding of the properties of such atomic-scale structures are prerequisites for the design and fabrication of atomic-scale devices. The group III–IV metals (Al, Ga, In, Sn, Pb) on Si(100) are a particularly interesting system because of their initial 1D anisotropic growth, which evolves into layer and island structures at higher coverage [58, 77–85, 87]. Previous low-energy electron diffraction (LEED) [77,78] photoemission spectroscopy [79,80] and STM studies [58, 81–85, 87, 88] have established not only the Stranski–Krastanov growth mode of these metals on Si(100), but also the dimerization of adsorbate metal atoms on a still-dimerized Si surface for coverage up to 0.5 ML [$1\,\mathrm{ML} = 6.8 \times 10^{14}\,\mathrm{atoms/cm^2}$, the surface Si density of nascent Si(100)-1×1, $a = 3.84$ Å]. Isolated metal ad-dimer chains are found to orient perpendicular to the underlying Si dimer rows. The ad-dimer configuration has also been determined both experimentally [58, 77–85, 87, 88] and theoretically [88–91] to be parallel to the chain direction with each atom triply bonded. Recent first-principles total-energy calculations of Pb on Si(100) point out further that the Pb ad-dimers are asymmetric [58, 91], in contrast with the symmetric ad-dimer structure for group III metals on Si(100) [88–90]. Since each group IV metal atom ($\mathrm{ns^2np^2}$) has one more valence electron than each group III atom ($\mathrm{ns^2np^1}$), different local bonding and electronic states are expected even though the geometrical adsorption structures appear similar on Si(100). STM is a powerful technique to detect such differences owing to its ability to image and probe local surface electronic states down to the atomic level. In this chapter, through selected examples of In and Pb on Si(100), we investigate their similarities and differences for the adsorption and tunneling behavior by STM and scanning tunneling spectroscopy (STS). Analyses of image contrasts on STM topographs allow us to discern unambiguously not only adsorbate structures from substrates but also whether the metal dimers are buckled. Furthermore, in an amazing analogy to the classical atom-selective imaging of the GaAs(110) surface via contrast shifts induced by charge transfer [92], we observe similar charge transfer within a buckled Pb dimer. On the other hand, since STS can provide spectroscopic information with a spatial resolution on the atomic scale (such site-specific information is not available by area-integrating surface techniques), we are able to find out (1) whether the isolated dimer chain is metallic and (2) whether the tunneling spectra for these chains are site- or length-dependent. Whereas the presence of a surface-state band gap is a common phenomenon for both In and Pb, the local surface states associated with adsorbate atoms show interesting differences in the frontier bands (orbitals) between In and Pb. The origin of nonmetallic behavior and image contrasts for metal dimer chains will also be discussed.

1.2.2 Results and Discussion

Are Metal Dimers Symmetric?

In order to answer this question, let us investigate the adsorption of In and Pb on Si(100) at low coverage. Figure 1.1 shows two STM topographs at 0.1 ML for In and 0.2 ML for Pb on Si(100). Each bright protrusion in the images corresponds to a metal ad-dimer and is clearly distinguishable from the underlying Si dimer rows. The 1D growth of both In and Pb on Si(100) is evident and is related to both highly strained Si dangling bonds induced by preexisting metal dimers and high mobility of single metal atoms and dimers on the surface [84, 89]. The chain polymerization is terminated by step edges or defects where the bonding requirements for metal atoms can hardly be satisfied.

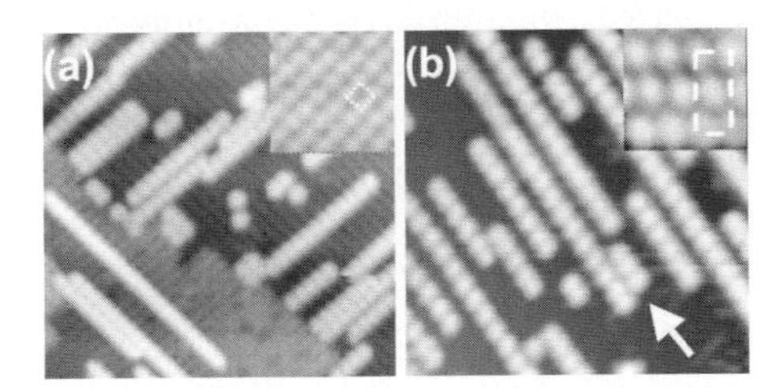

Fig. 1.1. Scanning tunneling microscopy (*STM*) topographs of 1D ad-dimer chains of (**a**) In on Si(100) at 0.1 monolayers (*ML*) ($23 \times 23\,\text{nm}^2$, +2.0 V, 0.1 nA, 295 K), with the *inset* showing the square Si(100)-2×2 In phase at 0.5 ML ($5 \times 5\text{nm}^2$, +4.0 V, 2.0 nA, 80 K) (**b**) Pb on Si(100) at 0.2 ML and 295 K ($14 \times 14\,\text{nm}^2$, 1.9 V, 0.5 nA), with the *inset* showing the rectangular Si(100)2- $\times$ 4 Pb phase at 0.5 ML ($3.1 \times 2.3\,\text{nm}^2$, 1.2 V, 0.5 nA) [76]

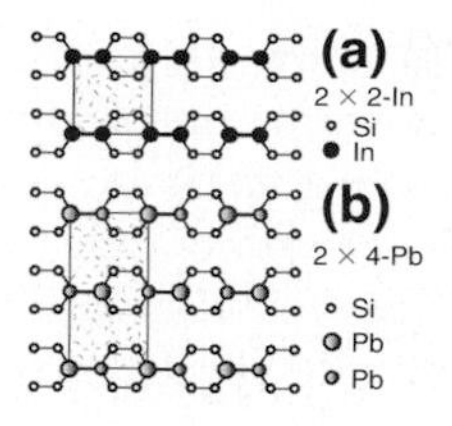

Fig. 1.2. The parallel ad-dimer adsorption structure and surface reconstruction at 0.5 ML. **(a)** In on Si(100) with the square unit cell *shaded* for the 2×2-In phase. The indium dimers are symmetric. **(b)** Pb on Si(100) with the rectangular unit cell *shaded* for the 2×4-Pb phase. The lead dimers are asymmetric, with the *large shaded atoms* buckled upward and the *small shaded atoms* buckled downward [76]

The dimerized configuration is illustrated in the parallel ad-dimer model in Figs. 1.2 with the metal dimer chain running perpendicular to the Si dimer rows. These chains are single-atom-type both in width and in height and thus represent a truly 1D system. Although we do not observe shifts of bright features in the neighboring chains spaced $2a$ apart for In on Si(100), we see apparent shifts of bright protrusions for Pb, as exemplified by the arrow in Fig. 1.1b. Such a buckling feature of Pb dimers is clearly resolved in the inset of Fig. 1.1b at 0.5 ML because we observe a rectangular surface reconstruction for Pb rather than the square structure shown in the inset of Fig. 1.1a for In. The symmetric 2×2-In phase is sketched as the shaded square in Fig. 1.2a and the asymmetric 2×4-Pb structure is shown in the shaded rectangle in Fig. 1.2b. The $2a$ spacing between the closest shifted neighbors of Pb chains further suggests that the buckling occurs often in opposite orientations in alternate chains, but in the same orientation within a chain.

Nevertheless, the 2×2 square structure is still applicable to In on Si(100) if the buckling of metal dimers is all oriented in the same direction, but this seems unlikely given the unprejudiced adsorption of In atoms on the same Si(100) surface. The perfect square pattern extended over the whole surface at 0.5 ML strongly signals a buckling-free structure for In dimers, whereas the rectangular surface structure of Pb on Si(100) clearly indicates a buckled structure for Pb dimers. We shall give more evidence for the buckling of Pb dimers in Sect. 1.2.2 through a charge transfer mechanism.

Image Contrasts Between Filled States and Empty States

Although both In and Pb show similar dimerized chain growth on Si(100), the difference in their electronic configurations should be reflected in STM imaging since tunneling from or to these chains is closely related to the surface electronic states associated with the adsorbate atoms. This is indeed the case, as illustrated in Fig. 1.3 through a sharp contrast for In and similar contrast for Pb between filled and empty states. For In on Si(100), the dimer chains

Fig. 1.3. Image contrasts between filled states and empty states for the same area at 80 K. (**a**) In on Si(100) for filled states (20 × 20 nm^2, 2.5 V, 0.2 nA), (**b**) In on Si(100) for empty states (+2.0 V, 0.2 nA), (**c**) filled state Pb on Si(100) (30 × 30 nm^2, −1.5 V, 0.2 nA), (**d**) empty-state Pb on Si(100) (+2.0 V, 0.2 nA) [76]

are very bright in the empty states, with an apparent height of 2.7–3.0 Å (which is bias-dependent), whereas in the filled states these chains are faintly noticeable, with an apparent height of only approximately 0.6 Å, as shown in Fig. 1.3a and b. It is worth noting that some indium chains have anomalously bright ends with an apparent height of 2.9–3.2 Å that are presumably related to single In atoms trapped by defects. On the other hand, for Pb, the chain feature appears equally bright, with an apparent height of approximately 2.3 Å for both filled and empty states (Fig. 1.3c, d). The additional bright dots in the empty states in comparison with the same area in the filled states arise from the C-type defects on the bare Si surface and they are thus irrelevant to Pb atoms. The actual geometrical height of these chains above the bare Si(100) surface is supposed to be slightly larger than the monoatomic height of Si(100) (1.4 Å) because of the larger atomic size of In and Pb with respect to that of Si.

The image contrasts displayed in the STM images of Fig. 1.3 can be traced to the local chemical bonding on the Si(100) surface through the four valence orbitals (1s + 3p) for each metal atom. As shown in Fig. (1.2a, each In atom (5s^2 5p^1) saturates all the unpaired dangling bonds after being bonded with two Si atoms and another In atom (yielding an ad-dimer). The highest occupied states are characterized by σ-type In–In bonding with mainly lying-down p_y-orbital interactions, as shown in Fig. 1.4 (which is a simplified picture of frontier orbitals associated with metal dimers). The lowest unoccupied states are actually the π-type bonding states between standing-up p_z orbitals that have a much better spatial extension toward the tip. In contrast, since each Pb atom (6s^2 6p^2) has one more valence electron than In, there is one electron

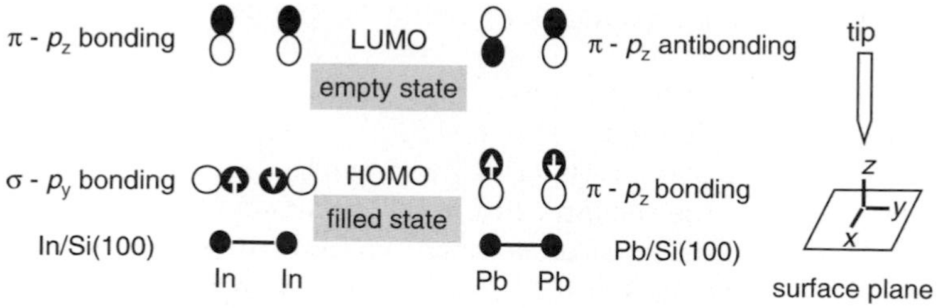

Fig. 1.4. Spatial extensions of the frontier orbitals related to the surface states of In and Pb on Si(100). The s–p orbital hybridization and the buckling of Pb dimers are not taken into account, but the spatial orbital extension picture remains essentially the same even if they are considered [76]

left on each Pb atom of the dimer after the formation of three similar bonds in the previously described In case, namely, two Pb–Si bonds and one σ-type Pb–Pb bond. This additional electron is located in the p_z-type orbital that is perpendicular to the surface and it is thus possible to form a π bond. The filled and empty states for Pb are therefore p_z–π bonding and antibonding states, respectively, as shown in Fig. 1.4. Both have a good spatial orbital extension into the vacuum. STM senses only the tails of those surface wave functions that protrude into the vacuum gap and overlap significantly with the tip electronic states. The filled states of In chains have only electronic states that are too close to the atomic cores. They have a poor spatial orbital extension and, as a result, are only weakly detected in spite of the topographic height of the metal atoms above the Si surface. In contrast, the empty states of In chains have spatially well extended p_z orbitals that can effectively overlap with the tip wave functions and thus offer sharp contrast between metal chains and substrate. The topographic height plays a major role in the contrast since it enables the metal p_z orbitals to approach the tip about 1.5 Å closer than the Si p_z orbitals (note that tunneling current has an exponential dependence on the gap distance). The same argument is applicable to both filled and empty states for Pb chains since both states involve well-extended standing-up p_z orbitals. As expected, we observe equally bright chain contrast for both bias polarities. Of course, the schematics in Fig. 1.4 is an oversimplified picture: it does not take into account the s–p orbital hybridization and the buckling of Pb dimers. However, even if these factors are considered, the spatial orbital extension picture remains essentially the same and the same argument still holds. Consistent with the observed image contrasts for different bias polarities and the previously given argument for such a contrast mechanism, the I–V data in Sect. 1.2.2 show a nearly symmetric curve for Pb on Si(100), but a very asymmetric curve for In chains, with the detected current of the filled states only about one seventh of that in the empty states.

Buckling and Charge Transfer

The asymmetric structure of Pb dimers versus that of symmetric In dimers also stems from the difference in the electronic configurations of the metal atoms. As described previously, each Pb atom has one half-filled dangling bond protruding into the vacuum, whereas this dangling bond is empty for In on Si(100). These half-filled dangling bonds are roughly parallel and are likely to form a π bond. However, both the predicted instability of symmetric dimers from theory [91] and the observed shifts of Pb dimer protrusions from STM images suggest that the formation of a regular double bond (σ plus π without polarization) between two Pb atoms in a dimer is not energetically favorable on the Si(100) surface. In order to lower the energy, the Pb dimers buckle at the expense of weakening π bonding, and accompanying this process there is a charge transfer from the Pb atom that moves down to the other one that moves up. All the dangling bonds are thus saturated and the surface becomes stable. (The buckling angle is approximately 12.5° according to the first-principles calculation in [91]. The dimerization and buckling phenomena are primarily determined by the local chemical bonding, but, to a lesser extent, might also pertain to the Peierls distortion mechanism for achieving a lower-energy configuration. STM is supposed to be sensitive to the valence states only, not to the element-specific core levels, and thus usually fails to offer chemical sensitivity. However, when there is a charge transfer between surface atoms, STM can readily disclose the difference in electronic states of these surface atoms through contrast mechanisms, e.g., the zigzag Si dimer chains for the buckled Si(100) surface, and in some cases, can even be used for atom-selective imaging, as elegantly demonstrated for the GaAs(110) surface [92]. In GaAs, STM can selectively image As atoms at negative sample bias and Ga atoms at positive bias because charge transfer from Ga to As results in filled states centered at As and empty states centered at Ga. The application of such an image-contrast mechanism to the present Pb on Si(100) system leads to the interesting observation shown in Fig. 1.5. The brightest features

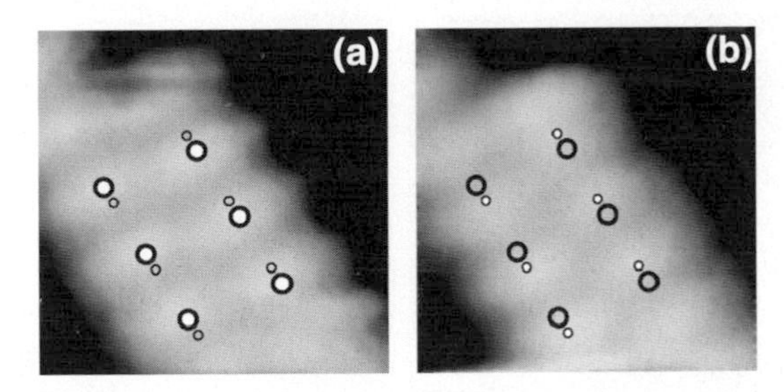

Fig. 1.5. Image contrast of Pb on Si(100) for the same $3.5 \times 3.5\text{-nm}^2$ area at 295 K between (**a**) the filled states (−1.8 V, 0.5 nA and (**b**) the empty states (+1.8 V, 0.5 nA). The buckled Pb dimers are marked as a guide for the eye. Charges are transferred from the buckled-downward atoms (*small circles*) to the buckled-upward atoms (*large circles*) [76]

in the filled-state image of Fig. 1.5a correspond to the Pb atoms that are buckled upward and negatively charged, whereas the brightest features in the empty-state image of Fig. 1.5b originate from the Pb atoms that are buckled downward and positively charged. The buckling of Pb dimers and the resultant charge transfer within a dimer or vice versa are evident. These observations are in good agreement with the theoretical simulation in [91], but the square pattern of bright protrusions simulated there contradicts our experimental observation of a rectangular structure. The difference lies in whether there is an alternation of buckling orientation in closest neighboring chains.

Tunneling Spectra: Are They Atomic Wires?

Are these metal dimer chains conducting? Do they show site- and length-dependent tunneling behavior? These are probably the first questions one would ask about their properties. In order to find out the electronic states of these metal dimer chains, I–V measurements were carried out via Current image tunneling spectroscopy (CITS) since the spectroscopy of differential conductance $\mathrm{d}I/\mathrm{d}V$ versus bias voltages is approximately related to the local density of states (LDOS) on the surface [96]. The tunneling density of states $(\mathrm{d}I/\mathrm{d}V)/(I/V)$ is normalized according to the simple offset technique of I/V replaced by $[(I/V)^2 + c]^{1/2}$, where c is a small constant, typically 0.01 for the data given here [97]. Figure 1.6 shows normalized tunneling spectra at specific sites for various lengths of metal dimer chains, and the averaged data are given in Fig. 1.7. The I–V characteristics for the bare Si(100) surface (the A sites in Fig. 1.6a, b) is in good agreement with the literature, with a surface-state band gap of approximately 1.1 eV [62, 94]. It is worth mentioning that the resemblance of the $\mathrm{dln}(I/V)\mathrm{ln}V - V$ curves to the electronic density of states is only limited to peak positions; the intensities of the peaks are not reliable and are dependent on the tip–substrate

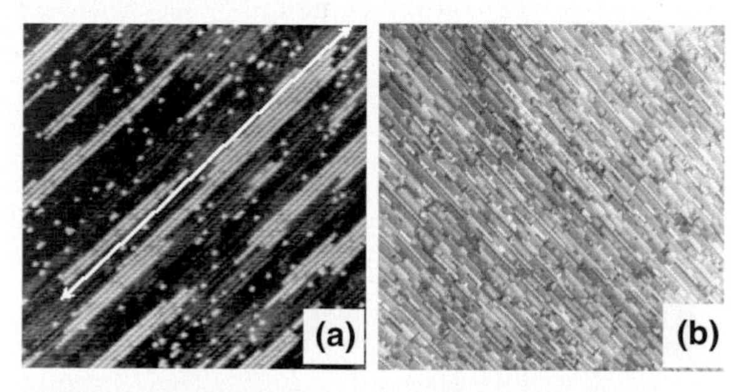

Fig. 1.6. (**a**) Pb chains of 80 nm on double-domain Si(100):75 × 75 nm^2, −1.85 V, 0.2 nA (**b**) Continuous Pb chains as long as 105 nm on single-domain Si(100): ~0.1 ML, 100 × 100 nm^2, −2.3 V, 0.2 nA, 295 K. The inset shows the clean single-domain Si(100) surface [102]

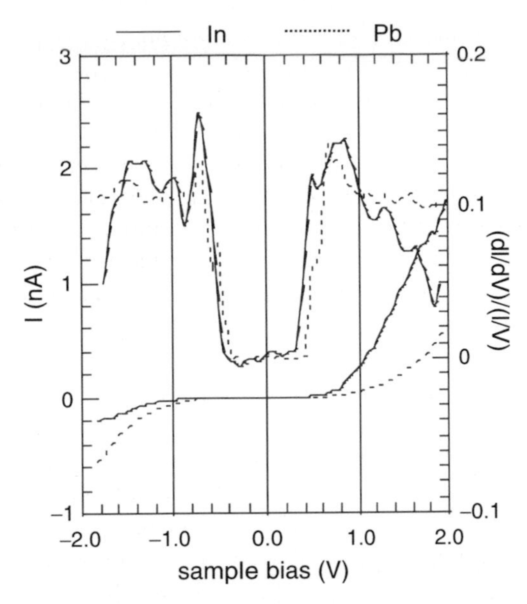

Fig. 1.7. Tunneling spectra averaged over various metal dimer chains for In (*solid lines*) and Pb (*dashed lines*) on Si(100). The *lower curves* are I–V characteristics and the *upper curves* are the normalized tunneling density of states. The difference in tunneling behavior and the presence of a surface-state band gap are evident [76]

configuration and normalization procedure. Before we proceed to a detailed analysis of the tunneling spectra, let us make an assumption that the metal dimer chains are metallic, and see what kind of tunneling spectra should be expected. In the simplest picture, the assumption of conductance means that electrons are confined within a chain (but delocalized along it) in a 1D square-well model. Quantum mechanics textbooks tell us immediately the quantized energy levels, the energy separation of neighboring states and the probability density distribution inside the well in the following three equations, respectively:

$$E_n = \frac{n^2 h^2}{8mL^2} = 0.63\left(\frac{n}{N}\right)^2 (\mathrm{eV}), n = 1, 2, ..., \tag{1.1}$$

$$E_{n+1} \mathrm{d} E_n = \frac{(2n+1)h^2}{8mL^2} = 0.63\frac{2n+1}{N^2}(\mathrm{eV}), \tag{1.2}$$

$$\psi_n^2 = \frac{2}{L}\sin^2\left(\frac{n\pi x}{L}\right) = \left(\frac{1}{Na}\right)\sin^2\left(\frac{n\pi x}{2Na}\right), \tag{1.3}$$

where n is the quantum number, L is the chain length ($2Na$, $a = 3.85$ Å), N is the number of dimers in a chain, h is the Planck constant and m is the electron mass. If we further assume that one electron per atom is filled into these energy states according to the Pauli exclusion principle, i.e., two electrons per state, then we have the energy gap as follows:

$$E_{\mathrm{g}} = E_{\mathrm{LUMO}} - E_{\mathrm{HOMO}} = E_{N/2+1} - E_{N/2} = \frac{(N+1)h^2}{8mL^2} = 0.63\frac{N+1}{N^2}\ (\mathrm{eV}). \tag{1.4}$$

Such a hypothesis results in the following implications:

1. The first energy state (or zero-point energy) should be roughly at 0.6, 0.2 and 0.1 eV for single-dimer, double-dimer and triple-dimer chains, respectively.
2. There should exist an energy gap of about 1.3, 0.5 and 0.3 eV for single-dimer, double-dimer and triple-dimer chains, respectively. The gap should be roughly inversely proportional to the number of dimers.
3. There should be a monotonic increase in energy-level spacing by about $1.3/N^2$ eV for higher-energy states.

In a word, the longer the chain is, the lower the first quantized energy state, and the smaller the energy gap and level separation. However, the observed tunneling spectra in Fig. 1.6 indicate no length-dependent trends at all for either energy position or level spacing; instead, they show a consistent surface-state band gap of about 1.2 eV with two LDOS maxima near the Fermi level (at $V = 0$), -0.7 V below and $+0.5$ V above E_{F} for In on Si(100), and -0.5 V below and $+0.7$ V above E_{F} for Pb on Si(100), respectively. In addition, the $\mathrm{d}I/\mathrm{d}V - V$ maps processed from the CITS data do not exhibit any corrugation variation of bright chain features that would be expected according to the probability density distribution of (1.3) for different bias voltages. Consequently, the assumption of electrons delocalized along the chain is not valid, the chain is thus semiconducting for both In and Pb on Si(100), and overall, no evident site or length dependency of tunneling spectra was observed, especially around the gap region. (It is interesting to see that the observed energy gap of about 1.2 eV agrees roughly with the quantum box model for single dimers. This is not a surprise since the electrons are indeed confined within a dimer.) However, site-related spectra show up in the following three situations:

1. Single dimers appear to indicate a slightly smaller surface-state band gap with an additional filled state at approximately -0.5 V for In (Fig. 1.6a, B) and -0.4 V for Pb (Fig. 1.6b, B), respectively.
2. Essentially different tunneling spectra are found at the anomalously bright ends of In chains shown in the filled-state images. They exhibit a metallic or small band-gap (less than 0.2 eV) behavior (Fig. 1.6a, G), whereas the ends of any Pb chains show a semiconducting property similar to that of any other sites (Fig. 1.6b, G). The origin of the spectral differences for single dimers and bright ends of indium chains (perhaps

single In atoms) is not yet well understood, but might be related to the defects that trap them.
3. Tunneling spectra on top of Pb chains other than the buckled-upward positions give generally the same shape of curves but without the LDOS maximum at −0.5 V, an indication of this peak being unique to the occupied state of buckled-upward Pb atoms.

The presence of the surface-state band gap and resultant semiconducting property can be rationalized by the local chemical bonding. Qualitatively and intuitively, the Si(100) surface is hungry for electrons owing to the presence of Si dangling bonds, so most of the valence electrons from the metal adsorbate atoms at a submonolayer coverage are simply sucked up by the substrate through the local covalent chemical bonding. For In on Si(100), each In atom saturates all the unpaired dangling bonds after being bonded with two Si atoms and another In atom and leaves no free electrons for metallic conduction. The one leftover valence electron for each Pb atom could possibly form a delocalized π-bonding system, but the Pb atoms are dimerized primarily owing to the local Si surface configuration and a minor effect, if any, of a Peierls dimerization mechanism, which would already have killed the conductivity. The buckling of Pb dimers results in further localization of electrons, thus leaving no free carriers for conduction among the adsorbate network. Now let us proceed to the correlation of tunneling spectra with the surface-state bands associated with metal–metal bonds and metal–silicon bonds. Tunneling spectra are sensitive to both filled and empty states. It can be viewed as a technique related to angle-integrated photoemission and inverse photoemission, and if combined with theoretical analyses, it can be used to determine electronic states via peak positions (but not the band structure owing to the lack of angle resolution). Figure 1.7 shows tunneling spectra averaged over various chains of In and Pb on Si(100), respectively. Indium chains show five LDOS maxima below E_F at, respectively, −1.7, −1.5, −1.3, −1.0 and −0.7 V, and five LDOS maxima above E_F at, respectively, +0.5, +0.8, +1.3, +1.6 and +1.8 V. There are five filled surface-state bands associated with each In dimer through one In–In bond and four In–Si backbonds with ten valence electrons available (from one dimer and four Si atoms). Comparison with theory [88–90] and photoemission studies [78, 99] allows us to assign the highest band, with a peak at −0.7 V, to the In–In σ bond and the remaining four to In–Si backbonds, by reference to the orbital interaction picture of Fig. 1.4 and the adsorption structure in Fig. 1.2. For the unoccupied surface states, we correlate the +0.5-V band to the π-type In–In bonding state, the +0.8-V band to the σ-type In–In antibonding state and the remaining three bands to the antibonding In–Si interactions.

The assignment of empty-state bands, especially the two bands below +1.0 V to metal–metal bonds, is also supported by the bias-dependent image shown in Fig. 1.8. This empty-state topograph exhibits dark grooves around

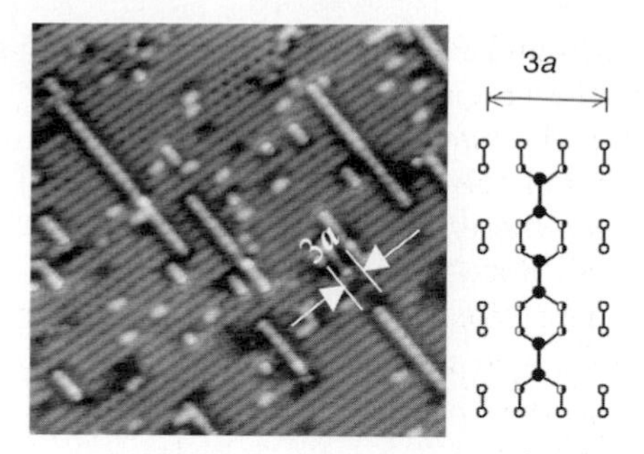

Fig. 1.8. Bias-dependent features of In chains on Si(100) at low bias voltages ($30 \times 30\,\mathrm{nm}^2$, +1.0 V, 0.2 nA, 80 K). The image contrast between the bright In chain and its depressed sides is related to the electronic states from different bonds, namely, the In–In bonds vs. In–Si bonds. The schematic on the *right* shows the local structure around a chain with a groove spacing of $3a$. The *solid circles* represent In atoms, the *hatched circles* are Si atoms bonded to In dimers and the *open circles* are Si atoms with one unsaturated dangling bond [76]

bright In chains at a bias of +1.0 V. Figure 1.8 explains why the dark grooves on the two sides of a chain are separated by $3a$. The In–Si states are not available at such a bias according to the tunneling spectra in Fig. 1.7, whereas the dangling bond states are readily detectable for both In dimers and unsaturated Si atoms. When bias voltages are above +1.3 V, In–Si empty states start to contribute to tunneling and groove-free normal images like those in Fig. 1.7 are observed. On the other hand, the tunneling spectra of Pb chains in Fig. 1.7 reveal six LDOS maxima below E_F at, respectively, −1.8, −1.5, −1.1, −0.9, −0.7 and −0.5 V, and five LDOS maxima above E_F at, respectively, +0.7, +1.1, +1.3, +1.6 and +1.8 V. There are six filled surface-state bands associated with each Pb dimer because of the additional π-type Pb–Pb bond (though weakened by buckling). In reference to the assignment of In chains, we correlate the first two bands to metal–metal bonds: the highest −0.5-V band to the π-type Pb–Pb bond, the −0.7-V band to σ-type Pb–Pb bond and the remaining four bands to the Pb–Si backbonds. In the empty states, the broad +0.7-V band is assigned to the unoccupied Pb–Pb states (including both σ- and π-antibonding interactions, and the remaining four bands are assigned to Pb–Si antibonding states. The small bump around +0.5 V is probably caused by the bare Si(100) surface, although it may be possibly related to the π-type Pb–Pb antibonding state. These tunneling spectra are in general agreement with the previous photoemission studies [78, 100].

1.2.3 Conclusion

We have investigated the similarities and differences of the adsorption behavior and tunneling spectra between In and Pb on Si(100) at low coverage by

STM and STS. Both In and Pb on Si(100) have been found to initially form 1D ad-dimer chains that register with the parallel ad-dimer configuration. Their differences lie in the image contrasts and buckling of metal ad-dimers. Indium dimers are symmetric on Si(100) but show sharp contrast between empty and filled states, whereas lead dimers give similar contrast for both polarities but are buckled. Both phenomena can be attributed to the difference in the electronic configurations of the metal atoms, namely, $3e^d$ In($5s^2$ $5p^1$) versus $4e^d$ Pb($5s^2$ $5p^2$), and resultant spatial extensions of atomic orbitals into the tunneling gap. Accompanying the buckling, there is a charge transfer from the buckled-downward Pb atom to the buckled-upward Pb atom within a dimer. The resultant difference in local electronic states is revealed through the contrast mechanism of STM topographs, with the negatively charged Pb atom detected in the filled states and the positively charged Pb atom in the empty states. The different electronic configurations of these two elements are also reflected in their tunneling spectra. Whereas the dimer chains for both metals exhibit a surface-state band gap of approximately 1.2 eV, the positions for the LDOS maxima around the band gap are different, with -0.7 to $+0.5$ V for In and -0.5 to $+0.7$ V for Pb. These two bands, plus another one at $+0.8$ V for In or $+0.7$ V for Pb, are assigned to the surface-state bands associated with metal–metal bonds, whereas the remaining bands farther away from the Fermi level are correlated with the metal–silicon backbonds. The presence of a surface-state band gap and the resultant semiconducting property arise from the complete saturation of dangling bonds of metal dimers upon the formation of local covalent chemical bonds and chain dimerization as well as buckling in the case of Pb dimers. As expected for a nonconducting line, no length dependency of tunneling behavior was observed. Nevertheless, since STS measures the conductivity perpendicular to the surface, it does not necessarily reflect the conductivity along these dimer chains. Lateral device-type measurements will help to clarify the issue, although experimental characterization of such atomic lines is difficult because they must be connected to macroscopic electric leads to perform transport measurements. On the other hand, the nonmetallic behavior observed for In and Pb chains may point out an intrinsic limitation for making a conducting atomic wire owing to the Peierls distortion mechanism in addition to the local chemical bonding requirement. According to Peierls [101], an equidistant 1D metal chain with one electron per atom can never be stable. The electron-to-phonon coupling will animate dimerization or bond alternation, and the atomic rearrangement is exactly such that an energy gap is opened up at the Fermi level. The solution to render an atomic-scale chain conducting may require a slight increase of the chain diameter by which the Peierls distortion effect can be ignored, e.g., in systems like carbon nanotubes, molecular wires and metal cluster wires.

1.3 Atomic-Scale Pb Chains on Si(100)

1.3.1 Introduction

An atomic chain, composed of a single row of atoms, is the ultimate limit in lateral miniaturization of a wire [102]. Owing to the atomic-scale dimension and boundary conditions imposed, the transport behavior of atomic chains is quite different from that in the bulk and could reveal a new intriguing aspect of physics. There is increasing interest in the fabrication of atomic-scale lines on semiconductor or metal surfaces for both fundamental research and application in nanoelectronics [103–109]. The approaches used include self-assembly [107–109] and atomic manipulation to form lines [103, 104, 109], or even dangling bond wires [105, 106, 110] on the surface. The adsorption of Pb on Si(100) is a particularly interesting system for the study of low-dimensional behavior owing to negligible mutual solubility at all temperatures [111]. Previous photoemission spectroscopy [111], LEED [112] and STM studies [58, 85] have established the Stranski–Krastanov growth mode of Pb on Si(100), analogous to that of the group III metals (Al, Ga, In) on the same surface [81, 83, 84, 108, 109, 116, 119]. Of particular interest is the initial 1D growth at a coverage far below 0.5 ML. Isolated metal ad-dimer chains are found to orient perpendicular to the underlying Si dimer rows. The ad-dimer configuration has also been determined both experimentally [58,81,83–85,111,112,116,119] and theoretically [88,89,91] to be parallel to the chain direction with each atom triply bonded. Recent first-principles total energy calculations on Pb/Si(100) [91] have pointed out further that the Pb ad-dimers are asymmetric, in contrast with the symmetric ad-dimer structure for group III metals on Si(100) [88, 89]. In addition to the varieties of chain structures themselves, the Pb/Si(100) system attracts our attention also because of the electronic properties. Our previous *I–V* data on In/Si(100) [84] have indicated a semiconducting property for such 1D structures, and this behavior can be readily understood in terms of the local chemical bonding of In atoms on the Si(100) surface. Each indium atom ($5s^2$ $5p^1$) saturates all the unpaired dangling bonds after being bonded with two Si atoms and another In atom (yielding an ad-dimer) and leaves no free electrons for metallic conduction. Since each lead atom has one more valence electron than indium, we are curious about the destiny of this electron for conduction under similar bonding geometry. Will it be delocalized over the chain, or localized by the dimerization and buckling? In this chapter, we report the STM images that support the buckling model of Pb dimers and investigate the chain evolution upon the increase of coverage up to 1.25 ML. An approach to obtaining long atomic lines is suggested in an effort to facilitate potential transport measurements. More significantly, we present tunneling spectra of the Pb dimer chains that show a semiconducting property. The shape of the curves and the origin of a surface-state band gap for the chain are briefly discussed.

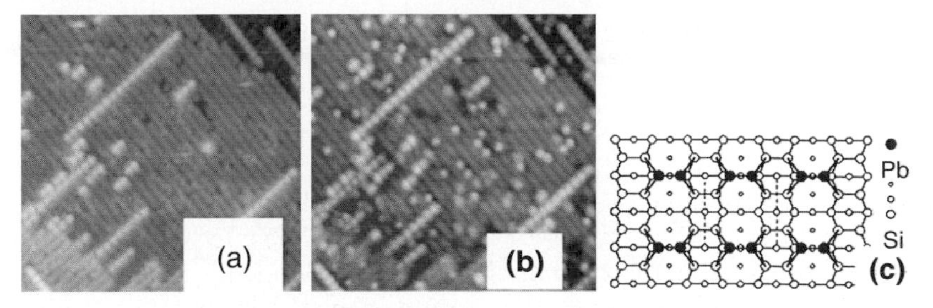

Fig. 1.9. STM topographs of Pb on the same area of Si(100) acquired at 80 K, 0.07 ML, 24 × 24 nm^2: (**a**) filled state, −1.5 V, 0.2 nA; (**b**) empty-state, +2.0 V, 0.2 nA; and (**c**) the parallel ad-dimer model without buckling [102]

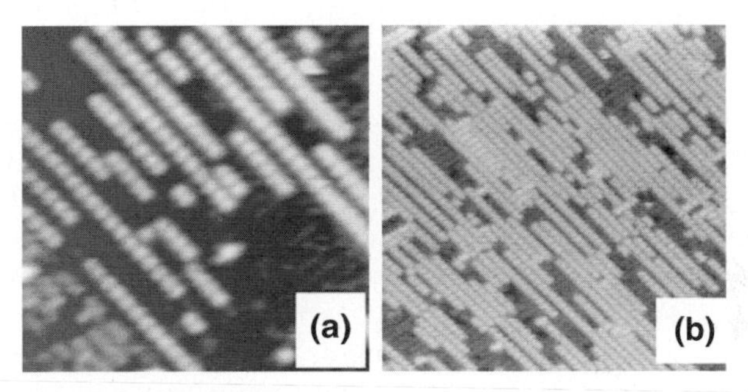

Fig. 1.10. Arrangement of Pb dimer chains upon the increase of coverage: (**a**) approximately 0.1 ML, 14 × 14 nm^2, −1.9 V, 0.5 nA, 295 K; (**b**) approximately 0.4 ML, 32×32 nm^2,−2 V,0.3 nA,80K [102]

STM Below 0.5 ML

Figure 1.9 shows two STM topographs at a 0.07-ML coverage of Pb on the same area of Si(100). Each bright protrusion in the filled-state image corresponds to a Pb ad-dimer and is clearly distinguishable from the underlying Si dimer rows. In the empty-state image, the Pb structures are almost as bright as in the filled state but the surface is adorned with additional bright dots arising from the C-type defects on the bare Si(100). The 1D growth of Pb on Si(100) is evident and the chain is aligned perpendicular to the Si dimer rows, as illustrated in the parallel ad-dimer model in Fig. 1.9c. These chains are single-atom-type both in width and height and thus represent a truly 1D system. As the coverage is increased, these chains become crowded on the surface and start to pack into 2×3 or 2×2 structures, as shown in Fig. 1.10. The 2×3 reconstruction at low coverage in Fig. 1.10a implies that there is a weak local repulsion between the metal dimer chains. A further increase of coverage up to 0.5 ML will extend the 2×2 structure observed in Fig. 1.9b over the whole surface. The 2×3 model is marked by the box in the structural model of Fig. 1.9c for a symmetric ad-dimer configuration. Nevertheless, the

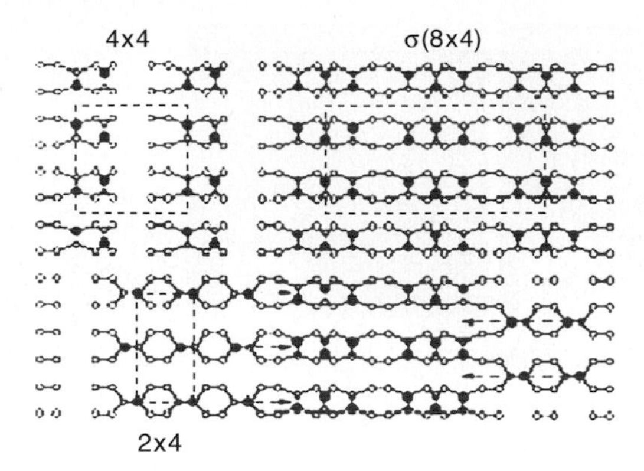

Fig. 1.11. Structural model of Pb on Si(100) below 0.75 ML. The *open circles* represent Si and the *shaded circles* are for Pb [102]

same 2×2 structure is still applicable if the buckling of Pb dimers is all oriented in the same direction. However, a careful check of Fig. 1.10 reveals the frequent shift of bright maxima in the neighboring chains spaced $2a$ apart. This presents clear evidence for the buckling of Pb dimers since symmetric dimers would give a perfect square pattern. The $2a$ spacing for the shifted neighbors further suggests that buckling occurs often in opposite orientations in alternate chains (but in the same orientation within a chain). The surface reconstruction of Pb in such an area is no longer 2×2 but changes to 2×4, as illustrated in the lower-left part of Fig. 1.11. The asymmetric structure of Pb dimers is related to the electronic configuration of Pb atoms. Each Pb atom has four valence electrons (or dangling bonds). When it is adsorbed onto the Si(100) surface, according to the parallel ad-dimer model, two of the electrons are used in the saturation of Si dangling bonds, and another electron is used in the formation of a σ-type Pb–Pb bond. There is a remaining electron in each Pb atom of the dimer that is located in the p-type orbital perpendicular to the surface and it is thus possible to form a π bond. However, both the instability of symmetric dimers predicted by theory [91] and the observed buckling of Pb dimers from STM images suggest that the formation of a double bond (σ plus π) between two Pb atoms in a dimer is not energetically favorable on the Si(100) surface.

In order to lower the energy, the Pb dimers buckle at the expense of weakening π bonding, and accompanying this process there is a charge transfer from the Pb atom that moves down to the one that moves up. All the dangling bonds are thus saturated and the surface is stable. Although we were not able to resolve the atomic structure within a Pb dimer, such buckling and charge transfer features are noticeable through the following observations: (1) as mentioned before, bright maxima often shift in neighboring chains (Figs. 1.10, 1.12) as modeled in the 2×4 structure of Fig. 1.11 instead of forming a perfect square pattern; (2) investigation of the bright maximum position with

Fig. 1.12. Coexistence of 2×2 or 2×4 (area A), $c(8\times4)$ (area B), and rare 4 ×4 (area C) phases on Si(100), approximately 0.6 ML: (**a**) 50 × 50 nm^2, −2.0 V, 0.3 nA; (**b**) 14 × 14 nm^2, −2.0 V, 0.3 nA, 80 K [102]

respect to the underlying Si dimer rows reveals that the brightest spots for Pb dimers are not located in the center of the grooves between Si dimer rows, but are shifted to the sides; (3) the high-resolution image shown in Fig. 1.12b indicates an asymmetric shape (and brightness distribution) for Pb dimers, in contrast to the otherwise ellipsoidal shape hypothesized for the buckling-free model; (4) careful investigation of both filled-state and empty-state images for the same area (Fig. 1.9) appears to show that bright maxima shift to one side of the groove in the filled-state image (buckled-upward atom) but to the other side for the empty-state image (buckled-downward atom) [91].

STM Above 0.5 ML

As the coverage is above 0.5 ML, a complicated surface structure with different reconstructions starts to form, as shown in Fig. 1.12 for 0.6-ML coverage. Area A corresponds to the 2×2 or 2×4 phase described before, area B corresponds to a more compact phase known as $c(8\times4)$ and area C to a new but minor 4×4 phase (upper left of Fig. 1.11). The high-resolution image in Fig. 1.12b clearly shows Pb chains composed of hexagonal rings with Pb dimers in very asymmetric shapes and thus buckled. Based on our STM observation and comparison with the theoretical calculation for the phase [124], a structural model of $c(8\times4)$ is presented in the upper right of Fig. 1.11, which is different from the one previously proposed by Itoh et al. [58] but similar to the one proposed for Sn/Si(100) by Baski et al. [87]. The ring chain in $c(8\times4)$ actually consists of three single-atom-wide Pb dimer chains with an interchain spacing of $1a$ ($a = 53.84$ Å). As more Pb atoms are deposited onto the surface, some of the Si–Si dimer bonds of the underlying Si surface have to be broken in order to accept additional Pb atoms into the first layer. The change of adsorption configuration from the parallel ad-dimer model to the orthogonal ad-dimer model implies a small difference in their total energy because half of the Si atoms are relaxed in this case. The $c(8\times4)$

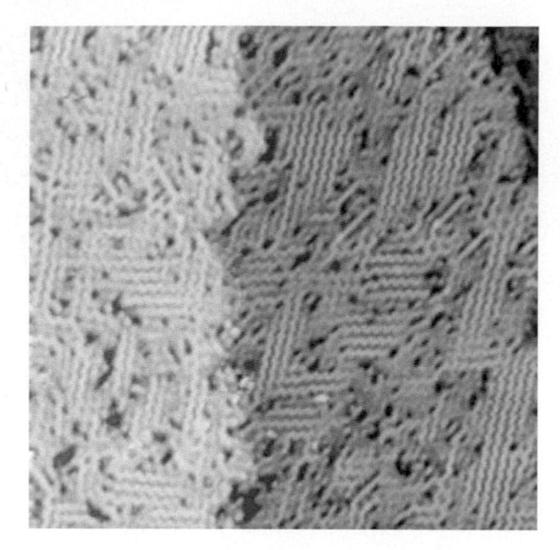

Fig. 1.13. Disordered $c(4\times4)$ phase of Pb on Si(100), approximately 1.25 ML, $60 \times 60\,\text{nm}^2$, -2.1 V, 1.0 nA, 80 K [102]

phase does not occur for either group III or group V elements on Si(100) and appears unique to the group IV metals for reasons related to the number of valence electrons and the resultant buckling of metal ad-dimers (symmetric versus asymmetric). The saturation coverage for this $c(8\times4)$ model is 0.75 ML because there are 24 Pb atoms per 32 Si atoms in a unit cell. When the coverage is above 0.75 ML, additional Pb atoms will have to reside on top of the unbroken Si dimers (which are then broken upon adsorption) and a 2×1 phase is formed for coverage close to 1 ML [58, 85]. A further increase of the coverage above 1 ML certainly leads to the formation of a second Pb layer, which was also reported to form buckled dimer structures, as illustrated in Fig. 1.13 for a phase related to the $c(4\times4)$ reconstructions [58, 85]. The $c(4\times4)$ phase consists of two orthogonal domains that reach saturation coverage of 1.25 ML for the two-layer model [85] and 1.33 ML for the three-layer model [58] and is found so far to be unique to Pb on Si(100).

Approaches to Long Chains

Although short chains may be potential functional units, a chain length on the order of 100 nm or longer is desirable for the chain to become a practical 1D conductor to bridge components.

Single Pb dimers are found to be mobile at room temperature since we observe occasional rearrangements of isolated dimers or even chains during scanning, although the mobility might be related not only to the thermal effect but also to the tip-induced electric field [126, 127]. The relatively long Pb chains in Fig. 1.9 imply that the mobile Pb atoms or dimers can sense

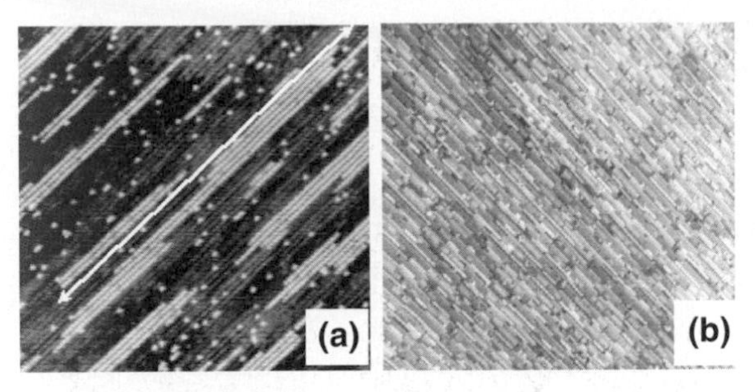

Fig. 1.14. (**a**) Atomically resolved 80-nm-long Pb chains on Si(100): $75 \times 75\,\text{nm}^2$, $-1.85\,\text{V}$, 0.2 nA. (**b**) Pb chains on single-domain Si(100): approximately 0.1 ML, $120 \times 100\,\text{nm}^2$, $-2.0\,\text{V}$, 0.5 nA, 295 K [102]

the attraction of the preexisting chains directly or indirectly for the chain polymerization mechanism proposed previously [84, 89]. The chain growth is terminated by steps or defects where the bonding requirements for Pb can hardly be satisfied. However, on a double-domain Si(100) surface, the typical maximum length of Pb chains is about 30 nm unless a surface with low defect density and wide long terraces can be prepared, in which case the longest chain atomically resolved is approximately 80-nm long, as shown in Fig. 1.14a. A better way to fabricate long atomic chains is to use single-domain vicinal Si(100) surfaces with a miscut angle greater than or equal to 4° [109] For single-domain Si(100), the monoatomic steps disappear in favor of double-atomic steps. More importantly, the step edge is almost straight and the surface is covered mostly by the 1×2 terraces in which the Si dimer rows are all aligned perpendicular to the step edges. This kind of substrate arrangement forces the metal ad-dimers to nucleate and grow selectively along the step edge direction because of the favorable parallel ad-dimer bonding configuration. An example is given in Fig. 1.14b. The chain could be much longer if appropriate heat treatment (outgassing, flashing and cooling) is applied to reduce the surface defect density.

STS Results

In order to probe the electronic states of the Pb dimer chains, tunneling spectra were measured on top of the Pb dimers for isolated Pb chains together with those on the bare Si(100) surface in the same image. The I–V data were acquired via the CITS technique [93] at a regulated position of $-2.0\,\text{V}$ and 0.5 nA at room temperature. The data shown in Fig. 1.15 are averaged over several dozen data points for a given type of position. CITS provides real-space imaging of surface electronic states with a perfect correspondence between the spectrum and the site measured. The brightest features along the Pb dimer chains are correlated with the buckled-upward atoms (filled states) according to the dimer buckling discussed before and these are the

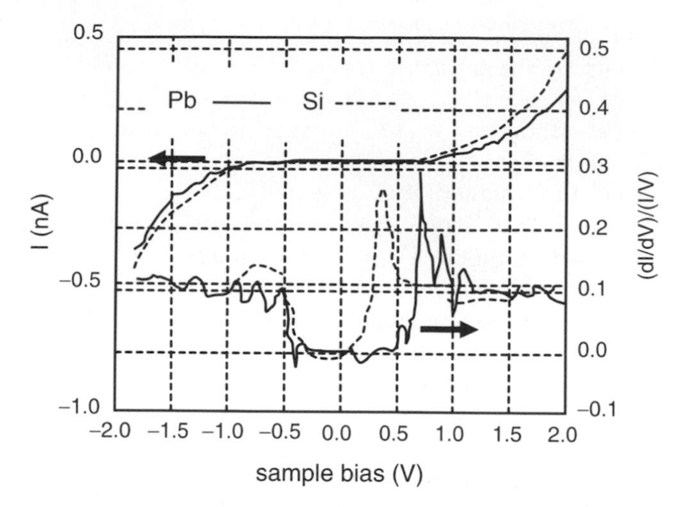

Fig. 1.15. Tunneling spectra of Pb dimers (specifically the buckled-upward positions) on Si(100) together with those on the bare Si surface. The *upper curves* are *I–V* characteristics and the *lower curves* are the normalized tunneling density of states. The presence of a surface-state band gap is evident [102]

particular sites measured. The tunneling density of states $(\mathrm{d}I/\mathrm{d}V)/(I/V)$ is normalized according to the simple offset technique of I/V replaced by $[(I/V)^2 + c]^{1/2}$, where c is a small constant, 0.005 in the present case [128]. The I–V characteristics for the bare Si(100) surface are in good agreement with the literature, with a surface-state band gap of approximately 1.1 eV (Fig. 1.15). The normalized tunneling spectra in Fig. 1.15 indicate that, on top of Pb dimers (in particular the buckled-upward atom), there are two LDOS maxima near the Fermi level (at $V = 0$), -0.5 V below and $+0.7$ V above E_F, respectively. This suggests a surface-state band gap of approximately 1.2 eV at the buckled-upward position, slightly larger than that of Si(100). The LDOS maximum at $+0.7$ V is likely related to the empty (antibonding) state of the Pb dimer, whereas the local maximum at -0.5 V is attributed to the filled state or a weakened bonding state of the Pb dimer.

There are four additional LDOS maxima below E_F, at, respectively, -0.8, -0.9, -1.3 and -1.7 V, which are presumably associated with the Pb–Si backbonds, although the peak at -0.8 V is likely to have some contribution from the bare Si surface. It is worth noting that tunneling spectra on top of Pb chains other than the brightest spots (buckled-upward positions) generally give the same-shaped curves but without the LDOS maximum at -0.5 V, an indication that this peak is unique to buckled-upward Pb atoms. These tunneling spectra are in general agreement with previous photoemission studies. The origin of the surface-state band gap for the Pb chains can be rationalized as follows. Qualitatively and intuitively, the Si(100) surface is hungry for electrons owing to the presence of Si dangling bonds, so most of the valence electrons from the metal adsorbate atoms at a submonolayer coverage are simply sucked up by the substrate through the local covalent chemical bonding. The leftover valence electrons are possible candidates for conduction, but the Pb atoms are dimerized primarily owing to the local Si surface configuration and a minor effect, if any, of a Peierls dimerization mechanism [101], which would kill the conductivity.

The buckling of Pb dimers further localizes electrons, thus leaving no free carriers for conduction among the adsorbate network. Our preliminary I–V measurements on the Pb chains of the $c(8\times4)$ and $c(4\times4)$ phases also indicate the presence of a surface-state band gap, although it slightly decreases upon the increase of coverage. A metallic feature starts to be observed only on top of 3D Pb islands in which the coverage is above 2 ML. The existence of a surface-state band gap implies a semiconducting property for the Pb dimer chain. Nevertheless, since STS measures the conductivity perpendicular to the surface, it does not necessarily reflect the conductivity along the surface (or chain). Theoretical calculation and lateral device-type measurements will help to clarify the issue, although experimental characterization of such atomic-scale wires is difficult because they must be connected to macroscopic electric leads for us to perform transport measurements.

1.3.2 Conclusion

Various Pb dimer chain structures have been investigated at different coverages on Si(100), ranging from the isolated Pb ad-dimer chains at a coverage far below 0.5 ML, to the hexagonal ring chains for $c(8\times4)$, and to the zigzag chains for $c(4\times4)$. The Pb dimers are found to be asymmetric in all cases. The selection of single-domain Si(100) is found to be an effective approach to obtaining long atomic chains because of straight step edges and the parallel ad-dimer configuration. Our tunneling spectra on top of isolated Pb dimer chains indicate the presence of a surface-state band gap around 1.2 eV. The local covalent chemical bonding, dimerization and buckling are responsible for the existence of the surface-state band gap and thus the semiconducting property of the Pb chains. Approaches to rendering these 1D structures metallic include electron doping into an empty band or artificial structuring of the atomic chains, increasing the temperature and pressure or, alternatively, the fabrication of larger-sized cluster chains or molecular wires rather than atomic lines.

1.4 Indium Ad-dimer Manipulation by a Scanning Tunneling Microscope Tip

1.4.1 Introduction

One of the ultimate goals of nanofabrication is directed at controlled positioning of individual atoms and molecules into designed functional structures [126]. Among all the approaches to atomic manipulation, STM is so far the simplest and the most general technique. This advantage arises from its ability to perform atomic-resolution imaging and manipulation using the same equipment. Whether a structure is imaged or modified depends on the strength of the interactions between the tip and the sample. In other words, by exploring different physical or chemical interaction mechanisms through which the tip and the sample "communicate," one can use STM to perform controlled manipulation of atoms or molecules through the control of location, magnitude, duration and polarity [132,133]. The tip–sample interactions can be classified into two broad categories: (1) forces that act on the adsorbate owing to the proximity of the tip, as illustrated in the elegant examples of Xe on Ni(110) [134] and the quantum corral of Fe on Cu(111) [135] via the atom sliding technique on metal surfaces and (2) effects caused by the electric field between the tip and the sample [138], or by the current flowing through the gap region, for example, in the atom extraction from semiconducting Si surfaces via chemically assisted field desorption [133, 137, 138] or by field induced evaporation alone [139]. In the tunneling gap region (0.5–1 nm), the electric field strength approaches values of the order of 10^8 V/cm when a voltage pulse is applied. Such fields are comparable to those experienced by valence electrons in atoms and are thus sufficient to break chemical bonds. Moreover, owing to the extremely small cross section of the scanning tunneling microscope beam (through one or a few tip-apex atoms) the current densities can be 10^{11}–10^{12} times higher than those of conventional electron beams. Local heating through inelastic tunneling is therefore highly possible and could promote vibrational or electronic excitations, facilitating the desorption of ad-atoms from the surface [142]. In this chapter, we report preliminary results on the manipulation of indium ad-dimers on Si(100) in an effort to make atomic-scale structures bridging the source and drain electrodes for subsequent transport measurements. The effects of voltage and current pulses on surface modification will be addressed.

1.4.2 Results and Discussion

How In Atoms Adsorb on the Semiconducting Si(100) Surface

Before we proceed to the manipulation, let us first take a look at how In atoms adsorb on the semiconducting Si(100) surface [84, 88, 141, 142]. Figure 1.16 shows a (430 Å)^2 image of double-domain Si(100)-2×1 with a 0.08-ML coverage of In, which was taken at room temperature with +2.0-V sample bias (empty states) and 0.1-nA tunneling current. The In atoms appear as well-separated bright protrusions in the topography that are clearly distinguishable from the underlying Si dimer structures. The most striking feature of the image is the 1D anisotropic growth of In chains on the surface, running perpendicular to the underlying Si dimer rows. Each bright protrusion within a chain, however, does not correspond to a single In atom but to two instead, as resolved in the inset showing a low-temperature

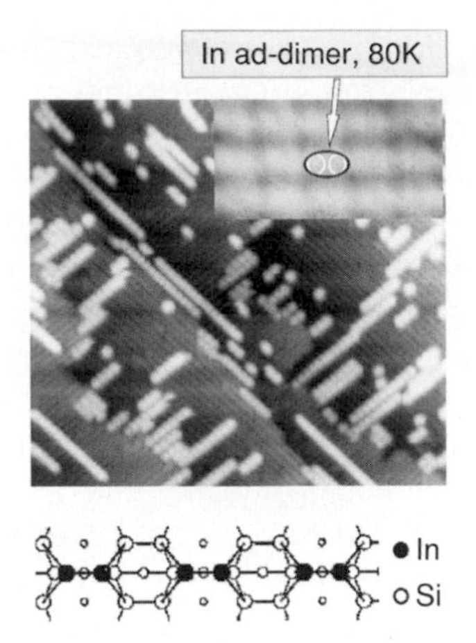

Fig. 1.16. *Top*: An image of (430 Å)^2 Si(100) with a 0.08-ML coverage of In, taken at room temperature with +2.0-V sample bias and 0.1-nA tunneling current. The image in the *inset* was taken at 80 K with +4.0 V and 2.0 nA for a 0.5-ML coverage. *Bottom*: A schematic view of the parallel ad-dimer configuration [126]

empty-state image. It is well established that each In atom is triply coordinated, making two bonds to two Si atoms adjacent within a dimer row and one bond to another In atom to form an ad-dimer with a spacing of approximately 3 Å, and the ad-dimer is located halfway between Si dimer rows and oriented parallel to the Si dimer bond (in other words, along the chain direction), as illustrated in the schematic drawing at the bottom of Fig. 1.16. The maximum length of isolated In chains on the double-domain Si(100) surface is only about 30 nm, under the experimental conditions. In order to increase the chain length to facilitate the connection between the source and drain electrodes, we want the In atoms to nucleate and grow selectively parallel to the step edges. A single-domain vicinal Si(100) surface is a wonderful candidate and is used for this purpose. Indeed, as shown in Fig. 1.17, the In chain length increases significantly to approximately 70 nm (composed of approximately 180 atoms, thanks to the chain periodicity of 0.77 nm). The inset shows a higher-resolution image which indicates that the Si dimer rows on the terrace are all aligned perpendicular to the step edges.

This kind of substrate arrangement forces the In ad-atoms to nucleate and grow only along the step edge direction because of the favorable parallel ad-dimer bonding configuration, as addressed previously [84,88,141,142,145]. Atomic manipulation is necessary for the fabrication of atomic-scale structures for at least two reasons in our case. First, the self-assembled In chains are still not long enough to bridge the electrodes, so repairing the (broken) gap or extending the chain can only be realized through artificial structuring. Second, we need manipulation to make structures with a desired pattern. Since In atoms are chemically bonded to the substrate Si atoms and the binding energy is as high as 2.8 eV (1.1 times larger than the enthalpy of for-

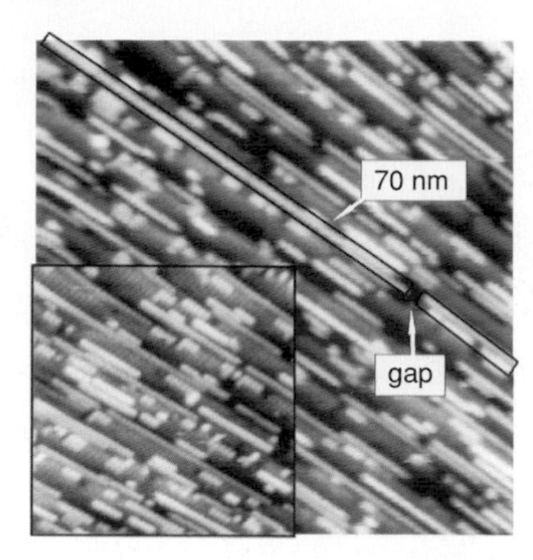

Fig. 1.17. A 70-nm^2 STM image showing the initial growth of In on the single-domain Si(100) surface at room temperature featuring an In ad-dimer chain as long as approximately 70 nm. The *inset* is a closeup of a 40-nm area which reveals the dimer row resolution. Both were taken at 0.1 ML, +2.0 V, 0.25 nA [126]

mation of bulk In, $\Delta H_{\mathrm{In}} = 2.52\,\mathrm{eV}$) [141], the vertical transport of In atoms between sample and tip involves bond breaking and thus requires significant activation. This can be realized by the application of high-voltage pulses or large currents through the gap region (details are given later). However, lateral transport of In ad-dimers through the atom sliding technique also appears possible according to the occasional rearrangements of In ad-dimers on the surface during scanning (at room temperature), as shown in Fig. 1.18. Two consecutive images of the same area were taken in a 5-min interval. The relatively high mobility of In ad-dimers is demonstrated by the occasional rearrangements marked within the boxes. One important observation is the presence of the horizontal scratching traces related to the In chains that agree with the scanning direction. This suggests that the redistributed In ad-dimers are probably dragged by the tip (through an electrostatic trap) under the imaging condition (+2.1 V and 0.1 nA). Nevertheless, it still remains unclear whether these rearrangements are caused by the electric field (approximately $2{\times}10^7$ V/cm for a 1-nm tunneling gap) or thermal diffusion (approximately 25 meV) or both. Data on the diffusion barrier of In on Si(100) are not available to our knowledge. However, Brocks et al. [89] reported a calculation on the adsorption of Al on Si(100), and since both In and Al belong to group III elements and have the same adsorption structure on Si(100), the values for Al/Si(100) can be used as a reference except that the bonding is weaker in the case of In, owing to its more diffuse orbitals.

For Al/Si(100), the calculated binding energy is 3.6 eV, the diffusion barriers of a single Al atom on the surface are 0.1 eV perpendicular to the Si dimer row and 0.3 eV parallel to the dimer row and the effective bond energy within an ad-dimer is 1.1 eV. Using $\Delta H_{\mathrm{In}}/\Delta H_{\mathrm{Al}}$ ($2.52/3.42 = 0.74$) as a

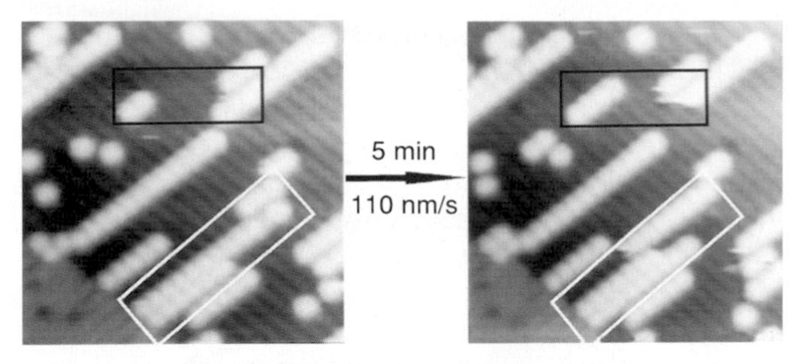

Fig. 1.18. In ad-dimer rearrangements on a Si(100) surface at 295 K after scanning, 20 nm^2, 0.08 ML, + 2.1 V, 0.10 nA [126]

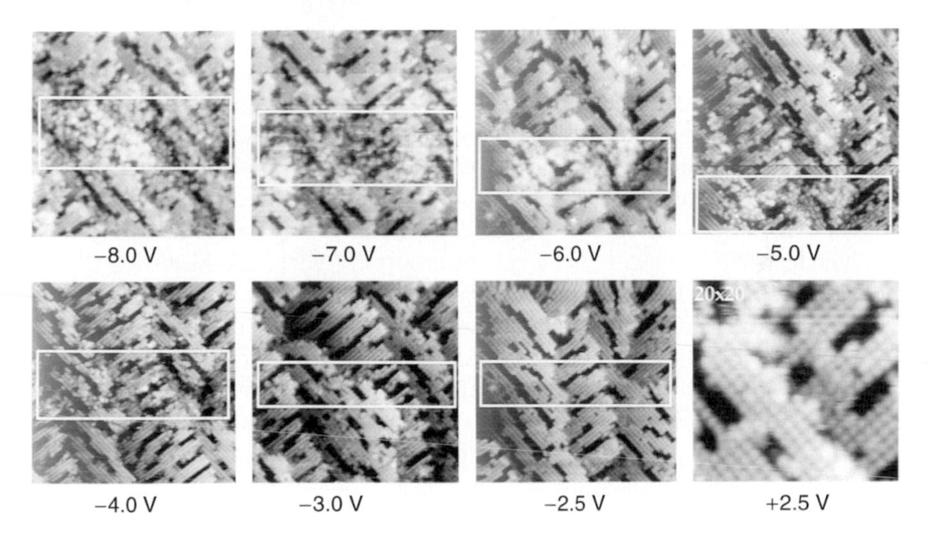

Fig. 1.19. Effects caused by the electric field from high-voltage pulses with the feedback loop on and a duration of 10 ms at 295 K on a 50-nm^2 Si(100) surface covered by 0.5-ML In. The *bottom-right image* shows the In-2×2 reconstruction. All image were taken at +2.5 V and 0.2 nA [126]

rough energy scale, we estimate that, for In/Si(100), the binding energy, diffusion barriers perpendicular and parallel to the dimer row and the effective bond energy within an In ad-dimer are 2.7, 0.1, 0.2 and 0.8 eV, respectively. The small barriers for diffusion mean that In atoms can be laterally moved on the surface in a controlled manner, particularly at low temperatures [84, 142]. Furthermore, the relatively larger bond energy per ad-dimer relative to the diffusion barrier suggests that the ad-dimer is likely to be moved as a unit. Unfortunately, we have not yet been able to test the atom sliding technique on In/Si(100) owing to software limitations. Now we come back to the effects of electric fields. Figure 1.19 shows a series of images that were taken after applying approximately 5 V pulses (feedback loop "on," −2.5 to 8 V, 10 ms, 0.2 nA) around a horizontal line through the center of the marked box on a surface fully covered by indium. The bottom-right image is a closeup of a 20-nm^2 area which exhibits the In-2×2 reconstruction upon a 0.5-ML coverage. The surface begins to be modified above −3 V but is badly damaged above −4 V as marked by the box. The higher the voltage pulse is, the worse the damage appears. Applying a positive bias pulse above 4 V causes similar damage. To be more specific, when a voltage pulse was applied on top of an isolated In ad-dimer (on a different lower coverage sample), the tip condition changed immediately in most cases (presumably related to the thermally activated probabilistic feature of field desorption). The image quality usually degraded but one could still see the effect of the pulse: the ad-dimer disappeared from the original spot. Placing the tip on top of an In chain followed by a pulse usually destroyed the chain structure, with individual In ad-dimers being scattered around. (One other possibility of modification is the deposition of tip contaminants or tip materials, but this is less likely here because the apex of the W tip was heated up to approximately 1273 K just before the experiment.) These observations suggest that with a high-voltage pulse, the missing In ad-dimers might sometimes jump to the tip. However, by using voltage pulses alone, we were not able to selectively extract an ad-dimer and deposit it back on the surface in a controlled way, most likely because the electric field strength falls off rather smoothly with lateral distance from the apex of the tip [147]. In order to improve the manipulation resolution, we used a different technique to manipulate In ad-dimers on Si(100) surfaces, i.e., through the combination of electric fields with large tunneling currents, with the speculation of employing the chemically assisted field desorption mechanism [133, 137]. Figure 1.20 shows four 10-nm^2 images taken at +1.8 V and 0.5 nA, in which the In ad-dimers were picked up from and deposited back on the surface on an almost one-by-one basis.

The manipulation process was as follows: approach the tip to the circular area by increasing the current to 10 nA with the feedback loop "on," keep it there for 10 ms while maintaining the same bias voltage, then move the tip to the rectangular area and apply the same current pulse again. After one repeat, we found that two ad-dimers inside the circle were missing but only one showed up in the rectangular area, so we applied the same current pulse to the rectangular area again. This more-or-less controlled manipulation is probably related to the following factors. The activation energy or critical field required for atom transfer decreases sharply as the tip–sample distance reduces. In addition, when a tip is brought very close to a surface (e.g., at an approximately 3 Å spacing), the tip could induce a chemical interaction between the selected adsorbate and the tip atoms. This interaction also reduces the activation energy required for desorption, so when a voltage bias (not a high-voltage pulse) is applied, the In ad-dimer can be selectively picked up sometimes. But this process is probabilistic, and it is still not yet understood why the same current pulse can put the ad-dimer back on the surface since the opposite bias polarity would make more sense (it worked also probabilistically). In summary, regarding the manipulation

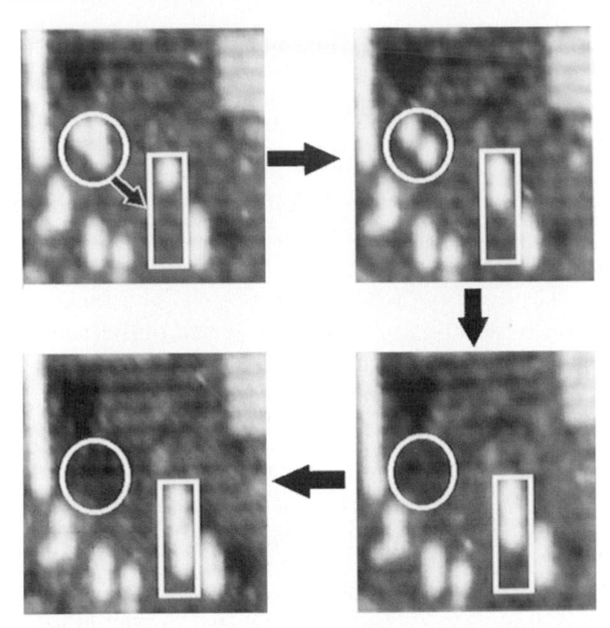

Fig. 1.20. STM images showing a step-by-step manipulation of In ad-dimers on 10-nm^2 Si(100) through the application of large currents. Current pulses +1.8 V, 10 nA, 10 ms, feedback loop "on"; imaging +1.8 V, 0.5 nA. See text for details [126]

of In ad-dimers on Si(100), voltage pulses above −4 V can easily modify the surface but the control of positioning is difficult owing to the smooth decay of the electric field along the lateral direction. In contrast, the application of large currents (current pulses) brings a sharp scanning tunneling microscope tip very close to the ad-dimer of interest and probably induces a chemical interaction between the selected ad-dimer and the tip. This interaction also reduces the activation energy required for the desorption of the adsorbate atoms and they can thus be manipulated at a relatively low bias voltage, which affords a better control over the positioning of the ad-dimers. It is to be noted that tip–sample interactions are complicated, and the present work is far from a completely controlled manipulation process. It remains a challenge to combine the self-assembly with manipulation to fabricate atomic chains long and stable enough to interface with the macroscopic world for subsequent property measurements.

1.5 Conductance Measurements Through a ML of Pb Films on Si(111)

In order to investigate the conducting behavior of adsorbed lead structures and the influence of step edges, we present in this section conductivity measurements on a ML lead film via in situ fabricated Au electrodes inside a UHV [148]. The coverage of lead was monitored by a quartz microbalance

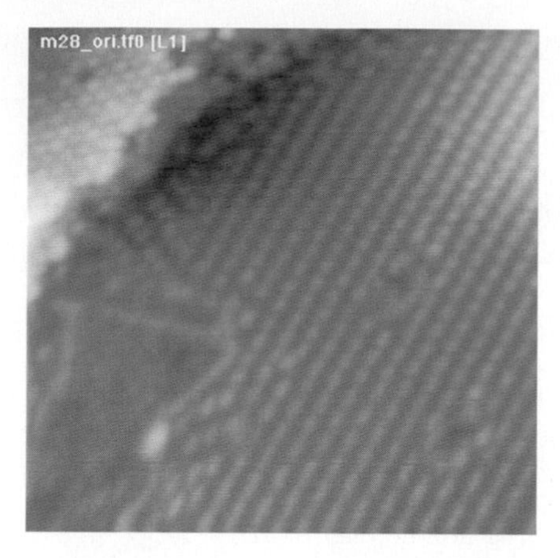

Fig. 1.21. STM image of lead on Si(111) at 43 K acquired at +1 V and 0.16 nA. Individual bright dots correspond to single lead atoms, which are aligned along $[1\hat{1}0]$, the step edge direction in this image [148]

in combination with the control of heating power. The Si(111) sample used in this experiment was n-type with a resistivity of 0.01 Ω cm. Au electrodes were fabricated onto the surface after lead deposition. The surface conductivity of the film was measured along the step edge direction (i.e., $[1\bar{1}0]$) and across step edges. Figure 1.21 shows a ML-covered lead film on Si(111) before three electrodes were formed on the surface. The individual bright dots in the bright ridges correspond to individual lead atoms. The reconstruction of adsorbed lead atoms has been modeled as Si(111)-$c(5\times\sqrt{3})$-Pb with each ridge composed of round single protrusions that are equally spaced, $\sqrt{3}a$ ($a = 0.384$ nm). The distance between neighboring ridges is $5a/2$. Although Fig. 1.21 shows that the direction of the ridges is parallel to the step edge, as a matter of fact, there are three equivalent directions owing to the threefold symmetry of the surface.

Three macroscopic Au pads were formed by the deposition of Au using a through-hole mask supported above the Pb-covered Si surface. The three electrodes were located in such a way that the first and third electrodes were aligned along step edges, whereas the second electrode was located several terraces away. The size of fine electrodes is 1 μm, with a gap distance of 1 μm between the ends of fine electrodes. These fine electrodes were connected to the 1-mm × 1-mm pads with which external probes make contact. The terrace width of this specific substrate is about 200 nm, so each electrode covers about five terraces. Au-coated probes were pushed gently toward the gold pads and the conductance was measured between each pair of electrodes through the lead layer. The three I–V curves in Fig. 1.22 show better conductance at room temperature (270 K) compared with that at 16 K, suggesting a semiconducting character. The conductance between electrodes 1 and 3 is larger than that for pairs 1 and 2, and 2 and 3. These results are reasonable

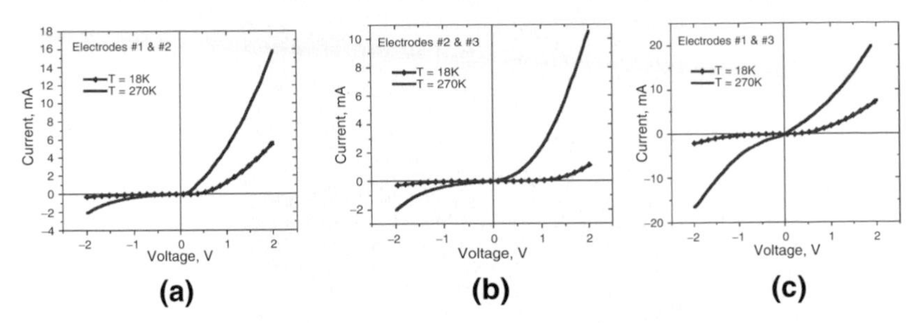

Fig. 1.22. I–V characteristics between three electrodes. The conductance between electrodes 1 and 2 (**a**) as well as that between electrodes 2 and 3 (**b**) is smaller than that between electrodes 1 and 3 (**c**). This effect is probably due to the electrode configuration. The conductance at 270 K (*solid line*) is greater than that at 16 K (*line with dots*), implying nonmetallic behavior. [148]

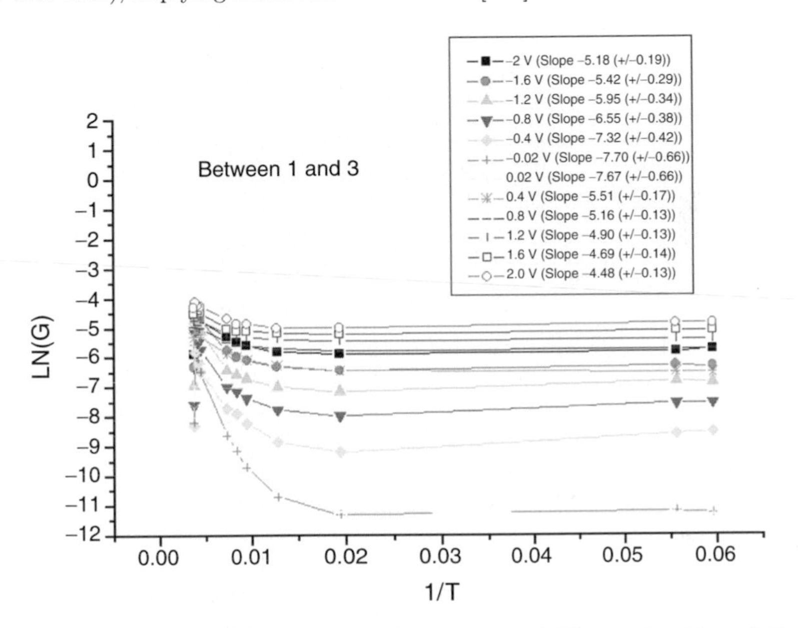

Fig. 1.23. Logarithmic differential conductance vs. $1/T$ as a function of the applied voltages between electrodes 1 and 2. The logarithmic conductance is almost constant when $1/T$ is large (below 50 K) and increases upon the increase of temperature with a strange sudden drop at the end. Such behavior is observed for all the applied voltages. Approximately, the smaller the applied voltage is, the greater the conductance varies across the temperature range [148]

according to the electrode configuration: conduction between electrodes 1 and 3 is along the step edges, whereas for the other it crosses step edges.

In order to study the contact nature between the Au electrodes and the sample, conductance versus $1/T$ was plotted as a function of given applied voltages between the electrodes (Fig. 1.23) using the measured values

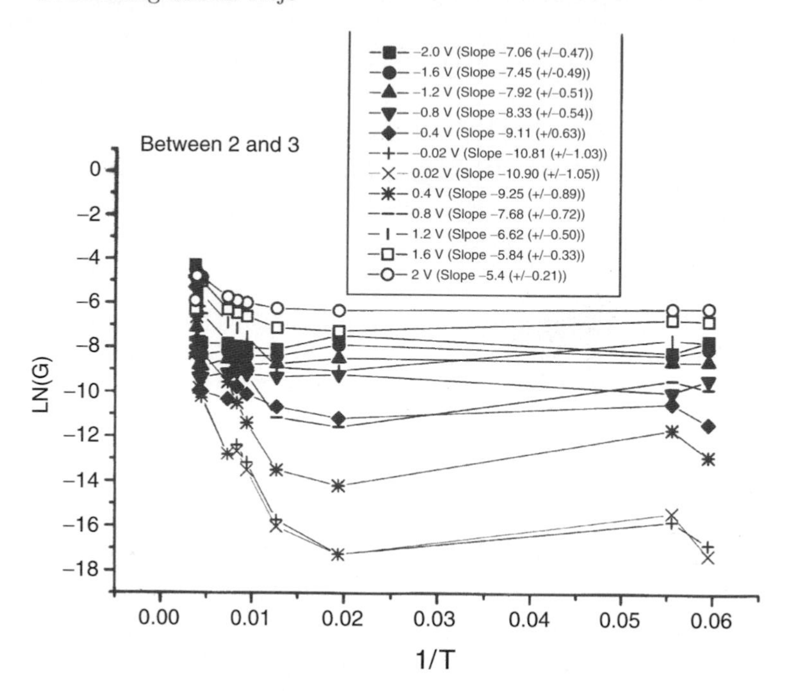

Fig. 1.24. Logarithmic differential conductance vs. $1/T$ as a function of the applied voltages between electrodes 2 and 3

(Fig. 1.22c) with additional data at other temperatures. The same was done for the other two configurations (Figs. 1.24, 1.25) The logarithmic conductance is almost constant when $1/T$ is large (below 50 K) and increases upon the increase of temperature with a strange sudden drop at the end. Such behavior is observed for all the applied voltages. Approximately, the smaller the applied voltage is, the greater the conductance varies across the temperature range. The nature of such behavior can be expressed as follows for a zero-bias voltage:

$$J_{\mathrm{sm}} = AT^2 \exp\left(- \frac{e\phi_{\mathrm{B}n}}{kT} \right), \tag{1.5}$$

where $A = 4\pi mek^2/h^3$ and $\phi_{\mathrm{B}n}$ is the Schottky barrier height. When the applied bias voltage is negative, current flows from semiconductor to metal, otherwise current flows from metal to semiconductor for the opposite polarity. Under a bias voltage of V, the current flux can be modified into the following form:

$$J_{\mathrm{ms}} = AT^2 \exp\left(- \frac{e\phi_{\mathrm{B}n}}{kT} \right) \exp\left(\frac{eV}{kT} \right) - 1. \tag{1.6}$$

The I–V curves are asymmetric between electrodes 1 and 2 and 2 and 3. Such a feature with a weak voltage dependence of current at the negative

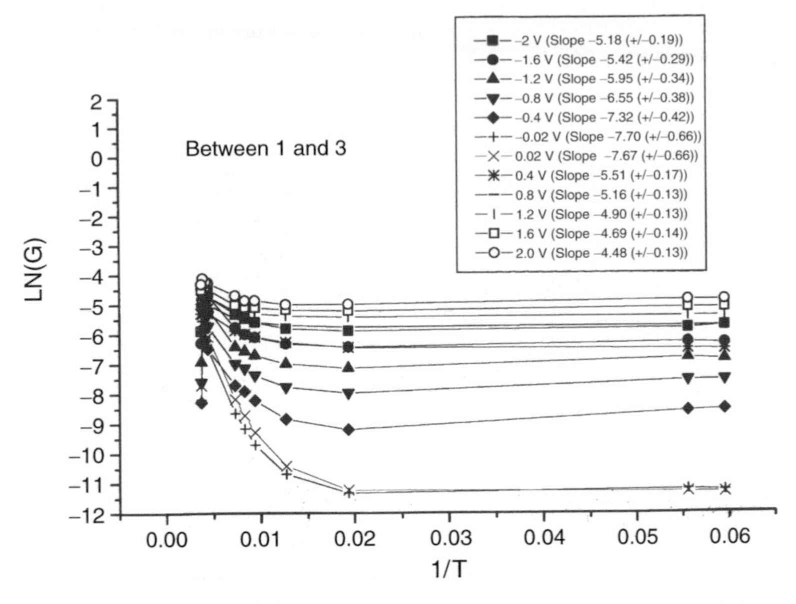

Fig. 1.25. Logarithmic differential conductance vs. $1/T$ as a function of the applied voltages between electrodes 1 and 3

polarity and a strong voltage dependence at the positive bias may be an indicator of the Schottky barrier that is formed between the lead layer and the Si(111) surface. However, between electrodes 1 and 3, the I–V curve is symmetric, and the current conduction behaves in the same way for both bias polarities. This does not agree with the previously given description of a Schottky barrier, and suggests that an Ohmiclike contact is formed at the interface of the electrodes with the sample. In other words, the barrier height is very low and the contact resistance is very small in comparison with that through the lead film. Therefore, the observed conductance feature is thought to arise from the conduction through the lead film, and is not a characteristic of the contacts. Nevertheless, the symmetric I–V curve for the conduction across the step edges needs to be studied further to clarify the effects of the step edges.

1.6 Transfer of Au Cusp Clusters on Si(111) 7×7 Surfaces from a Pure Au Tip of a Scanning Tunneling Microscope

1.6.1 Introduction

The possibility of transferring nanometer-sized Au clusters by applying voltage pulses between a Au tip and Au surfaces with a scanning tunneling microscope configuration was first demonstrated by Mamin et al. [150, 151].

It was recently confirmed by several groups with the same tip–sample configurations in air [152–154], in UHV [154] and in gases and a nonconducting liquid [155]. Recently, deposition of Au mounds from a gold tip onto Si(111)-7×7 surfaces was also performed in UHV by Köning et al. [156]. However, different mechanisms were proposed to explain the observed phenomena. The groups headed by Mamin [151] and Chang [154] attributed the Au-atom transfer to field evaporation. Chang et al. [154] further considered the heating of the tip apex by the field emission current to explain the sustained use of the tip for atom transfer, whereas Pascual et al. [152] concluded from their experiments that the creation of Au mounds on Au surfaces was the result of contact between the tip and sample surface. With regard to the question of whether a mound or a crater is created for a given voltage pulse, the experimental data are controversial [151–154]. The fact that better reproducibility of mound creation can be obtained in air than in UHV remains unexplained. It is also puzzling that the polarity of the voltage pulses has a much weaker effect on the creation of Au mounds in air than in UHV. The underlying mechanisms are still under debate [156–158]. Investigation of atom transfer in a clean UHV from a tip to an atomically well defined surface would be crucial for clarification and better understanding of the mechanisms, and thus better control of the STM technique in the nanostructure fabrications. In this chapter, we report an investigation in UHV of gold deposition onto well-known Si(111)-7×7 surfaces from a pure gold tip. Atomically resolved 7×7 STM images can be obtained with the same tip before and immediately after it was used to create Au mounds and craters. We focus on the effects of applying negative voltage pulses (levels and durations) to the tip on the creation of mounds and craters on Si(111) surfaces. All the mounds created appear as tiplike sharp protrusions. We demonstrate that Au mounds with similar sizes and shapes can be fabricated by tip–surface contact without applying a voltage pulse between the tip and surface, direct evidence of a nonfield evaporation mechanism. It is striking that the probability of mound creation reaches almost 70% for negative pulses of -4.8 to $-7.0\,$V with pulse durations from 10 to 100 ms. We find that repeated scans at low tunneling impedance over Au mounds can lead to strong modification, although they are stable upon subsequent in situ thermal annealing in UHV. Tip-enhanced diffusion of Au is proposed to explain such modifications. Furthermore, coalescence of Au mounds after thermal annealing is observed directly with STM.

1.6.2 Transfer of Au Cusp Clusters

Via voltage events, various voltage pulses were made between a pure Au tip and well-ordered Si(111)-7×7 terraces. Typical results made with 10-ms voltage pulses are displayed in Fig. 1.26. Only one pulse was made at each tip location. Parallel to the x-axis, three pulses are identical, whereas along the y-axis, from the bottom to the top (see inset), the pulse voltages are from

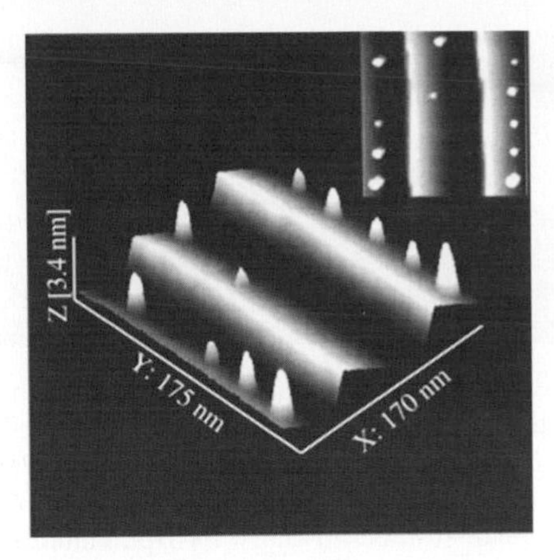

Fig. 1.26. Constant-current topograph (*CCT*) images ($V = 1.65\,\mathrm{V}$, $I = 0.3\,\mathrm{nA}$) taken after a series of 10-ms voltage pulses [150]

−5.0 to −6.0 V, with −0.2-V increment. The 3D image in Fig. 1.26 clearly reveals that all the mounds created appear as tiplike sharp protrusions. With the same tip, similar results were obtained by applying much longer (10-ms) voltage pulses. For a voltage event, no modifications on the Si(111)-7×7 surface were observed when the negative pulses was smaller than 4.4 V; therefore, for the gold tip on the Si(111) surface, the threshold negative voltage pulse for Au tip deposition in UHV is close to 4.6 V. It is slightly higher than the value obtained for a Au tip on a Au surface in air [151]. It is remarkable that above the voltage threshold, the average probability of mound creation almost reaches 70%. If craters are included, the probability of surface modification is almost 100%. The lengths at the base of mounds are in the range 5–20 nm, with typical mound heights of 2 nm, about 5 times the bulk gold lattice constant. With increasing pulse voltage, there is a tendency for accompanying craters or damage on the Si(111) surface, which appears as dark spots on the images. A change in the pulse duration by 3 orders of magnitude has hardly any effect on the threshold voltage and the probability of mound creation. There is no apparent dependence of the size of the mounds on the pulse voltages; however, the ratios of the height to the base length of the mounds is very close to 0.1. It is interesting to note that the shorter voltage pulses seem to create smaller and more uniform mounds. We observed that positive pulses usually give much less reproducible mound deposition and the size of the mounds is hardly controllable. The deposited mounds presumably are Au clusters, as similar results were obtained by using Au-coated tungsten tips. It is striking that the Au tip can well sustain itself for mound deposition after hundreds of pulses. The tip is very stable and sharp for obtaining atomically resolved images. This is illustrated in Fig. 1.27. The images were acquired after a 10-ms, −5.5-V pulse (generating a crater and several small clusters, Fig. 1.27a) and a 10-ms, −5.2-V pulse (generating a single mound,

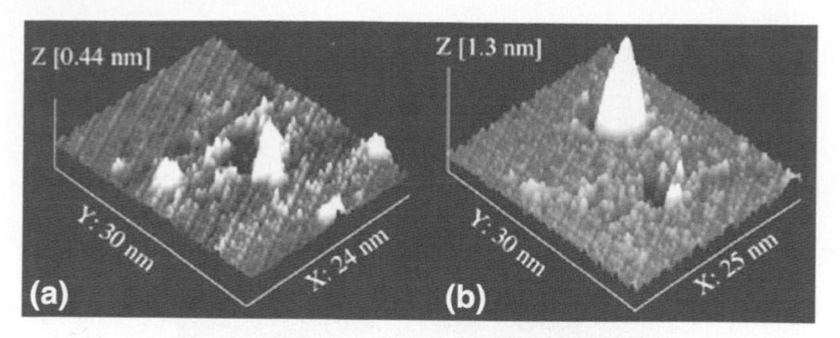

Fig. 1.27. CCT images ($V = 1.5\,\mathrm{V}$, $I = 0.5\,\mathrm{nA}$) taken immediately after two voltage pulses ranging from -5.0 to $-6.0\,\mathrm{V}$ (from bottom to top in the 10-ms voltage pulses of (**a**) $-5.5\,\mathrm{V}$ and (**b**) $-5.2\,\mathrm{V}$, which create a crater) had been applied between the Au tip and Si(111) surface. Very sharp tip and a mound, respectively. After pulsing, the tip is still stable and sharp for tiplike mounds are seen in the 3D atomically resolved images of Si(111)-7×7 terraces [150]

Fig. 1.27b, next to the crater). Sharp 7×7 reconstructed features, even in the area between the mound and the crater that are only approximately 5 nm apart (Fig. 1.27b), are resolved. These observations suggest that the tip pulses create only local modifications on the surface. Comparisons of Fig. 1.27a and b also reveal evolution of the crater and its accompanying clusters, and strong recovery of the 7×7 reconstruction. We take this as evidence that the clusters accompanied by the crater in Fig. 1.27a mainly consist of Si atoms. The tip-deposited mounds are not stable if scanned repeatedly by the scanning tunneling microscope with low tunneling impedance. The surface features before and after five successive scans (outlined part) at $V = 100\,\mathrm{mV}$, $I = 3\,\mathrm{nA}$ are illustrated in Fig. 1.28a and b. During those scans, we did not observe significant changes in the tip; therefore, the modifications are unlikely to be the result of scratching. One possible explanation would be the surface diffusion of Au atoms enhanced by the scanning tunneling microscope. In order to see whether the diffusion is thermally activated or not, Fig. 1.28c and d presents comparisons of Au mounds before and after thermal annealing in UHV at about 773 K. After the annealing, the main features remain unchanged, except that the sharp tiplike double Au mound in Fig. 1.28c has coalesced into a smooth, round, single mound (Fig. 1.28d). It is interesting to note that the height of the mounds is reduced by about 30% after annealing and coalescence, which is an indication of mound surface relaxation and possibly interdiffusion of Au into the Si substrate.

Therefore, we believe that the modification in Fig. 1.28b is the result of field-induced diffusion. In fact, strong tip-assisted Ag diffusion on Ag surfaces at room temperature has been reported by Li et al. [161]. Observations of fast Au diffusion on Au surfaces at room temperature [162] can also be attributed to a tip field-induced effect. Field evaporation alone cannot explain why it is easier to deposit Au mounds than to evaporate Si atoms on a Si surface when using voltage events, since the values of the critical fields for Au and Si are very close in both STM [159] and field ion microscopy [138]

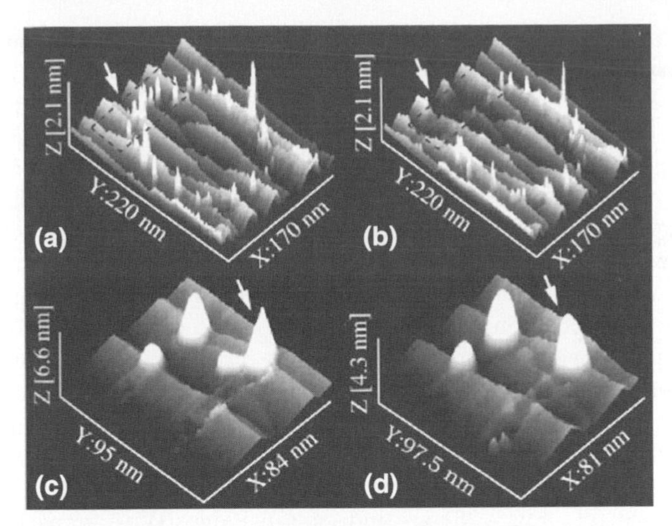

Fig. 1.28. Part of nanometer-scale patterns created on a Si(111)-7×7 surface by means of pure Au-tip deposition in a scanning tunneling microscope before modification (**a**) and after the outlined area in image **a** had been scanned five times at low tunneling impedance, i.e., $V = 100$ mV, $I = 3$ nA (**b**). As-created Au mounds on the Si(111) surface before annealing (**c**) and after annealing in UHV at about 773 K (**d**). Notice the coalescence of two Au mounds indicated with *arrows* in (**c**) occurred after thermal annealing. All CCT images were taken at $V = 1.55$ V and $I = 0.56$ nA [150]

regimes. It is also difficult to understand the sharp pointed shapes of the mounds. One would expect a smooth and round mound, as field-evaporated Au atoms will have enough energy to relax. Other field-induced effects, such as heating of the tip by an emission current, formation of protrusions [164], formation of a cusp-shaped liquid-metal cone owing to a field gradient [165] and mechanical deformation of the tip [166] can, of course, not be neglected.

Pascual et al. [152] offered a picture of nanometer-sized contact and neck formation to interpret the Au deposition on a gold surface. In our experiments using a voltage event (as in [154]), the Z piezo is forced to stay at the same position, because the feedback loop was turned off during the voltage pulsing. Thus, the distance between the tip and sample surface remained constant. As we noted earlier, the Au diffusion on a Au tip surface can be strongly enhanced by a high electric field. Such strong Au diffusion can make the apex of the gold tip behave like a liquid-metal source. Therefore, it is more likely that the initial contact is established through the formation of a cusp-shaped liquid-metal cone as proposed by Tsong [165]. The geometrical similarity of Au mounds observed by us and others is another argument. The contact between the tip and sample surface means that an abrupt, large transient current jump is expected to flow. This large current will melt the neck of the tip. The surface tension at both sides of the neck leads to further neck elongation and final disintegration. The ductile property of gold makes both broken ends of the neck very sharp. The deposited Au mounds thus appear as a sharp feature and, on the tip side, a sharp apex (maybe with single-atom sharpness) is formed. This argument is further supported by our observations that following the deposition of a tiplike sharp mound, atomically resolved images are often obtained. Therefore, the high reproducibility of Au-mound deposition and a very stable tip for atomic-resolution images immediately after pulsing can be attributed to such a tip self-shaping effect. This scenario is confirmed with tip and sample surface contact experiments. In order to eliminate possible field effects, both the tip bias voltage and the tunneling current were set to zero during the period (about 10 ms) of tip and sample surface contact. The mounds displayed in Fig. 1.29 were made by moving the tip from a tunneling distance (approximately 6 Å) toward the Si surface by a nominal step of 9 Å. Apparently, mounds created in this way and those with voltage pulses (cf., Fig. 1.26) are indistinguishable. With steps larger than 9 Å, the mound-creation probability reaches as high as 90% in a Z event. We believe that the tip and sample surface contact experiment is clearly direct evidence of nonfield evaporation mechanisms. After tip and sample surface contact the tip usually lost its atomic resolution. However such a tip can often be recovered by deposition of mounds by using negative voltage pulses (−6 to −8 V) at the tip. This suggests that the electric field at the tip end plays a role in shaping the tip. In the previous description, when there is no contact between the tip and the sample surface, no mound is expected. However, with a voltage event, the sample surface can still feel the strong field. Field evaporation from a substrate surface with a low threshold field is thus possible. Small clusters are also expected to accompany the crater, as seen in Fig. 1.27.

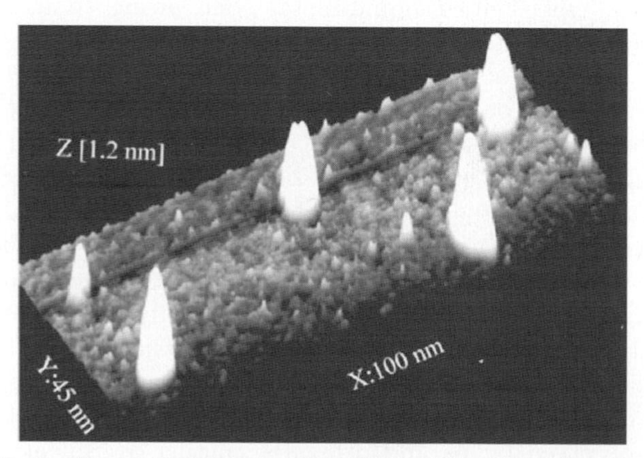

Fig. 1.29. 3D display of CCT image (2.0 V, 0.5 nA) of mounds created by direct tip surface contact. Both the tip bias and the tunneling current were set to zero during the period of contact [150]

1.7 Submicrometer Transmission Mask Fabrication

1.7.1 Introduction

As technology advances, people now desire to manipulate the nanometer-scale structure and even to design atomic- and molecular-scale arrays of ordered structures for information storage or small electrical devices and circuits [169–171]. Atomic manipulation with the use of a scanning tunneling microscope opens up the possibility of being able to construct artificial molecules, even atomic-scale electronic circuits through which current flows by electrons hopping between adjacent atoms. The scanning tunneling microscope can also be used to deposit metallic islands with dimensions of the order of a few tens of nanometers by a process akin to field emission. The technique has been demonstrated by Mamim et al. [172], who deposited gold mounds on a gold substrate. After the pulse a gold mound approximately 10 nm in diameter and 2-nm high was deposited beneath the scanning tunneling microscope gold tip. By repeating the process at different locations a pattern of a periodic 9316 array of dots was obtained. Recently, Fujita et al. [173] reported the fabrication of gold nanostructures on a vicinal Si(111)-7×7 surface by using a UHV scanning tunneling microscope. The results clearly suggest that the atom-transfer technique in UHV has proven to be a good candidate for fabricating nanometer-scale devices. Compared with manipulation of single atoms, the field emission approach is quite fast and the gold mounds would provide ideal charging islands for single-electron devices. The process of fabricating a single-electron transistor by field emission is shown schematically in Fig. 1.30. Although electron beam lithography and the lift-off process

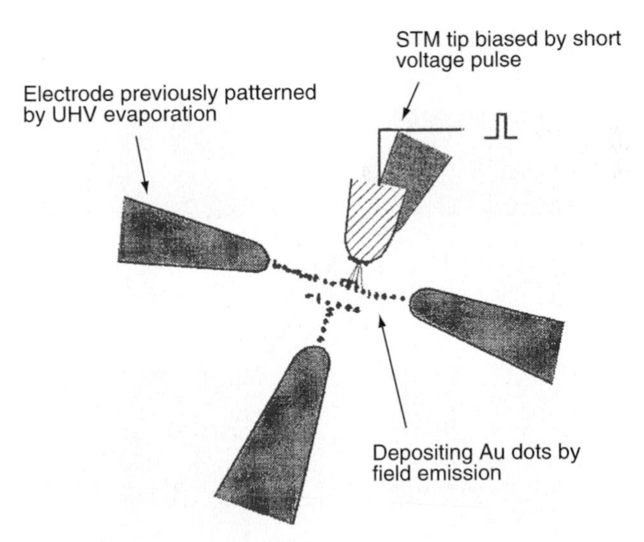

Fig. 1.30. A single-electron transistor fabricated by the field emission of gold islands from a scanning tunneling microscope tip [167]

can be used to realize the source, drain and gate patterns, after processing the substrate surface has been contaminated by the resist used for the electron beam or the optical lithography. In order to avoid contamination, all processes, including fabrication and characteristics measurements of the device, should be carried out in the UHV chamber, i.e., in situ process. In this case the submicrometer mask becomes essential for evaporating electrode patterns. On the other hand, for studying some physical phenomena such as single-electron transportation in single junctions and multi junctions, atomic clean semiconductor surfaces which should be obtained under UHV conditions can improve our understanding of electron transportation properties. For this, an evaporating method should be used to form contact patterns for these kinds of nanoscience and nanotechnology studies. Although a Si_3N_4 membrane mask can be made by electron beam lithography and reactive ion etching (RIE), a mask with a large area is difficult to fabricate because of the weakness of the Si_3N_4 membrane. Also, mask patterns, such as a long slit, cannot be made by the Si_3N_4 membrane method. In this chapter, we present a novel fabrication technique of a submicrometer-scale transmission mask. With use of this technique, an electrode pattern as large as $2.5 \times 2.5\,\mathrm{mm}^2$ with a gap as small as 0.8 mm has been made on a Si substrate.

1.7.2 Procedure for Fabrication

In order to use a four-probe test facility to measure dc and ac $I-V$ properties of the device fabricated by in situ UHV processes, the mask pattern

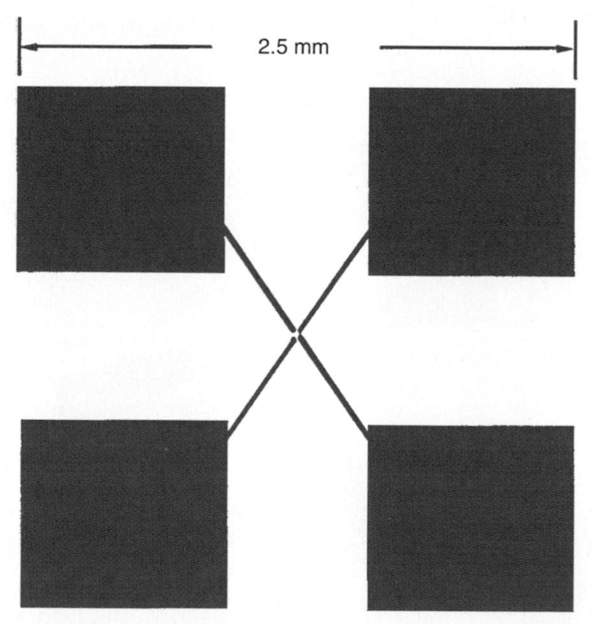

Fig. 1.31. Mask pattern for electron beam lithography [167]

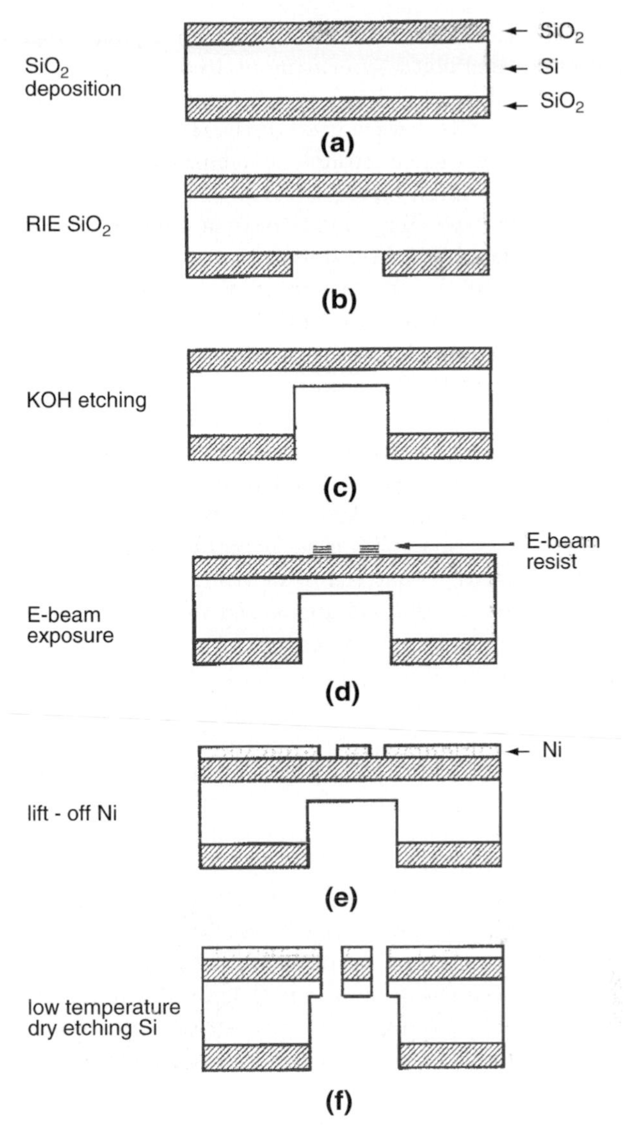

Fig. 1.32. Fabrication sequence of the transmission mask [167]

size was designed as $2.5 \times 2.5\ \text{mm}^2$ as shown in Fig. 1.31. The procedure for fabricating the Si transmission mask is presented in Fig. 1.32. The fabrication sequence is as follows. First, a two-side polished 250-mm-thick Si wafer was prepared. Then, 1 mm of SiO_2 was deposited on the front and back sides of the wafer, respectively (Fig. 1.32a). The wafer was cut to the required sizes, 4.5×8, 4.5×4.5 and $4 \times 7\ \text{mm}^2$. The sample was etched by RIE using SF_6 plasma to open a 3×3-mm^2 window in the SiO_2 surface of the back side

(Fig. 1.32b). In this case a metal mask was used to open the window. The RIE has a pair of parallel plates operating at 13.56-MHz radiofrequency (rf). SF_6 with 30-sccm gas flow, O_2 with 3-sccm gas flow, a rf power of 50 W and a chamber pressure of 50 mTorr were used during the etching process. The flow rate of the etching gases was controlled by a mass-flow controller. The etching time was 2 h and 15 min. The etch rates were about 11 and 160 nm/min for SiO_2 and Si, respectively. To ensure that the SiO_2 was completely etched away, overetching is necessary. In order to transfer the SiO_2 window pattern onto the underlying Si substrate, the sample was wet-etched in KOH solution (Fig. 1.32c) with 56% water and 44% KOH by weight for about 5.5 h at 333 K.

The Si etch rate was about 0.7 mm/min and the SiO_2 etch rate was 0.2 nm/min. After the KOH wet-etching process, Si about 20-mm thick was left in the window portion. A high-performance positive electron beam resist was used to image the electron beam exposure and to transfer the pattern into an underlying SiO_2 layer on the front side (Fig. 1.32d). The sample with a 0.3-mm-thick resist layer was baked at 443 K for 1.5 min. Exposure using a 30-keV electron beam was conducted using a modified scanning electron microscope equipped with a pattern generator. To determine the absolute electron dose, the beam current was measured using a Faraday cup attached to the sample holder. After exposure, the resist was developed in a developer, for 3 min at 295 K and rinsed for a few seconds. Then, a 17-nm-thick Ni layer was deposited at a deposition rate of 0.1 nm/s onto the front side by an electron beam evaporator at a pressure of 8×10^{-7} Torr. The sample was immersed in N,N-dimethylformamide for 30 min at room temperature for lift-off and was baked at 423 K for 20 min in order to increase the adhesion between the Ni and SiO_2 (Fig. 1.32e). The next step was to etch through the SiO_2 and Si parts which were uncovered with Ni using RIE (Fig. 1.32f). The etching was carried out at low temperature using liquid nitrogen. During etching, the sample temperature was about 73 K. SF_6 and O_2 of 30 and 3 sccm, respectively, and a rf power of 50 W were used for the etch. It took about 4 h to etch completely away about 0.4 mm of SiO_2 and 20 mm for Si. The last step was to mill the opaque part of the mask from 15 mm to a submicrometer level by using the focused ion beam (FIB) technique with a Ga ion source, 30-keV accelerating voltage and a 200-mm aperture size.

1.7.3 Submicrometer Transmission Mask

A finished mask is shown in Fig. 1.33. Figure 1.33a shows a SEM micrograph of the mask finished by the process used in Fig. 1.32f. Figure 1.33b shows the mask milled by a FIB. A high-magnification SEM micrograph of Fig. 1.33b is presented in Fig. 1.33c in which the about 0.8 mm opaque part between neighboring slits and the about 1.5 mm opaque part between opposite slits are very clear. We applied the Si transmission mask to evaporate a Au film pattern on the Si surface. A picture of a 30-nm-thick Au pattern evaporated on the Si substrate through the mask is shown in Fig. 1.33d in which the

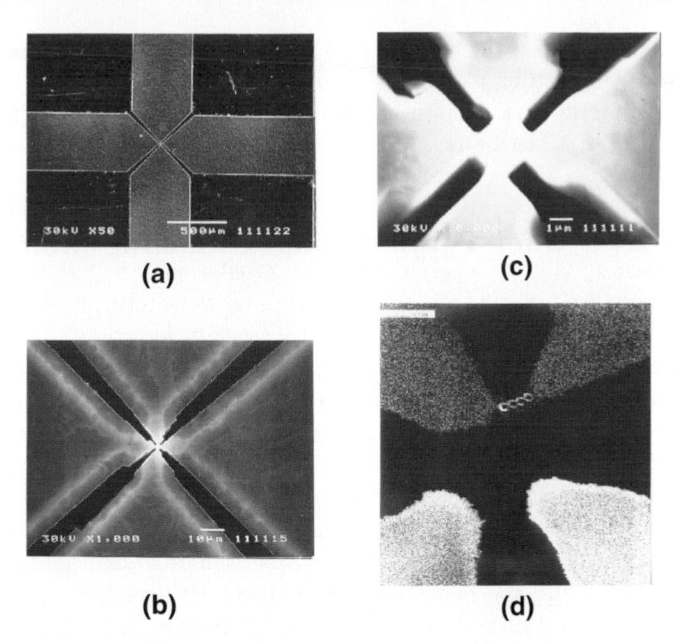

Fig. 1.33. Silicon transmission mask. (**a**) After low-temperature reactive ion etching. (**b**) After focused ion beam (*FIB*) milling. (**c**) Fine structure of the opaque part. (**d**) The 30-nm-thick Au pattern evaporated on the Si substrate through the Si transmission mask [167]

nanometer structures were fabricated in the gap between the electrodes. It is necessary to discuss the low-temperature RIE process further. Although SF_6 and O_2 gas mixtures in RIE plasma have been shown to yield high etch rates and anisotropic etching for Si, in order to satisfy anisotropy, a high etch rate and high selectivity at the same time, cryogenic deep RIE of silicon using SF_6 and O_2 gas has been studied [163, 176]. In our experiment, the better anisotropic etching profiles were obtained by adding O_2 to the etching gas. It is believed that oxygen contributes to sidewall passivation, therefore reducing the lateral etching. On the other hand, the etching rate can be increased by adding a few percent of oxygen to the etching gas. During the RIE process, the etching rate of Si was about 160 nm/min, which is much lower than in a previous study. One of the reasons may be attributed to the load effect because of the large etched area, i.e., the etching rate is the inversely proportional to the quantity of material being etched. The etching rate ratio for Si/SiO_2 is about 14. The Si/Ni selectivity is as high as about 3000 since a low rf power was used. The Ni is mainly etched by physical bombardment of incident ions. The result indicates that etching with low ion energy is effective for good mask selectivity. A large-capacity turbomolecular pump is necessary to realize low pressure with a high flow rate of etching gas. Otherwise, during etching the chamber pressure will increase and cause the machine to shut down. For the mask with the opaque part smaller than 3 mm, it was easier to break it down during etching owing to the large stress around the opaque part. This is why we use the FIB technique to mill the opaque part from 15 to 1 mm.

1.8 An UHV Dual-Tip Scanning Tunneling Microscope

1.8.1 Introduction

As electronic devices get smaller and smaller, it has become increasingly urgent to understand electronic properties of materials and structures at a nanometer scale [177]. The advent of the scanning tunneling microscope has lent us a powerful tool to investigate issues such as local electronic properties, together with UHV surface science, which enables us to create atomically well defined objects on surfaces. For instance, there have been studies of the single-atom rectification effect [178] and bistable I–V characteristics [179] by STM, establishing the fact that useful device properties can be realized in nanometer scale. In all these studies, the scanning tunneling microscope tip and sample form a two-terminal configuration like a diode. On the other hand, widespread active devices such as transistors and field-effect transistors have three terminals, which is essential for signal amplification and switching. To extend our investigation of nanoscale devices to a three-terminal configuration, it is necessary to place at least three terminals in proximity. A natural way to achieve this goal, besides lithographic methods [180,181], is to develop a dual-tip scanning tunneling microscope (DSTM) [182,183], whereby the two scanning tunneling microscope tips serve as two terminals while the sample serves as the third. The two tips can be manipulated independently and positioned as close as one desires limited only by their radius of curvature. It may appear that a DSTM is no more than an integration of two sets of scanning tunneling microscopes in one system. It turns out that the manipulation of the two tips in a safe and well-controlled manner is a serious challenge. The difficulty is compounded when the device is to be operated at cryogenic temperature, for which quantum properties of electrons are likely to be observed. In this chapter we describe the design and operation of our DSTM, which resides in a liquid helium cryostat and is integrated with an UHV specimen preparation and characterization chamber. There have been several theoretical proposals of using the cotunneling effect to obtain information on surface band structures [184, 185], the electronic structure of high-Tc superconductors [186] and shot-noise correlation in small coherent conductors [187]. The successful development of our DSTM should open new possibilities for testing these ideas as well as for three-terminal experiments at nanometer scale.

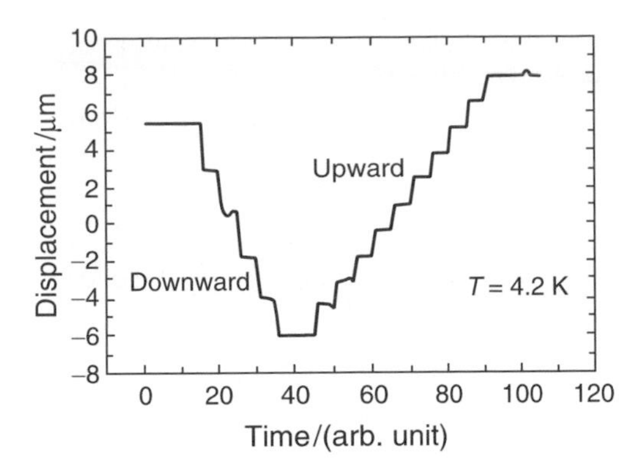

Fig. 1.34. The stepping motion curve of a stepper detected by a capacitance sensor at 4.2 K. Each step corresponds to 30 stick-and-slip motions [177]

1.8.2 Inertial Stepper Performance

Figure 1.34 shows helium-temperature performance of an isolated small stepper moving along the vertical direction (the y stepper) measured with the embedded position sensor. Although downward stepping is slightly faster than upward stepping (against gravity) reliable up/down motion is clearly demonstrated at 4.2 K. Each staircase-like feature corresponds to 30 steps. At room temperature each single step is detectable by the position sensor. The stepping efficiency is reduced by a factor of 20 at liquid helium temperature when compared with the room-temperature data.

1.8.3 Single-Tip Scanning Tunneling Microscope Performance

Figure 1.35 shows an image taken by the DSTM at room temperature. Despite the complexity of the DSTM and hence the compromised mechanical rigidity, a clear Si(111)-7×7 structure is seen. Remarkably, the mechanical drift of the images is only 2 Å/min even at room temperature, indicating the high stability of the DSTM mechanical structure. Mechanical resonant frequencies were measured to be approximately 200 Hz for both tips; therefore, it is better to the choose scanning speed in accordance with the size of the surface objects. For instance, the scanning speed was 0.8 Å/mS to obtain the atomic image in Fig. 1.35. For a large-area scan (several microns), we usually scan at a speed of 20 Å/mS. The resonant frequency was obtained by raising the feedback gain to generate signal oscillation. It is found that the oscillation frequency obtained this way depends on tip status and the preamplifier. In any case, the system resonant frequencies are not as high as those of a conventional scanning tunneling microscope but are nevertheless quite satisfactory considering the large number of degrees of freedom inherent in the DSTM.

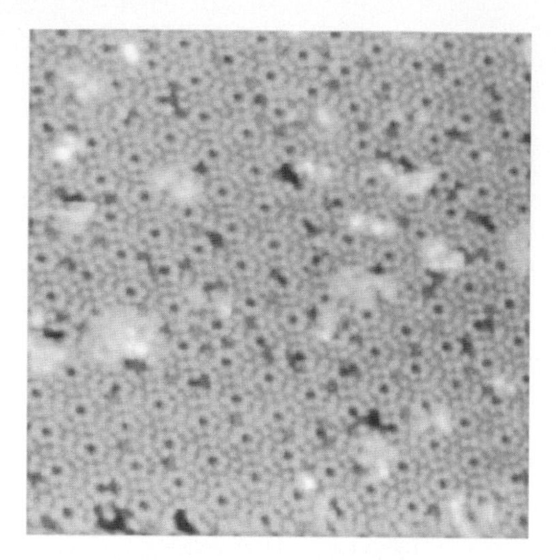

Fig. 1.35. An example of a Si(111)-7×7 image (tip voltage 2.2 V at 1 nA taken by one of the dual-tip scanning tunneling microscope systems) [177]

1.8.4 STM Performance at Low Temperatures

The helium-temperature performance is confirmed by using a Au/mica sample. Since the sample is on the stepper X which gives rise to a weak thermal contact to the helium bath, cooling the sample from room temperature to the liquid nitrogen temperature takes approximately 8 h, even though flexible braids are used to join the helium bath with the sample receptacle. But cooling from liquid nitrogen to liquid helium temperature takes just approximately 90 min as the heat capacity goes down as approximately T^3. The entire DSTM reaches precisely 4.2 K thanks to the enclosure of the scanning tunneling microscope inside the 4.2-K radiation shield.

1.8.5 DSTM Performance

Figure 1.36 shows a set of images of the same area on a Si surface obtained by the two tips, respectively. Prior to the insertion of the DSTM into the UHV chamber, the two tips were preset to a separation of a few millimeters to avoid crashing during handling. After reaching UHV pressure, the two tips were then set to a distance of 0.3 mm with the help of a small optical telescope approximately 0.7 m away. Finally, these tips were brought to within their scanning range following two iterations of the previously described alignment process using three mutually nonparallel planes. From here on, mutual tip adjustment is not necessary in principle unless the operating temperature is changed. In the rare event of losing the mutual tip position, as once occurred by a mechanical shock owing to the sample introduction onto the scanning

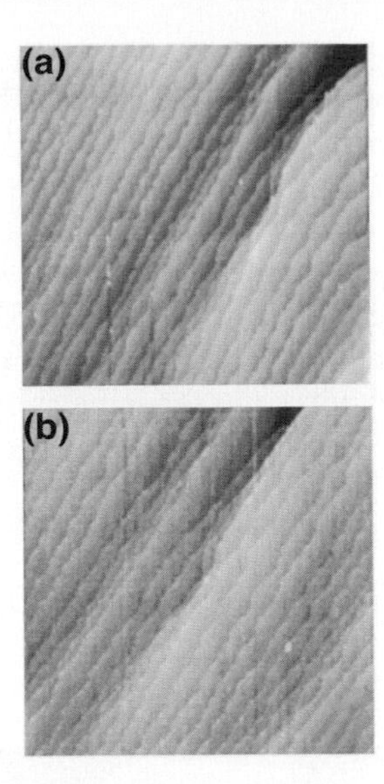

Fig. 1.36. Overlapped images of a Si(111)-7×7 surface taken by dual tips. Each tip approached each plane twice before taking this image. The image area is $2 \times 2\,\mathrm{mm}^2$ [177]

tunneling microscope unit, we can quickly readjust the relative tip positions thanks to the capacitive position sensor.

1.9 Fabrication and Lateral Electronic Transport Measurements of Au Nanowires

An important goal of microelectronics is the production of nanoscale devices [196]. Despite the success of atomic manipulation [197], lithographic methods [198, 199] and carbon nanotubes [200], problems of nanostructuring still persist or are generated when entering the nanoscale [201]. Part of the difficulty in fabricating artificial structures in nanometer scale is how to connect the nanostructures laterally to the outer world, so that their electrical properties can be measured and the device activated. Very promising toward solving the problems posed by nanowiring are the methods of self-assembly, where physicists and chemists have found a common research field and mutual progress has been made [109,202]. However, self-assembly growth of nanostructures is hard to control and is, as yet, barely used to produce integrated circuits. In this chapter we present a method to fabricate continuous gold wires with a width in the atomic range and a length of a few micrometers. The method is based on atomic force microscope technology using a commercially available piezoresistive cantilever [204–206] coated with gold. To measure the electronic properties, these wires are connected to large electrodes previously lithographically produced on the sample, providing the macroscopic pads for the four-point-probe electrodes. The ease of fabrication and the absolute shape control of such atomically thin metal wires are the main issue of this section and suggest that this method could play an important role in the production of the next generation of microprocessors.

The gold nanowire fabrication technique is rather simple. The tip of a piezoresistive cantilever is coated with 50 nm of gold (Fig. 1.37) using an electron beam evaporator technique. After inserting the cantilever into the atomic force microscope head, gold wires are made by simply bringing the cantilever in contact with the sample and drawing the wires by moving the tip on the surface of the sample. At room temperature, the Au migrates from the tip to the Si(111) surface (Fig. 1.38) without heating the cantilever. The wires are written at an average tip speed of 10 nm/s. After this procedure, the surface is imaged in noncontact AFM (nc-AFM) [207] mode using the very same cantilever.

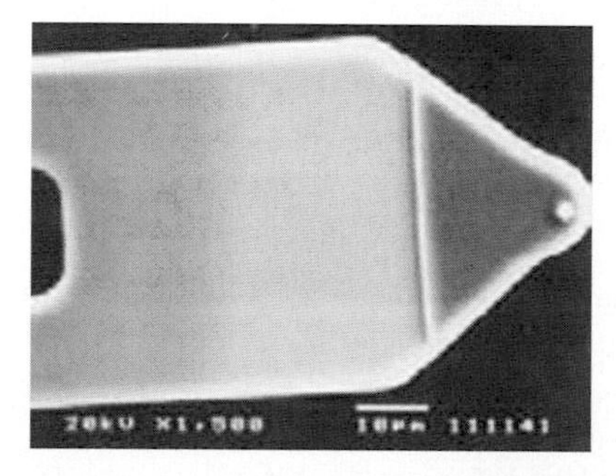

Fig. 1.37. Scanning electron microscope image of a Au-coated cantilever

Fig. 1.38. A 70 $\times$ 70-nm^2 STM image of a Au nanowire on Si(111) at atomic resolution

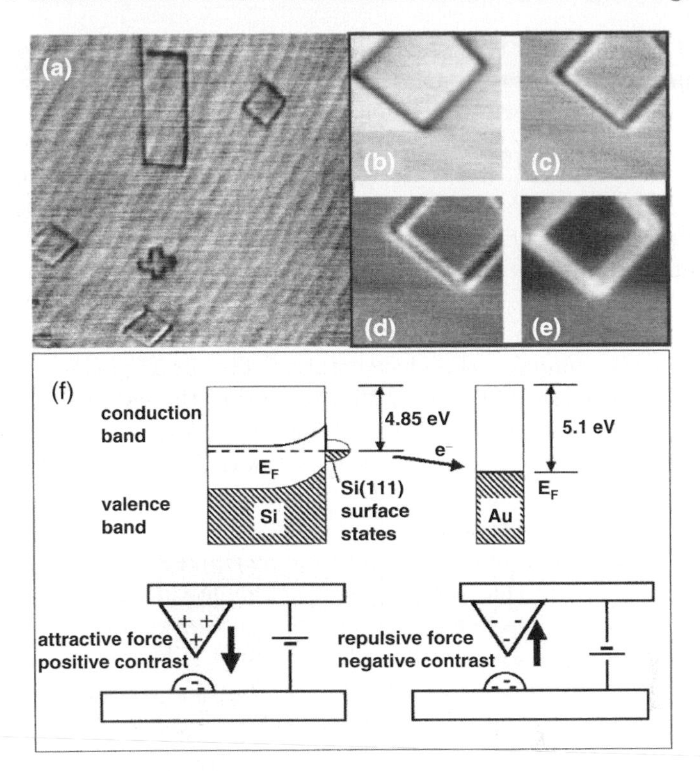

Fig. 1.39. (**a**) A 4-mm×4-mm noncontact atomic force microscopy (*nc-AFM*) image of various Au wires on Si(111). (**b**)–(**e**) Series of nc-AFM images (1 mm×1 mm) of the same gold wire taken with different tip bias voltages. The contrast changes by changing the tip voltage from negative (*black*) to positive (*white*): (**b**) $V_{\mathrm{tip}} = 21\,\mathrm{V}$, (**c**) $V_{\mathrm{tip}} = 0\,\mathrm{V}$, (**d**) $V_{\mathrm{tip}} = 1\,\mathrm{V}$, (**e**) $V_{\mathrm{tip}} = 1.5\,\mathrm{V}$. (**f**) The phenomenon of contrast change is explained by means of a drawing of the energy-band diagrams and the setup of nc-AFM. The different work functions of the Si(111) surface state and gold cause a negative charge transfer to the gold wires. If the cantilever tip is positively charged, the local electrostatic force between the tip and the gold wire produces a positive contrast in the nc-AFM image and vice versa [196]

Figure 1.39a shows a nc-AFM image of gold wires of different shape on Si(111) drawn with this technique. The negative contrast is caused by the different work functions of Si(111) and Au and depends on the tip bias field. A series of nc-AFM images of a square Au nanowire taken with different bias voltages are shown in Fig. 1.39b–e, clearly revealing the connection between contrast change and tip bias voltage. The transition of negative to positive contrast occurs at a bias voltage of roughly 10.5 V. This phenomenon can be explained by the energy-band diagram shown in Fig. 1.39f. A Fermi level difference of 0.25 eV causes an electron flow from the substrate via the surface states of Si(111)-7×7 [208, 209] to the gold wire until the Fermi levels reach a state of equilibrium. This results in a local negative charge on the Au wire. Depending on the applied tip bias field, the force between the cantilever tip and the Au wire is either attractive to a positively or repulsive to a negatively charged tip. Therefore, the gold wire appears with a positive or negative contrast, respectively. The contrast change produced by the changing tip bias voltage also indicates that the written nanowire is indeed a metal: semiconducting or insulating materials would not cause such a contrast change. This is proved by scratching an uncoated cantilever on a clean Si(111) surface as shown in Fig. 1.40a. A second pattern is drawn on the left-hand side of the scratched pattern, but is made with a gold-coated cantilever in order to directly compare the electrostatic behavior of these patterns. By changing the tip bias voltage from 13.5 to 23.5 V, the gold pattern changes the contrast, whereas the scratch maintains its positive contrast. The image shown in Fig. 1.40a was taken with a bias voltage of 23.5 V; hence, the Au nanowire appears black and the scratched pattern white. Several experiments were performed in order to reduce the lateral size of the wires. Basically, the line width is determined by the shape and sharpness of the cantilever tip. It does not depend on the contact force, i.e., deflection of the cantilever as shown in Fig. 1.40b. The fabrication speed is also negligible with regard to line width provided the speed is less than 50 nm/s. Higher speed could result in disconnected wires and/or huge gold dots at step edges. The amount of gold deposited can be controlled by resistively heating the piezoresistive cantilever, but has the undesirable side effect of thermal drift. At the lower limit, a gold wire of 4 nm in width was produced using this technique of wire fab-

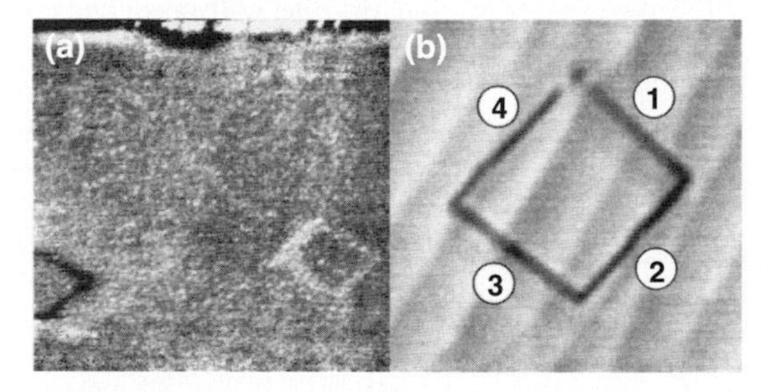

Fig. 1.40. (**a**) A nc-AFM image (3 mm × 3 mm) of two different types of nanowires on a Si(111) surface taken at a bias voltage of $V_{\text{bias}} = 23.5$ V. The wire on the *left* was fabricated with a gold-coated cantilever and shows a contrast change while changing from positive (*white lines* to negative (*black lines*) bias voltage, whereas the wire on the *right* was produced with a noncoated cantilever. The image contrast of this second wire is not affected by changing the bias voltage; it always remains as a white contrast to the surface. (**b**) A 1-mm ×1-mm nc-AFM image of a gold wire tests the influence of contact force on the line width. For each section (*1–4*) the contact force is increased by 50% starting with section *1*. No systematic variation of the line shape is observed, it seems to depend more on the tip shape than on the contact force [196]

rication. To measure the conductance of these nanowires, large macroscopic joint electrodes are needed to connect the nanowires to the contact electrodes of the four-point probe.

The four joint electrodes consist of large, millimeter-sized pads (contacts for the four-point probe) and 3-mm-wide probing electrodes. The ends of the probing electrodes are finally connected through the nanowire in order to measure the electrical transport properties of the wire. Molecular beam epitaxy with a through-hole mask placed between the sample and the evaporation source was used to fabricate the joint electrodes on the Si(111) surface (Fig. 1.43) [210]. Commercial FIB processing was used to produce the through-hole mask made of 3-mm-thick titanium foil (Figs. 1.41, 1.42). Once the joint electrodes have evaporated (Fig. 1.44) the sample is transferred to the AFM/STM stage where the electrodes are connected through AFM-produced Au nanowires. Figure 1.45a shows an AFM image of the previously evaporated Ag electrodes with gap distances of 5 and 10 mm,

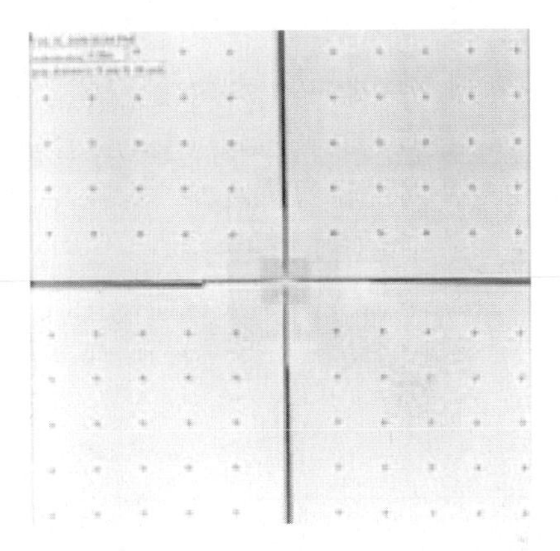

Fig. 1.41. FIB image of a through-hole mask. Electrodes and markers are visible

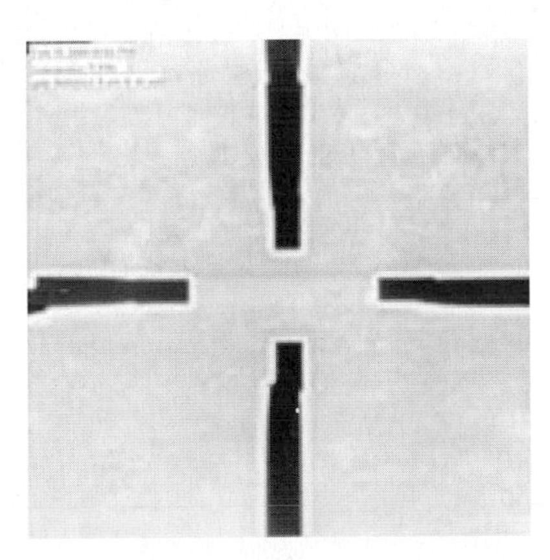

Fig. 1.42. FIB image of the gap region. Gap distance 5 and 10 μm

Fig. 1.43. 1000-nm × 1000-nm STM image of an evaporated Ag electrode on a Si(111) surface. During evaporation the mask was in contact with the sample. This results in dispersion of only a few nanometers

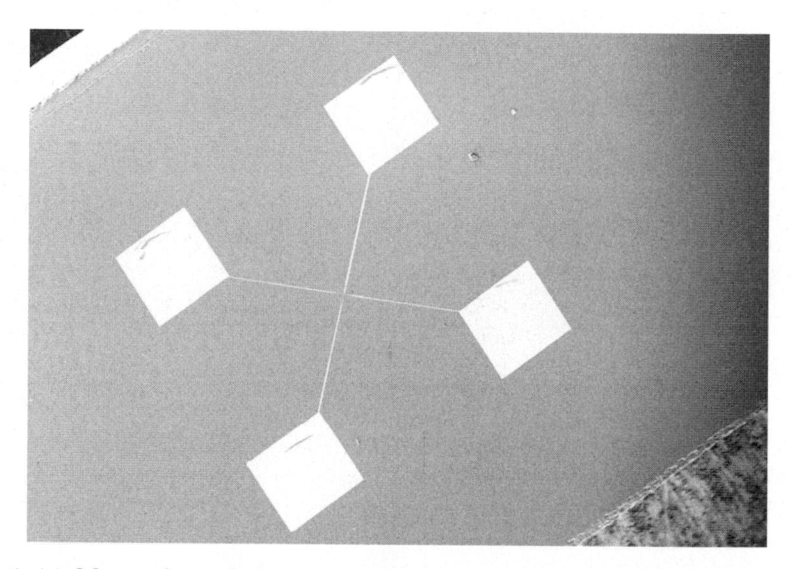

Fig. 1.44. Mounted on a helium cryostat (4 K at sample) Pad size (on the sample) $250 \times 250\,\mathrm{mm}^2$. Gap distance 3 mm

respectively, before the electrodes were connected through the nanowires. The thicknesses of the evaporated Ag electrodes were chosen between 5 and 10 nm in order to ensure proper contact to the electrodes of the four-point probe. Subsequently, electrodes 3 and 4 were connected through a Au wire using the very same Au-coated cantilever in AFM contact mode. A STM image of a fraction of the wire is shown in Fig. 1.45b and in detail in Figs. 1.46 and 1.47. The electrodes of the four-point probe approached the sample elec-

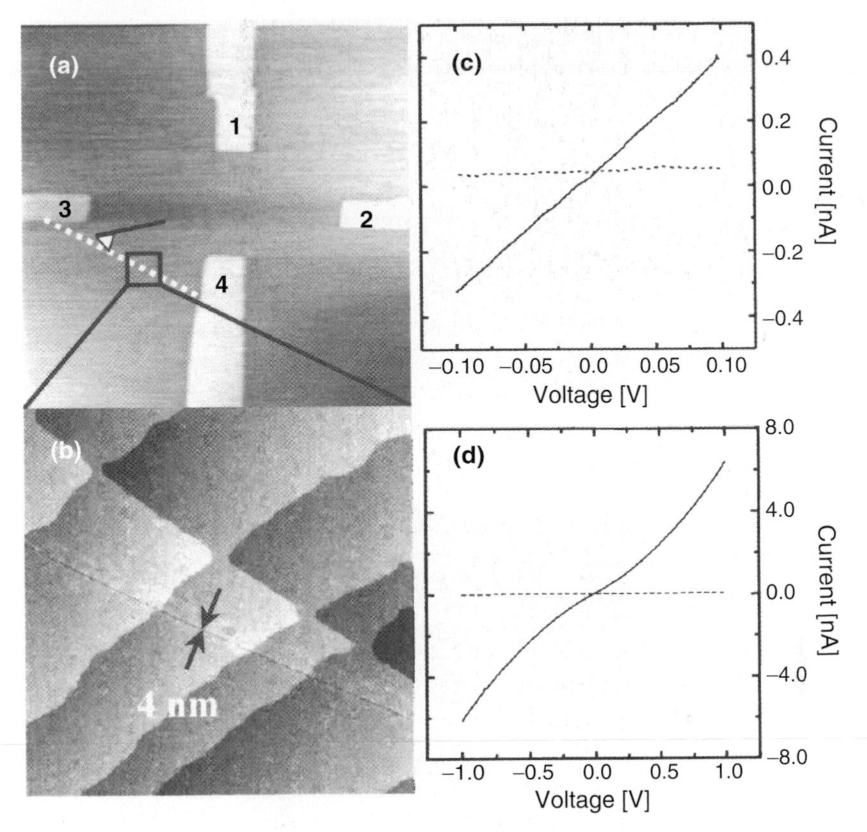

Fig. 1.45. (**a**) A 16-mm × 16-mm nc-AFM image of the evaporated Ag electrodes. The gap distances between the electrodes are 5 and 10 mm, respectively. Electrodes 3 and 4 are connected through a 4-nm wide Au wire (indicated by a *dashed line*), imaged in the 1-mm × 31-mm STM picture (**b**) Four-point-probe data taken at liquid helium temperature are shown in (**c**) and (**d**). Bias voltage I–V curve between 6100 mV for the connected electrodes 3 and 4 (**c**, *solid line*) reveals metallic behavior. For comparison, the data for the unconnected electrodes 1 and 2 are given (**c**, *dashed line*). At larger voltages (1000 mV) the Au wire shows nonlinear behavior (**d**, *solid line*). Data from a second experiment showing the I–V curve between the unconnected electrodes 3 and 4 are also presented (**d**, *dashed line*) [196]

trodes with inertial sliders guaranteeing a soft crash in order to avoid damage to the evaporated electrodes on the sample. The stage of the four-point probe is mounted on a liquid helium cryostat and can be cooled to 4.2 K. Four-point probe conductance measurements of a nanowire at liquid helium temperature are shown in Fig. 1.45c and d. The line width of the 7-mm-long Au nanowire averages 4 nm, whereas the thickness amounts to less than 1 nm. Bias voltage I–V curves were taken simultaneously across nanowire-connected electrodes (electrodes 3 and 4) and unconnected electrodes (electrodes 1 and

Fig. 1.46. A 1000-nm × 1000-nm STM image of a Au line on Si(111) made by AFM and evaporated Ag electrodes

Fig. 1.47. A 300-nm × 300-nm STM image of the Ag electrode edge

2; Fig. 1.45c). Across electrodes 3 and 4, the I–V curve for an applied bias voltage of $V_{\text{bias}} = 100\,\text{mV}$ is linear and thus the electrical behavior of the nanowire is metallic. For the larger applied bias voltage $V_{\text{bias}} = 1000\,\text{mV}$ seen in Fig. 1.45d, the I–V curve shows nonlinearity, for which the following two possible explanations are given:

1. A phenomenon of quantized conductance, observed in atomic-sized metallic contacts, gives rise to the nonlinearity of the I–V characteristics [211–214]. If the nanowire has some strictures, where only a few conductance channels exist, then the observed nonlinear effects in the I–V characteristics are caused by the lack of sufficient conductance channels [211]. Because of the instability of wire fabrication on silver electrodes, these strictures are most likely located at the contacts between the gold nanowire and the silver electrodes.
2. More plausibly, the nonlinearity can be explained by electrical conductance via the silicon sample. With an initially low applied bias voltage,

the metallic gold wire is mainly responsible for the transport of electrons. Therefore, the I–V curve is linear. However, while increasing the applied bias voltage, the electrical conductance via the sample starts to be the dominant electrical transport carrier and thus the I–V characteristics show semiconducting properties.

For comparison, I–V measurements of unconnected electrodes are given. In Fig. 1.45c (dashed line) the data from the measurements across electrodes 1 and 2 are presented, revealing no conductance between the electrodes. The same flat I–V characteristic was also measured between electrodes 3 and 4 before nanowire connection was made, shown in Fig. 1.45d (dashed line). Consequently, the resistance of the Si sample between the electrodes is orders of magnitude larger than that between the connected electrodes. In other words, real electron transport across the gold nanowire was measured. Notice that the electrical resistivity of the nanowire is 1.5×10^{-4} Ωm (wire dimension 7000 nm × 4 nm × 1 nm), which is relatively high compared with that of bulk Au at 1 K(2.2×10^{-1} Ωm) [215]. Scattering of electrons at impurities in the gold wire might be a possible interpretation of this discrepancy, all the more so in the initial stage of Au growth, as gold is absorbed by the Si(111)-7×7 surface [216]. Nevertheless the wire indisputably shows metallic behavior. In conclusion, we have demonstrated a rather simple technique to fabricate truly metallic gold nanowires on a clean Si(111) surface. In combination with other lithographic methods, it could be used to create devices on the nanometer scale, such as magnetic sensors. Developments toward arrays of cantilevers [217] could result in industrial relevance of this fabrication method for the fabrication of nanochips.

1.10 Conclusion

The conclusions from each of the methods are summarized as follows:

1. Surface conductivity measurements of lead nanostructures onto Si(in)-(1×1):H templates. Fabrication of lead nanowires together with macroscopic pads by applying high voltage after lead deposition on a hydrogen passivated surface worked very well. We could recognize the pattern with lead well. On the other hand, the method of lead deposition of hydrogen depassivation has not worked well so far. The tentative reason for the latter result is that either lead sticked to both the passivated and depassivated regions or the desorption of the lead from the passivated region was not enough.
2. Fabrication of nanowires using the tip contact technology. The gold nanowire was successfully connected to the macroscopic pads. It was confirmed that the line surely consists of gold not the scratch on the surface. The control of the line width will be continued further.

3. Connecting nanowires to macroscopic pads by placing a series of dots extracted from a scanning tunneling microscope tip. The macroscopic pads on the substrate were successfully fabricated and can withstand any mechanical stress by the four-point probe. Also, the series of gold dots was successfully formed in the gap and the wire was connected to the pads. Further, electric conductance of all three nanowires should be measured using the four-point probes.

We have presented some methods to fabricate metallic nanowires and even atomic chains:

1. Long atomic-scale lines of indium and lead atoms were produced through the selection of a vicinal single-domain Si(100) surface as the substrate via the combination of self-assembly with atomic manipulation.
2. Gold nanowires were formed from serial gold nanodots on a Si(111) substrate by the controlled field desorption and tip indentation techniques. The latter z-pulse technique has a better control over the size of nanodots and the deposition probability, and hence is a potential means to bridge atomic-scale structures with macroscopic pads.
3. Gold nanowires were fabricated by transfer from the atomic force microscope cantilever tip to substrate surface. This technique is rather simple minded but straightforward, since the transfer probability is relatively controllable.

Also in situ confimation is possible, using the same cantilever to fabricate the nanowires. This has the great advantage that we can eliminate the difficulty of finding the positions where the nanowires were fabricated beforehand. On the other hand, we have developed the pattern transfer techniques for four-probe measurements through either in situ UHV evaporation of metal atoms through a closely attached shadow mask or conventional electron beam lithography and a lift-off process. The technique and pattern structure can be selected and designed to meet different experimental requirements. In addition, electrical measurements on the ML lead film on Si(111) and the 1.3-ML lead film on Si(100) suggest that the contact between the pad and surface is Ohmiclike with a low Schottky barrier height. The former is semiconducting, whereas the latter is metallic in character.

Direct measurements of the intrinsic resistivity of nanowires and even atomic chains still remain as a challenging task because of the technical difficulty in connecting the macroscopic electrodes to such ultrafine structures. To avoid possible ambiguities owing to poor contacts, four-probe measurements are generally required.

Unfortunately, it is not easy at all to position four probes onto a nanowire using the present technology. When the contact between the wire and the electrode is not perfect or the wire itself is not defect-free, two or more tunneling junctions are formed, and thus the effect one measures is usually the single-electron tunneling featured by Coulomb blockade or staircase. The technical issues for staying away from this problem involve (1) how to reduce the coupling between the nanostructure and the dissipative external measuring device and (2) how to suppress thermal smearing as well as $1/f$ voltage (or charge) noise. Even if good contacts are assumed, there are other factors to consider, e.g., (1) the mechanical strength of the nanowire upon a small static discharge and current passing (current-carrying capability), both of which could result in fatal damage or melting of the structure, and (2) the separation of the signal from the supposed-to-be insulating substrate. In other words, how do we make sure that transport is through the nanowire?

Nanotechnology is certainly coming of age. The progress and future of the field depend on, firstly, how well and how quickly one can fabricate well-defined nanostructures in an ordered array via self- assembly and artificial structuring, secondly, how precisely and how rapidly one can measure the electronic properties of these nanostructures and, finally, how nanostructures such as nanodots, wires and sheets can be engineered into an integrated nanocomponent. A lot more work remains to be done before we can integrate different nanostructures into a nanodevice.

References

1. N. Agrait, A. L. Yeyati, J. M. van Ruitenbeek: Phys. Rep. **377**, 81 (2003)
2. H. W. Yeom et al.: Phys. Rev. Lett. **82**, 4898 (1999)
3. D. F. Kyser, R. Pyle: IBM J. Res. Develop. **24**, 426 (1980)
4. C. Drukan, M. A. Schneider, M. E. Welland: J. Appl. Phys. **86**, 1280 (1999)
5. C. Drukan, M. E. Welland: Phys. Rev. B **61**, 14215 (2000)
6. S. Yasin, D. G. Hasko, H. Ahmed: Appl. Phys. Lett **78**, 2760 (2001)
7. C. Drukan, M. E. Welland: Ultramicroscopy **82**, 125 (2000)
8. C. P. Collier, E. W. Wong, M. Belohradsky, F. M. Raymo, J. F. Stodarrt, P. J. Kuekes, R. S. Williams, J. R. Heath: Science **285**, 391 (1999)
9. C. P. Collier, G. Mattersteig, E. W. Wong, Y. Luo, K. Bevery, J. Sampaio, F. M. Raymo, J. F. Stodarrt, J. R. Heath: Science **289**, 1172 (2000)
10. Y. Huang, X. Duan, Y. Cui, L. J. Lauhon, K. H. Kim, C. M. Lieber: Science **294**, 1313 (2000)
11. R. J. Behm, N. Garcia, H. Roher: *Scanning Tunneling Microscopy and Related Methods*, (Kluwer Academic Publishers, Dordrecht 1990)
12. C. J. Chen: J. Vacuum. Sci. Technol. A **9**, 440 (1991)
13. J. E. Demuth, U. Koehler, R. J. Hamers: Journal of Microscopy **152**, 299 (1988)
14. T. Fujisawa, T. Saku, Y. Hiroyama et al: Appl. Phys. Lett. **63**, 51 (1993)
15. S. Tarucha, T. Saku, Y. Tokura et al: Phys. Rev. B **47**, 4064 (1993)
16. T. Honda, S. Tarucha, T. Saku et al: Jpn. J. Appl. Phys. **34**, L72 (1995)
17. S. Tarucha, T. Honada, T. Saku et al: Solid State Communications **94**, 413 (1995)
18. T. Fujisawa, S. Tarucha: Appl. Phys. Lett. **68**, 526 (1996)
19. T. H. Wang, S. Tarucha: Appl. Phys. Lett. **71**, 2499 (1997)
20. Ch. Würsch, C. Stamm, S. Egger et al: Nature **389**, 937 (1997)

21. A. Yacoby, M. Heiblum, V. Umansky et al: Phys. Rev. Lett. **73**, 3149 (1994)
22. X. Ying, J. P. Lu, J. J. Heremnas et al: Appl. Phys. Lett. **65**, 1154 (1994)
23. V. V. Moshchalkov, L. Gielen, M. Dhallé, C. Van Haesendonck Y. Bruynseraede: Nature **361**, 617 (1993)
24. H. Grabert: Z. Phys. B-Condensed Matter **85**, 319 (1991)
25. P. Lafarge, H. Pothier, E. R. Williams et al: Z. Phys. B-Condensed Matter **85**, 327 (1991)
26. S. T. Ruggiero, J. B. Barner: Z. Phys. B-Condensed Matter **85**, 333 (1991)
27. D. B. Haviland, L. S. Kuzmin, P. Delsing et al: Z. Phys. B-Condensed Matter **85**, 339 (1991)
28. L. J. Geligs, S. M. Verbrugh, P. Hadley et al: Z. Phys. B-Condensed Matter **85**, 349 (1991)
29. U. Meirav, P. L. McEuen, M. A. Kastner et al: Z. Phys. B-Condensed Matter **85**, 355 (1991)
30. L. P. Kouwenhoven, N. C. van der Vaart, A. T. Johnson et al: Z. Phys. B-Condensed Matter **85**, 367 (1991)
31. D. C. Glattli, C. Pasquier, U. Meirav et al: Z. Phys. B-Condensed Matter **85**, 375 (1991)
32. A. Ramdane, G. Faini, H. Launois et al: Z. Phys. B-Condensed Matter **85**, 405 (1991)
33. M. Ueda, F. Guinea: Z. Phys. B-Condensed Matter **85**, 413 (1991)
34. W. Zwerger, M. Scharpf: Z. Phys. B-Condensed Matter **85**, 421 (1991)
35. W. Haüsler, B. Kramer, J. Masěk: Z. Phys. B-Condensed Matter **85**, 435 (1991)
36. G.-L. Ingold, P. Wyrowski, H. Grabert: Z. Phys. B-Condensed Matter **85**, 443 (1991)
37. G. Falci, V. Bubanja, G. Schön: Z. Phys. B-Condensed Matter **85**, 451 (1991)
38. A. M. van der Brink, A. A. Odntsov, P. A. Bobbert et al: Z. Phys. B-Condensed Matter **85**, 459 (1991)
39. M. H. Devoret, R. J. Schoelkopf: Nature **406**, 1039 (2000)
40. K. K. Likharev: IEEE Circuits and Devices **16**, 16 (2000)
41. W. Liang, M. Backrath, D. Bozovic, J. H. Hafner, M. Tinkham, H. Park: Nature **411**, 665 (2001)
42. J. Wang, H. Guo, J. L. Mozos et al: Phys. Rev. Lett. **80**, 4277 (1998)
43. N. R. Arista, M. A. Fuentes: Phys. Rev. B **63**, 165401 (2001)
44. Sang-Kee Eah, Wonhho Jhe, T. Saiki, M. Ohtsu: Optical Review **3**, 450 (1996)
45. K. Jang W. Jhe: Optics Letters **21**, 236 (1996)
46. H. Ito, T. Nakata, K. Sakaki, M. Ohtsu, K. I. Lee W. Jhe: Phys. Rev. Lett. **76**, 4500 (1996)
47. Ki Hyun Kim, Sang-Kee Eah, Byougho Lee, Chang Ho Cho, Wonho Jhe: Rev. Sci. Instrum. **68**, 2783 (1997)
48. Changyeon Won, Seung Hyup Yoo, Kyunghwan Oh, Un-Chul Paek, Wonho Jhe: Optics Communications **161**, 25 (1999)
49. Seung Hyup Yoo, Changyeon Won, Jong-An Kim, Kihwan Kim, Unyob Shim, Kyunghwan Oh, Un-Chul Paek Wonho Jhe: J. Opt. B: Quantum Semiclass. Opt. **1**, 364 (1999)
50. Myong R. Kim, June-Hyoung Park Wonho Jhe: Jpn. J. Appl. Phys. **39**, 984 (2000)
51. June-Hyoung Park, Myong R. Kim Whonho Jhe: Optics Letters **25**, 628 (2000)

52. R. M. Feenstra, W. A. Thompson, A. P. Fein: J. Vacuum. Sci. Technol. A **4**, 1315 (1986)
53. J. A. Strocio, R. M. Feenstra, A. P. Fein: Phys. Rev. Lett. **57**, 2579 (1986)
54. R. S. Becker, T. Klitsner, J. S. Vickers: Phys. Rev. B **38**, 3537 (1988)
55. R. S. Becker, G. S. Higashi, Y. J. Chabal et al: Phys. Rev. Lett. **65**, 1917 (1990)
56. J. Nara, T. Sasaki, T. Ohno: Phys. Rev. Lett. **79**, 4421 (1997)
57. R. M. Feenstra, W. A. Thompson, A. P. Fein: Phys. Rev. Lett. **56**, 608 (1986)
58. H. Itoh, H. Tanabe, D. Winau et al: J. Vacuum. Sci. Technol. B **12**, 2086 (1994)
59. J. Slezák, M. Ondréjcék, Z. Chvoj et al: Phys. Rev. B **61**, 16121 (2000)
60. J. Slezák, P. Mutombo, V. Cháb: Surface Science **454-456**, 584 (2000)
61. J. Slezák, V. Cháb, Z. Chvoj et al: J. Vac. Sci. Technol. B **18**, 1151 (2000)
62. R. J. Hamers, Ph. Avouris, F. Bozso: Phys. Rev. Lett. **59**, 2071 (1987)
63. Mitch Jacoby: Science/Technology **78**, 42 (2000)
64. G. P. Lopinski, D. D. M. Wayner, R. A. Wolkow: Nature **406**, 48 (2000)
65. R. L. Schult, H. W. Wyld, D. G. Ravenhall: Phys. Rev. B **41**, 12771 (1990)
66. X. Duan, Yu. Huang, J. Wang et al: Nature **409**, 66 (2001)
67. AM. Morales, CM. Lieber: Science **279**, 208 (1998)
68. B. Kim, S. L. Tripp, A. Wei: J. Am. Chem. Soc. **123**, 7955 (2001)
69. A. Drury, S. Maier, A. P. Davery et al: Synthetic Metals **119**, 151 (2001)
70. A. Drury, S. Maier, A. P. Davery et al: Synthetic Metals **119**, 151 (2001)
71. H. Park, A. K. L. Lim, A. P. Alvisatos et al: Appl. Phys. Lett. **75**, 301 (1999)
72. L. W. Swocnson, A. E. Bell: Liquid Metal ion Sources In: *The Physics and Technology of ion sources*, ed by I. G. Brown (John Wiley & Sons, New York 1996) pp. 313–330
73. M. Komuro, O. Kitamura, S. Okayama et al: Microelectronic engineering **9**, 285 (1989)
74. S. Okayama, M. Komuro, W. Mizutani et al: J. Vacuum. Sci. Technol. A **6**, 440 (1988)
75. S. Hasegawa et al: Progress in Surface Science **60**, 89 (1999)
76. Z.-C. Dong, D. Fujita, H. Nejoh: Phys. Rev. B **63**, 5402 (2001)
77. J. Knall, S. A. Barnett, J.-E. Sundgren, J. E. Greene: Surf. Sci. **209**, 314 (1989)
78. R. G. Zhao, J. F. Jia, W. S. Yang: Surf. Sci. Lett. **274**, L519 (1992)
79. H. W. Yeom, T. Abukawa, M. Nakamura, S. Suzuki, S. Sato, K. Sakamoto, T. Sakamoto, K. Kono: Surf. Sci. **341**, 328 (1995)
80. G. Le Lay, K. Hricovini, J. E. Bonnet: Phys. Rev. B **39**, 3927 (1989)
81. J. Nogami, in Atomic Molecular Wires, Vol. 341 of NATO Advanced Studies Institute Series E: Applied Sciences, edited by C. Joachim S. Roth. Kluwer, Dordrecht, 1997, p. 11.
82. J. Nogami, S.-I. Park, C. F. Quate: Appl. Phys. Lett. **53**, 2086 (1988)
83. A. A. Baski, J. Nogami, C. F. Quate: Phys. Rev. B **43**, 9316 (1991)
84. Z.-C. Dong, T. Yakabe, D. Fujita, Q. D. Jiang, H. Nejoh: Surf. Sci. **380**, 23 (1997)
85. L. Li, C. Koziol, K. Wurm, Y. Hong, E. Bauer, I. S. T. Tsong: Phys. Rev. B **50**, 10 834 (1994)
86. H. Itoh, H. Tanabe, D. Winau, A. K. Schmid, T. Ichinokawa: J. Vac. Sci. Technol. B **12**, 2086 (1994)

87. A. A. Baski, C. F. Quate, J. Nogami: Phys. Rev. B **44**, 11 167 (1991)
88. J. E. Northrup, M. C. Schabel, C. J. Karlsson, R. I. G. Uhrberg: Phys. Rev. B **44**, 13799 (1991)
89. G. Brocks, P. J. Kelly, R. Car: Phys. Rev. Lett. **70**, 2786 (1993)
90. T. Yamasaki, M. Ikeda, Y. Morikawa, K. Terakura: Mater. Res. Soc. Symp. Proc. **318**, 257 (1994)
91. M. E. Gonzalez-Mendez N. Takeuchi: Phys. Rev. B **58**, 16172 (1998)
92. R. M. Feenstra, J. A. Stroscio, J. Tersoff: A. P. Fein, Phys. Rev. Lett. **58**, 1192 (1987)
93. R. J. Hamers, R. M. Tromp, J. E. Demuth: Phys. Rev. Lett. **56**, 1972 (1986)
94. F. J. Himpsel, Th. Fauster: J. Vac. Sci. Technol. A **2**, 815 (1984)
95. R. J. Hamers, Ph. Avouris, F. Bozso: Phys. Rev. Lett. **59**, 2071 (1987)
96. J. Tersoff, D. R. Hamann: Phys. Rev. B **31**, 805 (1985)
97. M. Prietsch, A. Samsavar, R. Ludeke: Phys. Rev. B **43**, 11850 (1991)
98. H. W. Yeom, T. Abukawa, Y. Takakuwa, Y. Mori, T. Shimatani, A. Kakizaki: S. Kono: Phys. Rev. B **53**, 1948 (1996)
99. H. W. Yeom, T. Abukawa, Y. Takakuwa, Y. Mori, T. Shimatani, A. Kakizaki, S. Kono: Phys. Rev. B **55**, 15 669 (1997)
100. S. Odasso, M. Gothelid, V. Y. Aristov, G. Lelay, H. J. Kim, T. Buslaps, R. L. Johnson: Surf. Rev. Lett. **5**, 5 (1998)
101. R. E. Peierls: *Quantum Theory of Solids*, 1st ed. (Oxford University Press, Oxford, Engl, 1955, p. 108)
102. Zhen Chao Dong, Daisuke Fujita, Taro Yakabe, Hanyyi Sheng, Hitoshi Nejoh: J. Vac. Sci. Technol. B **18**, 2371 (2000)
103. M. F. Crommie, C. P. Lutz, D. M. Eigler: Science **262**, 218 (1993)
104. L. J. Whitman, J. A. Stroscio, R. A. Dragost, R. J. Celotta: Science **251**, 1206 (1991)
105. T. C. Shen, C. Wang, G. C. Abaln, J. G. Tacker, J. W. Lyding, Ph. Avouris, R. E. Walkup: Science **268**, 1590 (1995)
106. T. Hashizume, S. Heike, M. Lutwyche, S. Watanabe, K. Nakajima, T. Nishi, T. Wada: Jpn. J. Appl. Phys., Part 2 **35**, L1085 (1996)
107. P. Soukiassian, F. Semond, A. Mayne, G. Dujardin: Phys. Rev. Lett. **79**, 2498 (1997)
108. J. Nogami, A. A. Baski, C. F. Quate: Phys. Rev. B **44**, 1415 (1991)
109. Z.-C. Dong, T. Yakabe, D. Fujita, T. Ohgi, D. Rogers, H. Nejoh: Jpn. J. Appl. Phys.: Part 1 **37**, 807 (1998)
110. P. Doumergue, L. Pizzagalli, C. Joachim, A. Altibelli, A. Baratof: Phys. Rev. B **59**, 15910 (1999)
111. G. Le Lay, K. Hricovini, J. E. Bonnet: Phys. Rev. B **39**, 3927 (1989)
112. R. G. Zhao, J. Jia, W. S. Yang: Surf. Sci. Lett. **274**, L519 (1992)
113. L. Li, C. Koziol, K. Wurm, Y. Hong, E. Bauer, I. S. T. Tsong: Phys. Rev. B **50**, 10834 (1994)
114. H. Itoh, H. Tanabe, D. Winau, A. K. Schmid, T. Ichinokawa: J. Vac. Sci. Technol. B **12**, 2086 (1994)
115. J. Nogami: In: *Atomic Molecular Wires*, NATO ASI Series E: Applied Sciences, Vol. 341, ed by C. Joachim, S. Roth (Kluwer, Dordrecht, 1997) p. 11
116. J. Nogami, S.-I. Park, C. F. Quate: Appl. Phys. Lett. **415**, 301 (1998)
117. A. A. Baski, J. Nogami, C. F. Quate: Phys. Rev. B **43**, 9316 (1991)
118. Z.-C. Dong, T. Yakabe, D. Fujita, Q. D. Jiang, H. Nejoh: Surf. Sci. **380**, 23 (1997)

119. A. A. Baski, J. Nogami, C. F. Quate: Phys. Rev. B **50**, 10834 (1994)
120. J. E. Northrup, M. C. Schabel, C. J. Karlsson, R. I. G. Uhrberg: Phys. Rev. B **44**, 13799 (1991)
121. G. Brocks, P. J. Kelly, R. Car: Phys. Rev. Lett. **70**, 2786 (1993)
122. M. E. Gonzalez-Mendez N. Takeuchi: Phys. Rev. B **58**, 16172 (1998)
123. R. J. Hamers, R. M. Tromp, J. E. Demuth: Phys. Rev. Lett. **56**, 1972 (1986)
124. N. Takeuchi: Phys. Rev. B **58**, R7504 (1998)
125. A. A. Baski, C. F. Quate, J. Nogami: Phys. Rev. B **44**, 11167 (1991)
126. Z.-C. Dong, T. Yakabe, D. Fujita, H. Nejoh: Ultramicroscopy **73**, 169 (1998)
127. J.-Y. Veuillen, J.-M. Gomez-Rodriguez, R. C. Cinti: J. Vac. Sci. Technol. B **12**, 1010 (1996)
128. M. Prietsch, A. Samsavar, R. Ludeke: Phys. Rev. B **43**, 11850 (1991)
129. R. J. Hamers, Ph. Avouris, F. Bozso: Phys. Rev. Lett. **59**, 2071 (1987)
130. S. Odasso, M. Gothelid, V. Y. Aristov, G. Lelay, H. J. Kim, T. Buslaps, R. L. Johnson: Surf. Rev. Lett. **5**, 5 (1998)
131. Z.-C. Dong, T. yakabe, D. Fujita, H. Nejoh: Ultramicroscopy **73**, 169 (1998)
132. Ph. Avouris: (Ed.) *Atomic Nanometer Scale Modication of Materials: Fundamentals Applications* (Kluwer Academic Publishers, Dordrecht 1993)
133. Ph. Avouris: Acc. Chem. Res. **28**, 95 (1995)
134. D. M. Eigler, E. K. Schweizer: Nature **344**, 524 (1990)
135. M. F. Crommie, C. P. Lutz, D. M. Eigler: Science **262**, 218 (1993)
136. T. T. Tsong: *Atom-Probe Field Ion Microscopy* (Cambridge University Press, Cambridge, 1990)
137. I.-W. Lyo, Ph. Avouris: Science **253**, 173 (1991)
138. T. C. Shen, C. Wang, G. C. Abeln, J. R. Tucker, J. W. Lyding, Ph. Avouris, R. E. Walkup: Science **268**, 1590 (1995)
139. M. Aono, A. Kobayashi, F. Grey, H. Uchida, D.-H. Huang: Jpn. J. Appl. Phys. **32**, 1470 (1993)
140. R. E. Walkup, D. M. Newns, Ph. Avouris: Phys. Rev. B **352**, 600 (1993)
141. J. Knall, S. A. Barnett, J.-E. Sundgren, J. E. Greene: Surf. Sci. **209**, 314 (1989)
142. A. A. Baski, J. Nogami, C. F. Quate: J. Vac. Sci. Technol. A **9**, 1946 (1991)
143. Z.-C. Dong, T. Yakabe, D. Fujita, Q. D. Jiang, H. Nejoh: Surf. Sci. **380**, 23 (1997)
144. J. E. Northrup, M. C. Schabel, C. J. Karlsson, R. I. G. Uhrberg: Phys. Rev. B **44**, 13799 (1991)
145. B. E. Steele, L. Li, J. L. Stevens, I. S. T. Tsong: Phys. Rev. B **47**, 9925 (1993)
146. G. Brocks, P. J. Kelly, R. Car: Phys. Rev. Lett. **70**, 2786 (1993)
147. C. Girard, C. Joachim, C. Chavy, P. Sautet: Surf. Sci. **282**, 400 (1993)
148. H. Nejo, D. Fujita, Z.-C. Dong, S. Odasso, D. Rogers, J. Slezak : Fabrication of nanowires and conductance measurements In: *Nanometer Sclae Science and Technology*, ed by M. Allegrini, N. Garcia, O. Marti (IOS Press, Amsterdam, Oxford, Tokyo, Washington DC 2001) pp. 331–352
149. L. Langer, V. Bayot, J.-P.: Phys. Rev. Lett. **76**, 479 (1996)
150. Q. D. Jiang, D. Fujita, H. Y. Sheng, Z. C. Dong, H. Nejoh: Appl. Phys. A **64**, 619 (1997)
151. H. J. Mamin, P. H. Guethner, D. Rugar: Phys. Rev. Lett. **65**, 2418 (1990)
152. J. I. Pascual, J. Méndez, J. GómezHerrero, A. M. Baró, N. Garcia: Phys. Rev. Lett. **71**, 1852 (1993)

153. K. Bessho, S. Hashimoto: Appl. Phys. Lett. **65**, 2142 (1994)
154. C. S. Chang, W. B. Su, Tien T. Tsong: Phys, Rev. Lett. **72**, 574 (1994)
155. T. C. Chang, C. S. Chang, H. N. Lin, Tien T. Tsong: Appl. Phys. Lett. **67**, 903 (1995)
156. R. Kning, O. Jusko, L. Koenders, A. Schlachetzki: J. Vac. Sci. Technol. B **14**, 48 (1996)
157. H. J. Mamin, D. Rugar: Phys. Rev. Lett. **72**, 1128 (1994)
158. J. I. Pascual, J. Méndez, J. GómezHerrero, A. M. Baró, N. Garcia: Phys. Rev. Lett. **72**, 1129 (1994)
159. A. Kobayashi, F. Grey, R. S. Williams, M. Aono: Science **259**, 1724 (1993)
160. H. Uchida, D. H. Huang, J. Yoshinobu, M. Aono: Surf. Sci. **287**, 1056 (1993)
161. Jiutao Li, Richard Berndt, WolfDieter Schneider: Phys. Rev. Lett. **76**, 1888 (1996)
162. R. Emch, J. Nogami, M. M. Dovek, C. A. Lang, C. F. Quate: J. Appl. Phys. **65**, 79 (1989)
163. T. T. Tsong: *Atom-Probe Field Ion Microscopy* (Cambridge University Press, Cambridge, 1990)
164. Vu Thien Binh, N. Garcia: Ultramicroscopy **42-44**, 80 (1992)
165. Tien T. Tsong: Phys. Rev. B **44**, 13703 (1991)
166. C. X. Guo, D. J. Thomson: Ultramicroscopy **42-44**, 1452 (1992)
167. H. Y. Sheng, D. Fujita, T. ohgi, H. okamoto, H. Nejoh: J. Vac. Sci. Technol. B **16**, 2982 (1998)
168. C. G. Smith: Rep. Prog. Phys. **59**, 235 (1996)
169. W. M. Tolls: Nanotechnology **7**, 59 (1996)
170. H. J. Mamim, P. H. Guethner, D. Rugar: Phys. Rev. Lett. **65**, 2418 (1990)
171. D. Fujita, Q. Jiang, H. Nejoh: J. Vac. Sci. Technol. B **14**, 3413 (1996)
172. K. Ono, H. Shimada, S. Kobayashi, Y. Ootuka: Jpn. J. Appl. Phys., Part 1 **35**, 2369 (1996)
173. D. A. Darbyshire, C. W. Pitt, A. A. Stride: J. Vac. Sci. Technol. B **5**, 575 (1987)
174. C. P. D. Emic, K. K. Chan, J. Blum: J. Vac. Sci. Technol. B **10**, 1105 (1992)
175. M. Esashi, M. Takinami, Y. Wakabayashi, K. Minami: J. Micromech. Microeng. **5**, 5 (1995)
176. I. W. Rangelow, H. Loschner: J. Vac. Sci. Technol. B **13**, 2394 (1995)
177. Hiroshi Okamoto, DongminCheng: Rev. Sci. Instrum. **72**, 4398 (2001)
178. P. J. Bedrossian, D. M. Chen, K. Mortensen, J. A. Golovchenko: Nature **342**, 258 (1989)
179. I. B. Altfeder, D. M. Chen: Phys. Rev. Lett. **84**, 1284 (2000)
180. E. E. Ehrichs, W. F. Smith, A. L. DeLozanne: J. Microsc. **42.44**, 1438 (1992)
181. U. Ramsperger, T. Uchihashi, H. Nejoh: Appl. Phys. Lett. **78**, 85 (2001)
182. S. Tsukamoto, B. Siu, N. Nakagiri: Rev. Sci. Instrum. **62**, 1767 (1991)
183. H. Watanabe, C. Manabe, T. Shigematsu, M. Shimizu: Appl. Phys. Lett. **78**, 2928 (2001)
184. Q. Niu, M. C. Chang, C. K. Shih: Phys. Rev. B **51**, 5502 (1995)
185. Y. S. Chan, E. J. Heller: Phys. Rev. Lett. **78**, 2570 (1997)
186. J. M. Byers, M. E. Flatte: Phys. Rev. Lett. **74**, 306 (1995)
187. T. Gramespacher, M. Buttiker: Phys. Rev. Lett. **81**, 2763 (1998)
188. H. Okamoto, D. M. Chen: Rev. Sci. Instrum. **72**, 1510 (2001)
189. S. T. Smith, D. G. Chetwynd: *Foundations of Ultraprecision Mechanism Design* (Gordon and Breach, Amsterdam, 1992)

190. D. W. Pohl: Rev. Sci. Instrum. **58**, 54 (1986)
191. D. W. Pohl: Surf. Sci. **181**, 174 (1987)
192. I. B. Altfeder, A. P. Volodin: Rev. Sci. Instrum. **64**, 3157 (1993)
193. R. C. Richardson, E. N. Smith: *Experimental Techniques in Condensed Matter Physics at Low Temperatures* (Addison-Wesley, Reading, MA, 1988)
194. H. Okamoto, D. M. Chen: J. Vac. Sci. Technol. A **19**, 1822 (2001)
195. K. Reichelt, H. O. Lutz: J. Cryst. Growth **10**, 103 (1971)
196. U. Ramsperger, T. Uchihashi, H. Nejoh: Appl. Phys. Lett. **78**, 85 (2001)
197. H. C. Manoharan, C. P. Lutz, D. M. Eigler: Nature **403**, 512 (2000)
198. S. Hong, J. Zin, C. A. Mirkin: Science **286**, 523 (1999)
199. K. Wilder, C. F. Quate, D. Adderton, R. Bernstein, V. Elings: Appl. Phys. Lett. **73**, 2527 (1998)
200. S. J. Tans, M. H. Devoret, H. Dai, A. Thess, R. E. Smalley, L. J. Geerligs, C. Dekker: Nature **386**, 474 (1997)
201. P. A. Serena, N. Garciá (Ed.) *Nanowires*, NATO ASI Ser. E, Vol. 340 (Kluwer Academic, Dordrecht 1997)
202. J. Weckesser, J. V. Barth, C. Cai, B. Muller, K. Kern: Surf. Sci. **431**, 168 (1999)
203. Z. C. Dong, T. Yakabe, D. Fujita, T. Ohgi, D. Rogers, H. Nejoh: Jpn. J. Appl. Phys., Part 1 **37**, 807 (1998)
204. Park Scientific Instruments, ThermoMicroscopes, 1171 Borregas Ave., Sunnyvale, CA 94089.
205. H. J. Mamin: Appl. Phys. Lett. **69**, 433 (1996)
206. M. Tortonese, R. C. Barett, C. F. Quate: Appl. Phys. Lett. **62**, 834 (1993)
207. T. R. Albrecht, P. Grutter, D. Horne, D. Rugar: J. Appl. Phys. **69**, 668 (1991)
208. S. Heike, S. Watanabe, Y. Wada, T. Hashizume: Phys. Rev. Lett. **81**, 890 (1998)
209. G. V. Hansson, R. I. G. Uhrberg: Surf. Sci. Rep. **9**, 197 (1988)
210. C. Stamm, F. Marty, A. Vaterlaus, V. Weich, S. Egger, U. Maier, U. Ramsperger, H. Fuhrmann, D. Pescia: Science **282**, 449 (1998)
211. J. I. Pascual, J. Mendez, J. Gomez-Herrero, A. M. Baro, N. Garcia, Uzi Landman, W. D. Luedtke, E. N. Bogachek, H.-P. Cheng: Science **267**, 1793 (1995)
212. J. Abellan, R. Chicon, A. Arenas: Surf. Sci. **418**, 493 (1998)
213. J. L. Costa-Kramer, N. Garcia, P. Garcia-Mochales, P. A. Serena, M. I. Marques, A. Correia: Phys. Rev. B **55**, 5416 (1997)
214. H. Ohnishi, Y. Kondo, K. Takayanagi: Nature **395**, 780 (1998)
215. D. R. Lide: (Ed.) *Handbook of Chemistry and Physics* 73rd ed., (CRC Press, New York 1992)
216. I. Chizhov, G. Lee, R. F. Willis: Phys. Rev. B **56**, 12316 (1997)
217. P. Vettiger, M. Despont, U. Drechsler, U. Durig, W. Haberle, M. I. Lutwyche, H. E. Rothuizen, R. Stutz, R. Widmer, G. K. Binnig: IBM J. Res. Dev. **44**, 323 (2000)

2 Nanolithography

Duncan Rogers

2.1 Introduction

Nanolithography is the fabrication of nanometer scale patterns on surfaces. The method of nanolithography described in this chapter consists of template formation and adsorption, followed by reactions that lead to the imprinting of the template onto a surface. Three kinds of templates are considered. The first template is formed by depositing a resist onto a surface and exposing selected areas of the resist to a beam of electrons. The resulting structure acts as a template if atoms or molecules adsorb preferentially on either the exposed or unexposed regions of the surface. The second template is not fabricated by resist exposure; instead, molecule and surface are selected such that a natural template forms on adsorption. For this case, the self-assembled pattern of molecules is the template. Instead of forming the template on the surface, the template can be formed on the molecule by designing the molecule to have desired atomic groups at specific locations. The third template is the pattern of atomic groups in the molecule.

For the template formed by resist exposure, nanometer scale patterning requires that the resist be no thicker than a single layer of atoms. This requirement minimizes the scattering of electrons within the resist. The single atomic layer of hydrogen atoms on the hydrogen passivated Si surface is a suitable resist for nanolithography. To form a template on the hydrogen passivated surface, hydrogen atoms are removed from selected areas of the surface by tracing the tip of a Scanning Tunneling Microscope (STM) across the surface in the pattern of the template. Atoms and molecules bond to the bare-Si regions of the surface in preference to the hydrogen passivated regions, allowing the structure to act as a template.

After template formation and adsorbate deposition, the template is imprinted onto the surface by inducing reactions between the adsorbates and surface. The reactions can be induced by heating, electron irradiation or photon irradiation. The outcomes of these surface treatments include the desorption, diffusion, and dissociation of the adsorbates. For the case of the dissociation of an adsorbed molecule, a desired atom, or molecular fragment, is deposited onto the surface at a nearby atomic site. This process transfers the pattern of molecules into a pattern of atoms, completing the fabrication of the nanostructure.

Before describing nanostructure fabrication in detail, I present an overview of the effects of electrons, photons, and electric fields on adsorbate-covered surfaces. Since many of the processes are induced by the current and electric field between a surface and an STM tip in close proximity to the surface, the additional effect of the STM tip is also presented.

2.2 Adsorbate Manipulation with Electrons, Photons and Electric Fields

Atoms and molecules adsorbed on a surface can be induced to desorb, diffuse or dissociate by means of photon irradiation, electron irradiation or the application of an electric field. The outcomes of these processes can be applied to the fabrication of nanostructures on the surface. Many of the processes can be induced by the current and electric field between an adsorbate and an STM tip in close proximity to the adsorbate. Before discussing the role of the STM tip, I describe simple, one-dimensional models of electron and photon excitation of adsorbates leading to desorption, diffusion and dissociation.

2.2.1 Adsorbate Excitation

Irradiation of a surface with electrons or photons can lead directly to the electronic, vibrational, or rotational excitation of an adsorbate. If an incoming electron attaches to an adsorbate, an electron attachment reaction can be initiated. Excitation of an adsorbate can also occur indirectly through substrate carriers excited by substrate photon-absorption. The excited substrate carriers can be thermalized by carrier-carrier collisions, carrier-phonon collisions, or electron-hole recombination followed by Auger electron excitation [1]. The result is surface heating and possible thermal excitation of an adsorbate leading to a reaction. If an excited substrate carrier diffuses to the surface, it can attach to an adsorbate, also initiating a reaction.

The reactions can be understood by qualitative models. The Menzel-Gomer-Redhead (MGR) model [2,3] for desorption is illustrated in Fig. 2.1(a). The lower curve is the potential energy of an adsorbate, A, in its ground electronic state on a surface, S. The substrate-adsorbate distance is plotted along the x-axis. The upper curve is the potential energy of the electronically excited adsorbate, A*. In its ground electronic state the adsorbate can have different vibrational excitations which are indicated by the lines parallel to the x-axis. Electronic excitation of an adsorbate on the ground state at position 1 results in a vertical transition to position 2 on the excited-state potential energy curve. According to the Franck-Condon principle there is no change in adsorbate geometry and no nuclear motions during an electronic excitation since the greater mass of the nucleus compared to the electron renders the nucleus effectively stationary during the transition. Thus only vertical transitions are allowed.

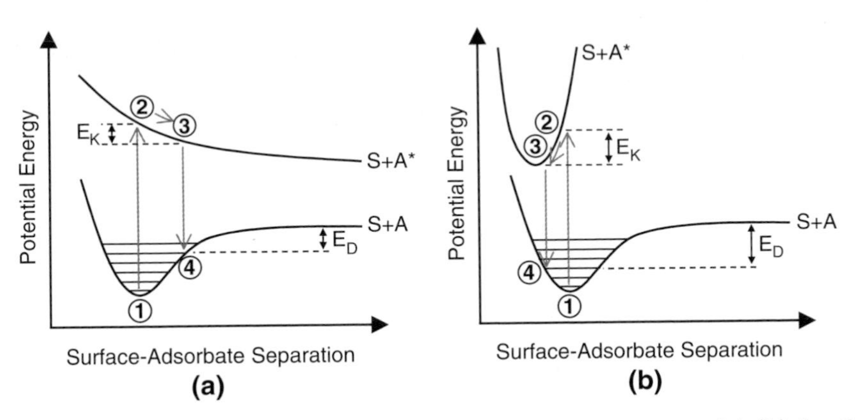

Fig. 2.1. Menzel-Gomer-Redhead model (**a**) and Antoniewicz model (**b**) for the desorption of an adsorbate, A, on a surface, S. For both models, the adsorbate is excited from position 1 on the ground electronic state to position 2 on the upper electronic state. The adsorbate moves away from the surface in (**a**) and towards the surface in (**b**). When the adsorbate reaches position 3 on the upper electronic state it is quenched to position 4 on the ground electronic state. If the kinetic energy E_K gained on the upper state is greater than the desorption barrier E_D on the ground state, the adsorbate can desorb from the surface

In the MGR model the upper curve is a repulsive state. The adsorbate at position 2 on the upper curve moves away from the surface. The time spent by the adsorbate on the upper curve is governed by the rate of quenching to the lower curve. If the rate of quenching is low, the adsorbate remains on the upper curve long enough for desorption to occur. For more rapid quenching the adsorbate eventually is quenched to the lower curve. Figure 2.1(a) shows an adsorbate that moved from position 2 to position 3 on the upper curve before being quenched to the ground state at position 4. The longer the adsorbate remains in the excited state, the greater is its degree of excitation after falling to the ground state. If the kinetic energy gained on the upper curve, E_K, is greater than the desorption barrier on the lower curve at position 4, E_D, the adsorbate can desorb from the surface. If the energy E_K is less than E_D, the adsorbate dissipates its excess vibrational energy, eventually falling to the lowest energy vibrational state.

In the Antoniewicz model [4] of Fig. 2.1(b), the upper electronic state is attractive instead of repulsive. This can occur if, for instance, a negatively charged excited state is attracted to the surface by its image charge. Electronic excitation excites the adsorbate into the upper electronic state via a vertical Frank-Condon transition (1 to 2 in Fig. 2.1(b)). The attractive potential pulls the adsorbate towards the surface (to position 3 in Fig. 2.1(b)). Eventually the adsorbate is quenched to the repulsive wall of the ground state in a vertical Franck-Condon transition (shown as transition 3 to 4 in Fig. 2.1(b)). For a negatively charged adsorbate the quenching could be due to an electron tunneling from the adsorbate to the substrate. If the kinetic energy, E_K, gained on the upper curve is greater than the desorption barrier on the lower curve at position 4, E_D, the adsorbate can desorb from the surface.

The models discussed in Fig. 2.1 can be extended to the dissociation and diffusion of adsorbates. Figure 2.2 shows the case for the dissociation of an adsorbate AB. The adsorbate is excited to the (repulsive) excited electronic state. If the adsorbate moves far enough along the upper curve it can be quenched into the dissociative ground state (sequence 1-2-3-4 in Fig. 2.2). The mechanism for diffusion of an adsorbate entails adsorbate excitation leading to the severance of a substrate-adsorbate bond and the formation of a substrate-adsorbate bond at an adjacent site. This mechanism was invoked to explain the electron- and photon-induced diffusion of benzene on Si(111)7×7 [5]. It is consistent with the observation that the energy barrier to diffusion of benzene on Si(111)7×7 is approximately equal to the energy barrier to desorption [6].

An alternative mechanism of excitation is multiple vibrational excitation of an adsorbate in the ground electronic state [7,8]. If the adsorbate receives enough vibrational energy it can cross the potential energy barrier to desorption. Desorption by this mechanism is illustrated in Fig. 2.3. Each arrow between the vibrational levels of the ground electronic state indicates excitation by electrons or photons. Multiple vibrational excitation provides enough energy for the adsorbate to cross the potential energy barrier leading to desorption.

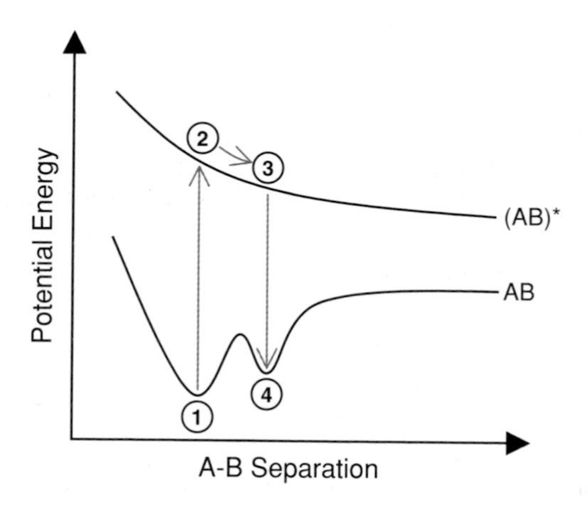

Fig. 2.2. Model for the dissociation of an adsorbate AB. The adsorbate at position 1 on the ground electronic state is excited to position 2 on the upper electronic state. The adsorbate moves to position 3 on the upper state where it is quenched to the dissociative ground state at position 4

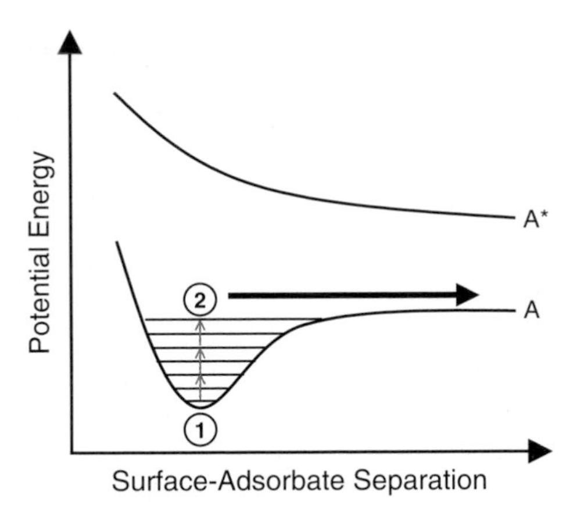

Fig. 2.3. Model for desorption by multiple vibrational excitation. The curve labeled A is the ground electronic state of the adsorbate, and the curve labeled A* is the excited electronic state. The horizontal lines indicate vibrational levels of the adsorbate in the ground electronic state. Electron or photon irradiation can vibrationally excite the adsorbate from position 1 to position 2 leading to desorption as indicated by the horizontal arrow at position 2

2.2.2 Adsorbate Manipulation with the STM Tip

The surface-adsorbate reactions described in the previous section can be induced by the tip of an STM. To operate the STM, its sharp metal tip is placed in close proximity (~10 Å) to the surface of a conductor or semiconductor. A bias voltage applied between the tip and surface causes electrons to flow between the tip and surface, and also results in an electric field between the tip and surface. The electrons and electric field can modify the surface and any surface adsorbates. The narrow radius of the STM tip leads to a narrow electron beam and minimizes the area of the surface that experiences the STM-induced electric field, resulting in the confinement of the STM-induced modifications to a narrow region of the surface. The confinement of the modifications, as well as the capability of accurately positioning the STM tip to within a few Å, are the main advantages of the STM as a tool for nanolithography.

The electrons that flow through the tip-surface junction of the STM must cross the potential energy barrier that separates the tip and surface. The barrier is crossed in one of two ways, depending on the magnitude of the bias voltage. For bias voltages of low magnitude (<5 V), the electrons tunnel through the potential energy barrier. The tunneling current is inversely proportional to the exponential of the distance z separating the tip and surface:

$$I \propto \exp(-\alpha z),$$

where I is the current, α is a constant, and the bias voltage is fixed [9–11]. For bias voltages of greater magnitude, the electrons are field emitted over the barrier. The field emission current is given by the Fowler-Nordheim equation:

$$I = AE^2 \exp(-B/E),$$

where A and B are constants that depend on the tip work function, and E is the electric field [12]. The field emission current increases exponentially with increasing electric field. Both the tunneling and field emission currents increase as the magnitude of the bias voltage increases since for increasing magnitude more substrate-adsorbate electronic states become available for the current to flow into or out of.

STM Parameters

Nanolithography executed by STM is influenced by several STM parameters including the shape of the tip, the distance between the tip and surface, the bias voltage and the current. To use the STM as a device for nanolithography requires that the electrons irradiate an area of the surface which is as small as possible. The diameter of the electron beam at the surface depends in part on the shape of the tip. Since the electric field at the tip is inversely proportional to the square of the radius of curvature of the tip, the electric field, and hence the current, are maximum at the region of the tip with the minimum radius of curvature. Reducing the radius of curvature narrows the region of the tip that emits electrons, leading to a smaller electron beam diameter on the surface. A typical W STM tip has a radius of curvature of $\sim$1000 nm in scanning electron microscopy images, but the electrons most likely are emitted from an asperity at the end of the tip. The beam diameter can change during nanolithography if the structure of the tip apex is altered by contact between tip and surface, or by electron and electric field induced rearrangement of the atoms at the tip apex.

The diameter of the electron beam at the surface also depends on the distance between the tip and surface. As shown in Fig. 2.4, the electric field lines at the surface become more closely spaced as the distance between the tip and surface decreases. Since the electrons follow the field lines, the beam diameter can be reduced by moving the tip closer to the surface.

During nanolithography, the STM operator controls the distance between the tip and surface by turning on or off the feedback loop of the STM. When the feedback loop is on and the bias voltage is fixed, as in the normal imaging mode, the feedback loop adjusts the distance between tip and surface in order to maintain a constant tunneling current. If the bias voltage is increased, the requirement for constant current forces the feedback loop to move the tip away from the surface. This increases the electron beam diameter at the surface. It also increases the area of the surface that experiences the high electric field due to the increased bias voltage. If the feedback loop is off and the bias voltage is raised, the separation between tip and sample remains constant resulting in an increased current. The smaller tip-surface distance with the feedback loop off localizes the electron beam and electric field to a

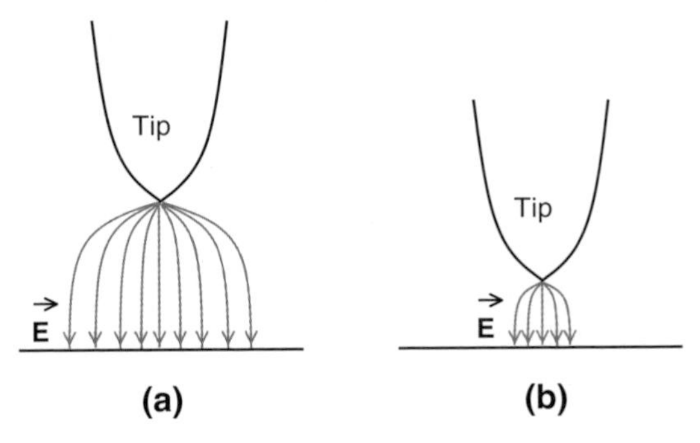

Fig. 2.4. Electric field lines between an STM tip and a surface. The electric field lines at the surface (**a**) become more closely spaced if the tip is moved closer to the surface (**b**)

smaller area of the surface. Compared to the case where the feedback loop is on, turning the feedback loop off when the bias voltage is increased results in an electric field of greater magnitude at the surface. The small tip-sample separation required for narrow linewidth patterns is achieved by keeping the bias voltage low or by turning off the feedback loop if the bias voltage needs to be raised to a high value.

In addition to controlling the feedback loop, the operator of the STM controls the current and the bias voltage. The energy of the electron is set by the bias voltage applied between the tip and surface. If the electron has sufficient energy it can excite an adsorbate by the mechanisms described in Sect. 2.2.1. For example, the STM-induced dissociation and desorption of molecular oxygen adsorbed on Si(111)7×7 has been ascribed to electron capture by the adsorbate followed by excitation to either a repulsive or an attractive electronically excited state [13]. In addition, the STM-induced desorption of H atoms from the hydrogen passivated Si surface was postulated to occur through electronic excitations at high bias voltages (>6 V) [7].

To increase the likelihood for excitation, the operator can increase the current. If the bias voltage is low and the current is increased, multiple vibrational excitations of adsorbates can be induced. The mechanism of multiple vibrational excitation explains the STM-induced dissociation of molecular oxygen on Pt(111) [14], and the low bias voltage STM-induced desorption of H atoms from the hydrogen passivated Si surface [7].

Tip-Adsorbate Chemical Interaction

The potential energy diagrams discussed in Sect. 2.2.1 are modified by the presence of an STM tip. One effect is a lowering of the barrier for the removal of an adsorbate from the surface. The barrier lowering occurs because at close proximity a degree of chemical bonding is established between the tip and adsorbate. Instead of removing the adsorbate from the surface by moving it into the vacuum far from the surface, the adsorbate is removed from the surface by bonding the adsorbate to the STM tip [15]. Figure 2.5(a) is the potential energy diagram of an adsorbate, A, on a surface, S, with and without an STM tip near the adsorbate. The energy barrier to desorption without an STM tip nearby (E_D) becomes, in the presence of an STM tip, a lower barrier (E_T) for removal of the adsorbate to the STM tip. To maintain a degree of chemical bonding between the tip and adsorbate it is essential that the tip remain close to the surface when the bias voltage is raised to the value appropriate to overcome the potential energy barrier to reaction. This often requires that the feedback loop be turned off when the bias voltage is raised. Evidence for barrier lowering by chemical interaction is provided by comparing the threshold electric fields required for Si atom removal in the field ion microscope and the STM. Removing a Si atom from a W STM tip in close proximity to a Si surface requires a threshold field of 1 V/Å [16], while removing a Si atom from a field-ion-microscope W tip, located relatively far from any surface, requires a higher threshold of 4 V/Å [17].

The energy barrier for removal to the STM tip is further reduced by the electric field between the tip and surface [18, 19]. The effect is analogous to field evaporation in the field ion microscope [17, 20]. The adsorbate can be ionized by electric field ionization or by a tunneling electron. After ionization, the electric field between the tip and surface lowers the potential energy barrier as shown in Fig. 2.5(b). The barrier to removal of the adsorbate to the STM tip is lowered from E_T to E'_T after ionization of the adsorbate. This mechanism accounts for the STM-induced decomposition of ferrocene on Si(111)7×7 [21] and the STM-induced removal of Si atoms from the Si(111)7×7 [16] and Si(100)2×1 [22] surfaces.

Stark Effect

In addition to assisting in the lowering of activation barriers, the electric field can assist in the localization of adsorbate excitations, leading to narrow-linewidth patterns. Delocalization of an adsorbate excitation occurs through energy transfer to surrounding adsorbates. Through the Stark effect the electric field under the STM tip can shift the resonant vibrational frequency of the adsorbate [23]. This shift in frequency can prevent energy transfer to neighbouring adsorbates, thereby localizing the excitation to the adsorbate under the STM tip. The localization increases the likelihood for multiple vibrational excitation. This mechanism was first proposed to account for the narrow linewidths observed in a study of H-desorption by multiple vibrational excitation of hydrogen passivated Si(100) [7].

Summary

In Sect. 2.2.1 I showed how an adsorbate on a surface can be electronically and vibrationally excited by electron and photon irradiation of the surface.

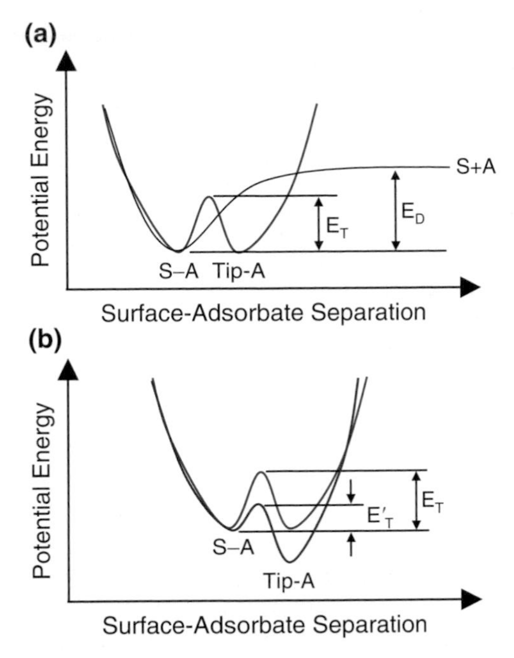

Fig. 2.5. Potential energy diagrams of an adsorbate A on a surface S showing the lowering of energy barriers by (**a**) chemical bonding to a nearby STM tip and (**b**) adsorbate ionization and the electric field between an STM tip and the surface. The black curve in (**a**) is the potential energy without an STM tip near the adsorbate and the red curve is the potential energy when an STM tip is near the adsorbate. The energy barrier for desorption (E_D) becomes, in the presence of an STM tip, a lower barrier (E_T) for removal of the adsorbate to the STM tip. The red curve in (**b**) is the potential energy when an STM tip is near the adsorbate and the blue curve is the potential energy after adsorbate ionization. The electric field between the tip and surface lowers the energy barrier for removal of the adsorbate to the STM tip from E_T to E'_T after adsorbate ionization

The excitation can lead to the desorption, diffusion or dissociation of the adsorbate. It can be induced by macroscopic beams of electrons and photons; it can also be induced by the microscopic beam of electrons that flows between the surface and an STM tip in close proximity to the surface, as shown in this section. The sharp tip of the STM minimizes the diameter of the electron beam at the surface. This leads to the confinement of the electron-induced reactions to a small area of the surface, thereby allowing the fabrication of nanometer-scale patterns. Confinement of reactions is also achieved by maintaining a small tip-sample separation. If necessary, the STM feedback loop can be turned off during a bias voltage increase in order to maintain a small tip-sample separation. The STM also lowers the barrier to reaction by tip-adsorbate chemical interactions and by the electric field between the tip and surface. The localization of adsorbate excitations is assisted by the Stark effect induced by the electric field between the tip and surface.

The next section describes how the adsorbate excitations are utilized in the fabrication of nanostructures on surfaces. Since the substrate for that discussion is the surface of a Si crystal, I first discuss the structure of the Si surface.

2.3 Methods of Nanolithography

2.3.1 Silicon: the Substrate for Nanolithography

Crystalline silicon has the diamond structure. In this structure the three p orbitals and one s orbital in the outer shell of each silicon atom hybridize into four sp^3 orbitals which bond to the atom's four nearest neighbours in a tetrahedral configuration, with a nearest-neighbour separation of 2.35 Å. The termination of the bulk crystalline structure at the surface of the crystal has the result that each Si atom at the surface does not have enough neighbouring atoms with which to bond and fully saturate its four sp^3 orbitals. The non-bonding sp^3 orbitals are referred to as 'dangling bonds'. There is an energy cost associated with the dangling bonds. The atoms at the surface reconstruct into a non-ideal geometry in order to reduce the number of dangling bonds, and hence lower the surface energy.

Si(111) Surface

At the ideal Si(111) surface each Si atom is bound to the crystal by covalent bonds between three of its sp^3 orbitals and three subsurface Si atoms. This leaves one dangling bond which is oriented perpendicular to the surface, pointing away from the crystal. The Si atoms in the plane of the surface are separated from one another by 3.84 Å, and are arrayed in the structure with the diamond shaped 1×1 unit cell shown in Fig. 2.6(a). This structure does not spontaneously form on the surface at room temperature. The actual structure that is formed depends on how the surface is prepared.

The surface prepared by cleaving a Si crystal in ultra high vacuum (UHV) has the 2×1 reconstruction as seen by low energy electron diffraction [24] and STM [25]. The model which best describes the 2×1 surface is the π-bonded chain model proposed by Pandey [26]. In this model, the 2×1 surface is formed from the ideal 1×1 surface by a drastic change in the bonding topology of the surface atoms. To form the 2×1 surface, half of the first layer atoms become second layer atoms by moving inward towards the crystal bulk, eliminating their dangling bonds, and bonding to atoms in the third layer. In addition, half of the second layer atoms become first layer atoms, each with a dangling bond, after their bonds to third layer atoms are severed and they move outward away from the crystal bulk. The resulting surface, as shown

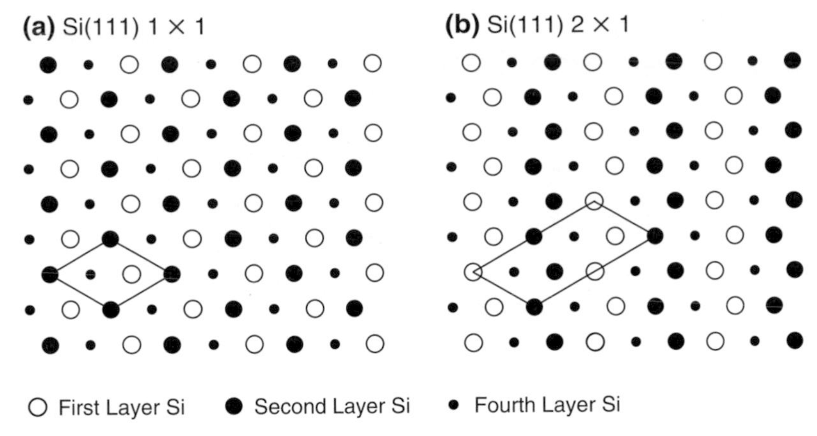

Fig. 2.6. The ideal Si(111) surface (**a**) and the reconstructed Si(111)2 × 1 surface (**b**). The 1 × 1 and 2 × 1 unit cells are outlined in (**a**) and (**b**) respectively. The 1 × 1 unit cell has sides 3.84 Å in length

in Fig. 2.6(b), consists of parallel rows of first layer atoms with neighbouring rows separated by 6.66 Å. Each of the first layer atoms has a dangling bond. The dangling bonds are on nearest neighbour sites allowing them to significantly interact by forming a chain of π bonds, thus lowering the surface energy. In contrast, there is no significant interaction between the dangling bonds on the 1 × 1 surface because of their larger separation on next nearest neighbour sites.

The standard method of preparing a Si(111) surface for surface science studies in UHV is to heat a Si(111) crystal to 1250°C in UHV and then cool to room temperature. The high temperature is required to remove carbon contamination from the surface. The surface has the 1 × 1 structure above 860°C but it transforms to a 7 × 7 structure at 860°C [27]. The 7 × 7 structure is the stable room temperature structure [28]; heating the 2 × 1 surface to 600°C and cooling to room temperature leaves the surface in the 7 × 7 configuration [29].

The 7 × 7 unit cell is diamond shaped with sides of length 26.9 Å. The structure of the 7 × 7 surface is described by the dimer-adatom-stacking fault (DAS) model [30]. In the DAS model one triangular half of the unit cell contains a stacking fault; the other triangular half has no stacking fault. Separating the two triangular halves are rows of Si-Si dimers. Twelve Si adatoms bond on top of the 7 × 7 unit cell, six on each triangular half. The formation of this complicated structure is driven by the reduction of dangling bonds. The 49 dangling bonds in the ideal 7 × 7 unit cell is reduced to 43 after the formation of the stacking fault and the bonding together of pairs of Si atoms to form the Si dimers. The number of dangling bonds per 7 × 7 unit cell is further reduced to 19 when twelve Si adatoms bond to each 7 × 7 unit cell. The Si adatom bonds to three surface Si atoms, eliminating three

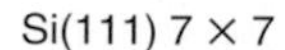

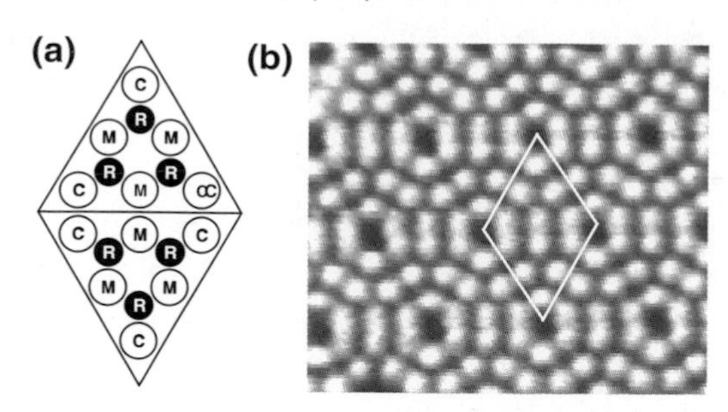

Fig. 2.7. Diagram of the Si(111) 7 × 7 unit cell (**a**) and an STM image of the Si(111)7 × 7 surface (**b**). The diamond shaped 7 × 7 unit cell has sides 26.9 Å in length. The twelve Si adatoms in the unit cell are divided into six corner adatoms (labeled C) at the corners of each triangular half of the unit cell and six middle adatoms (labeled M) between the corner adatoms. The six rest atoms in each unit cell (labeled R) are in the first atomic layer below the adatom layer. The twelve Si adatoms can be seen within the outlined 7 × 7 unit cell of the −1 V tip-bias STM image in (**b**)

dangling bonds at the cost of one dangling bond on the adatom itself. The 19 dangling bonds in the 7 × 7 unit cell consist of twelve dangling bonds on the twelve Si adatoms; six dangling bonds on six Si atoms called 'rest atoms' which are located between the adatoms one layer below the adatom layer; and one dangling bond on the Si atom at the corner of the unit cell, three layers below the adatom layer. The dangling bonds are the sites for the adsorption of atoms and molecules.

The diagram in Fig. 2.7(a) shows the twelve Si adatoms and the six Si rest atoms within a 7 × 7 unit cell. The six Si adatoms at the corners of each triangular half of the unit cell are called corner adatoms, and the six Si adatoms between the corner adatoms are called middle adatoms. In the STM image of the surface shown in Fig. 2.7(b), the outlined 7 × 7 unit cell contains twelve bright protrusions, each corresponding to one of the twelve Si adatoms of the unit cell. The six Si rest atoms of the unit cell are not observed in this image which was made with a bias voltage of −1 V on the tip, but are observed in images made with positive bias voltages. The faulted and unfaulted halves of the unit cell can be distinguished by imaging with a positive bias voltage on the tip; for a positive bias voltage the faulted half is brighter than the unfaulted half, while for a negative bias both halves have the same brightness.

Si(100) Surface

At the ideal Si(100) surface, each surface atom has two dangling bonds which protrude away from the crystal, making a 35.3° angle to the surface. As seen

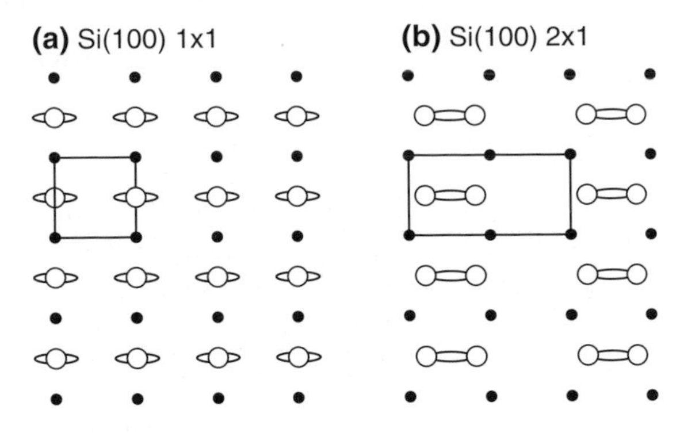

Fig. 2.8. The ideal Si(100)1×1 surface (**a**) and the reconstructed Si(100)2×1 surface (**b**). Each first layer Si atom of the ideal surface has two dangling bonds indicated by the lobes on the open circles in (**a**). On the 2×1 surface in (**b**) neighbouring first layer Si atoms have bonded together to form parallel rows of Si dimers. The 1×1 and 2×1 unit cells are outlined in (**a**) and (**b**) respectively. The 1×1 unit cell has sides 3.84 Å in length

in Fig. 2.8(a), the surface components of the two dangling bonds on each surface atom are directed in opposite directions toward neighbouring surface atoms. To reduce the surface energy, pairs of neighbouring Si atoms bond together to form Si dimers. The Si dimer is held together by a double bond: the inward facing dangling bonds form a σ-bond, and the outward facing dangling bonds form a weaker π-bond. The surface after dimer formation has a 2×1 reconstruction, and consists of parallel rows of Si dimers, as shown in Fig. 2.8(b). Neighbouring dimer rows are separated by 7.68 Å, and neighbouring dimers within a row are separated by 3.84 Å. The Si-Si dimer bond length is 2.4 Å.

The Si dimer is tilted with one Si atom of the dimer displaced towards the crystal bulk and the other displaced away from the crystal bulk. The tilting of the dimer is due to the weakness of the π bond. The diamond (100) surface also has a 2×1 reconstruction with pairs of surface atoms bonded together in dimers. The strong π bonding in the C dimer results in a symmetric dimer. The tilting of the Si dimer is observed at low temperature by STM, but at room temperature the dimers appear symmetric by STM because of thermally-induced rapid changes in the direction of tilting [32]. Atoms and molecules can adsorb onto the surface by breaking the π bond of the Si dimer. In Sect. 2.3.4 it is shown how the Si dimer is analogous to the alkene group of organic chemistry, leading to the attachment of organic molecules to the dimers in [2 + 2] and [4 + 2] cycloaddition reactions.

2.3.2 Hydrogen Passivated Si

The hydrogen passivated Si surface is a Si surface that is covered by a monolayer of hydrogen atoms. It has no dangling bonds since all dangling bonds have been eliminated by bonding to hydrogen atoms. On the Si(111)7×7 surface, hydrogen passivation eliminates the 7×7 reconstruction in favour of a 1×1 surface with all dangling bonds terminated by hydrogen atoms. On the Si(100)2×1 surface, hydrogen passivation eliminates the Si-dimer π bonds. Since the dangling bonds and π bonds are the preferential sites for adsorption, incident atoms and molecules are much less likely to bond to the hydrogen passivated surface than to the unpassivated surface. The monolayer of hydrogen atoms can be used as a resist for the fabrication of a nanoscale template on the surface. In this process, the STM tip is both the resist exposure tool and developing agent. The advantages of the hydrogen atom resist over the conventional polymer resist are the elimination of electron scattering within the resist and the elimination of resist polymerization.

There are two ways to form the hydrogen passivated surface: wet chemistry, and atomic hydrogen exposure in vacuum. The wet chemical method begins with a standard RCA clean [33] consisting of organic contamination and particle removal in an aqueous ammonia and hydrogen peroxide solution (SC-1 step), and metal contamination removal in a hydrochloric acid and hydrogen peroxide solution (SC-2 step). After the RCA clean, the surface contains a chemically grown oxide. To hydrogen passivate the surface, the oxide must be removed and the Si surface must be terminated by a monolayer of hydrogen atoms. This is accomplished by immersing the Si substrate in an HF solution. For both Si(111) and Si(100), immersion in 49% to 10% HF (pH = 1.0–2.0) results in a surface that is atomically rough. This has been observed by STM, and inferred by infrared absorption spectroscopy experiments which indicate surface Si in the monohydride, dihydride and trihydride configurations after HF etching [34].

The tendency for roughening on Si(111) can be counteracted by raising the pH of the etching solution. If the Si(111) substrate is immersed in an etching solution that has had its pH raised to 9–10 by the addition of NH_4OH, an atomically smooth surface is obtained. The surface smoothing was indicated by an STM study which showed that the atomic corrugation on Si(111) terraces was reduced from 3.7 Å after etching in the low pH solution to only 0.07 Å after etching in the high pH solution [35]. Further evidence for the elimination of surface roughening was provided by the appearance of only the monohydride configuration in the infrared absorption spectrum [34]. Scanning tunneling microscopy images show that the hydrogen passivated Si(111) surface prepared by wet chemistry has the ideal 1×1 structure with nearest-neighbour Si atoms in the surface plane separated by 3.8 Å [35].

Immersion of Si(100) in the higher pH solution does not result in an atomically smooth surface. The surface roughening is a result of the increased concentration of OH^- which accelerates the etching of (100) planes in preference to (111). Since (111) planes are preferentially formed, the surface is rough. This is confirmed by infrared absorption spectroscopy which indicates an increased concentration of Si in the monohydride configuration with the Si-H bond tilted away from the surface normal [34]. The wet chemical method is thus not recommended for hydrogen passivation of Si(100). Exposure to hydrogen atoms in vacuum is a better alternative for Si(100).

It is important for the reduction of surface contamination that the sample be rinsed thoroughly in deionized water after etching in the HF solution [35]. The hydrogen passivation protects the surface from oxidation, but it does not provide good protection from hydrocarbon uptake and the resulting formation of SiC defects [36]. Exposure of the hydrogen passivated surface to hydrocarbon containing ambients must be minimized. Thus the sample should be introduced into UHV through a loadlock that has been carefully prepared by previous baking and venting with filtered nitrogen. The loadlock should also be pumped with oil free pumps. These procedures have resulted in hydrogen passivated surfaces which have only 1% contamination [35].

The standard method of preparing clean Si surfaces for surface science studies in UHV is to heat the surfaces to 1250°C in UHV. This creates a 7×7 structure on Si(111) and a 2×1 structure on Si(100). Both of these surfaces can be hydrogen passivated by exposure to atomic hydrogen in UHV. The hydrogen passivation procedure is the same for both surfaces. Atomic hydrogen is produced by leaking hydrogen gas into the UHV chamber and dissociating the gas on a tungsten filament heated to 1900°C. During hydrogen exposure the Si sample faces the filament. Typical exposures are 4×10^{-5} Torr H_2 for 100 s (4000 L; 1L $= 10^{-6}$ Torr s) with the Si sample located about 1.5 cm from the filament. For a Si surface at room temperature, continuous exposure to atomic hydrogen results eventually in surface etching. To prevent surface etching the Si sample is heated to 300–400°C during hydrogen exposure [38, 39].

On Si(100)2×1 the π bond of a Si dimer can be severed if each of the its sp^3 orbitals bonds to a hydrogen atom. At the lowest exposures the π bonds are severed, but the Si-dimer σ bonds remain intact and the surface maintains the 2×1 periodicity (Fig. 2.9(a)). Since the sp^3 orbitals which form the π bonds are the sites for the adsorption of atoms and molecules, the surface is passivated after removal of all π bonds by the bonding of hydrogen atoms. For higher exposures, the σ bond of the dimer is also severed by the hydrogen atoms. This completely severs the Si dimer bond resulting in a 1×1 surface periodicity (Fig. 2.9(b)) [40].

Passivation of the Si(111) surface is more complicated since there are adatoms and stacking faulted regions which must be removed in order to create a 1×1 hydrogen-passivated surface [39]. Hydrogen atoms, for the lowest exposures, bond to the dangling bonds of the surface. As the exposure continues, additional hydrogen atoms break the backbonds between the adatoms and the substrate. Eventually the adatoms are removed from the surface as

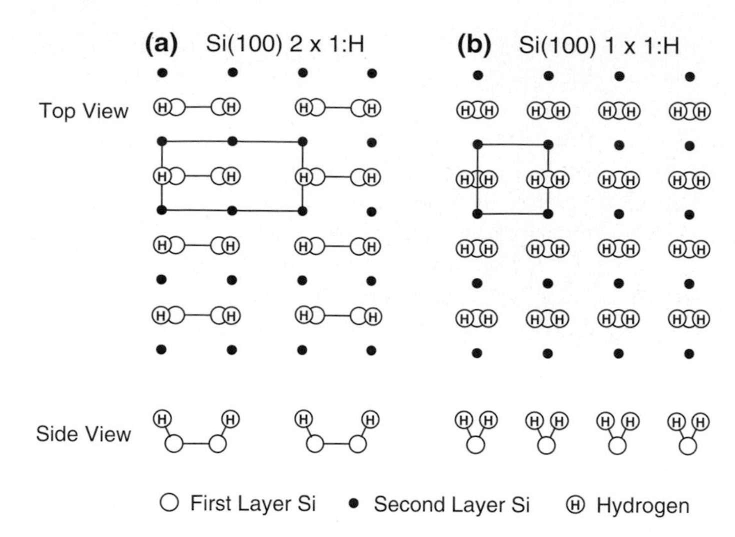

Fig. 2.9. The hydrogen passivated Si(100)2×1 surface with one hydrogen atom bonded to each Si atom (**a**), and the hydrogen passivated Si(100)1×1 surface with two hydrogen atoms bonded to each Si atom (**b**). The 2×1 and 1×1 unit cells are outlined in (**a**) and (**b**)respectively. The 1×1 unit cell has sides 3.84 Å in length

SiH_4, or they diffuse on the surface and assist in eliminating some of the stacking faulted regions and dimer rows. There are more Si adatoms on the 7×7 surface than is required to remove all of the stacking faults and dimer rows. Since not all of the excess Si adatoms are removed from the surface as SiH_4, the excess adatoms collect into islands on the surface. The islands of extra Si adatoms, and a few remnants of the stacking faulted regions and dimer rows, are observed in STM images of the hydrogen passivated Si(111) surface [39].

2.3.3 Template Formation on Hydrogen Passivated Si

A template on the hydrogen passivated Si surface can be formed by an STM tip that traces out the pattern of the template while it emits electrons of energy sufficient to desorb hydrogen atoms from the surface. The early depassivation work was done on Si surfaces in air [41]. The technique was later extended to Si(111) [42–46] and Si(100) [7,47–49] surfaces in vacuum.

An example of STM-induced depassivation of Si(111) in UHV is shown in Fig. 2.10 [45]. The bright square area in Fig. 2.10(a) is a 330 nm × 330 nm region that was depassivated by the STM tip; the rest of the surface remained hydrogen passivated. The removal of a hydrogen atom from a Si atom results in the reappearance of the Si atom dangling bond and the associated increase in electronic surface state density. Because of the higher surface state density on the bare Si atom compared to that on the hydrogen passivated Si atom, the

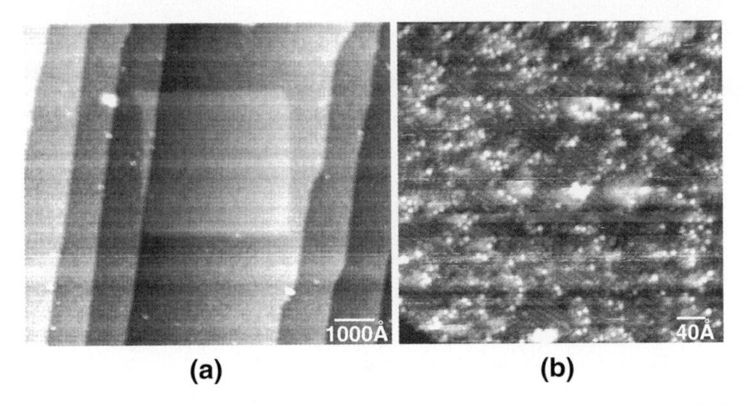

Fig. 2.10. STM images of a hydrogen passivated Si(111) surface after STM-induced depassivation. The bright square in (**a**) is a 330 nm × 330 nm area of depassivation. The lines running from the top to bottom of (**a**) are atomic steps. The high magnification image in (**b**) shows the π-bonded chains of the 2 × 1 reconstruction within the area of depassivation. Additional Si atoms can be seen on top of the π-bonded chains in (**b**). (Reproduced from [45].)

feedback loop, in order to maintain a constant tunneling current, increases the distance between tip and surface when the tip is over a bare Si atom and decreases the distance when the tip is over a hydrogen passivated Si atom. Thus, in the STM image of Fig. 2.10(a) the square region of bare Si atoms appears brighter than the rest of the surface.

To remove the hydrogen atoms from the square area, a −8 V bias was applied to the tip, the current was set to 10 nA, and the tip was scanned across 1024 equally-spaced horizontal lines within the square region. The feedback loop remained on during depassivation. The spacing between traces was 3.3 Å, and the injected charge density over the scan area was 31 pC/nm^2. In addition to the square area of depassivation, a narrow linewidth pattern was formed by limiting the STM tip to just a few traces during depassivation [45]. High magnification STM images of Si(111) show that the surface forms the 2 × 1 reconstruction after STM-induced depassivation [42, 45]. The π-bonded chains of the 2 × 1 reconstruction are evident in Fig. 2.10(b) which is a high magnification image of the area of depassivation in Fig. 2.10(a). The bright protrusions on top of the π-bonded chains in Fig. 2.10(b) are most likely Si atoms or clusters of Si atoms.

A threshold voltage of about 6 V is required to induce desorption of hydrogen from hydrogen passivated Si(100) [7, 49] and Si(111) [42, 44, 46]. The threshold voltage is evident in Fig. 2.11 which is a plot of desorption yield versus bias voltage for hydrogen passivated Si(100) [49]. Above the threshold voltage, the yield is independent of voltage. The mechanism for desorption has been postulated to be electronic excitation of Si-H from the bonding state σ to the antibonding state σ^* [7]. The antibonding state is

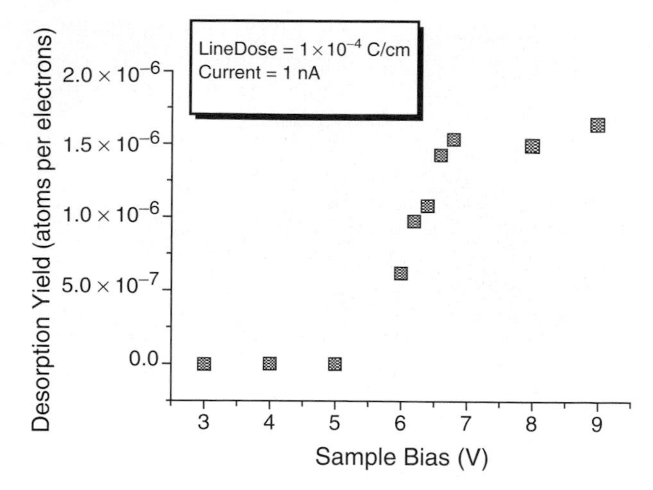

Fig. 2.11. Desorption yield of hydrogen from a hydrogen passivated Si(100) surface as a function of STM bias voltage. (Reproduced from [49].)

repulsive leading to desorption. Theoretical calculations of H on Si(111) find a separation between the bonding and antibonding levels of 5.8 eV, close to the observed threshold for desorption [50].

Desorption can also occur at bias voltages below the threshold voltage. The lower bias voltage minimizes the tip-sample separation resulting in the narrowest linewidths. Lines as narrow as the width of a single dimer row, or 7 Å, have been formed on the Si(100) surface by using a bias voltage of 3 V during depassivation [7, 51]. An STM image of one of the narrow linewidth patterns on Si(100) is shown in Fig. 2.12. Desorption at low bias voltages is explained by multiple vibrational excitation. For this mechanism the desorption yield increases with the tunneling current.

In the course of nanostructure fabrication, heating of the surface after pattern formation may be necessary. Temperature programmed desorption experiments for hydrogen atoms on Si surfaces show that H is desorbed from SiH_2 and SiH_3 species at about 420°C and from the SiH species at about 540°C [52]. Any heat treatments must be below the desorption temperature from the monohydride, and must not allow diffusion of hydrogen atoms from the passivated region to the bare Si region. One study of the diffusion of hydrogen atoms on partially-hydrogenated Si(111)7 × 7 showed that hydrogen atoms diffuse preferentially at 385°C and 460°C from the rest atoms to the adatoms [53]. The same study also showed a preferential diffusion of hydrogen from the corner adatoms to the middle adatoms at 385°C and 460°C, but the maximum distance that a hydrogen atom can diffuse was not obtained.

To show that a template on a hydrogen passivated surface remains unchanged after a heat treatment, the pattern in Fig. 2.10(a) underwent a series of 30 s heat treatments at 300°C, 350°C and 375°C [45]. The pattern was

Fig. 2.12. Narrow linewidth pattern of depassivation fabricated by STM manipulation of a hydrogen passivated Si(100)2 × 1 surface. The hydrogen-passivated Si-dimer rows are dark and the depassivated rows are bright. The distance between neighbouring dimer rows is 7.68 Å. (Reproduced from [7].)

unchanged even after the highest temperature heat treatment. In the higher magnification STM images of the bare Si region after the 300°C heating, the disordered array of Si atoms and clusters seen before the heat treatment (Fig. 2.10(b)) assumed a more ordered arrangement (Fig. 2.13), but the overall 2 × 1 periodicity of the surface did not change.

2.3.4 Adsorption on the Surface Template

Atoms and molecules, in general, bond to the Si dangling bonds within the template in preference to the surrounding hydrogen passivated Si. If atoms are deposited onto the surface or if the surface is exposed to molecules, the preference for adsorption on the bare Si atoms of the template transfers the pattern of depassivation into a pattern of adsorbed atoms or molecules. Examples of nanostructure fabrication by adsorption on the template are described in this section.

Oxide Lines

Oxide lines were the first nanostructures fabricated by STM manipulation of a hydrogen passivated Si surface [41]. The lines were fabricated on a hydrogen

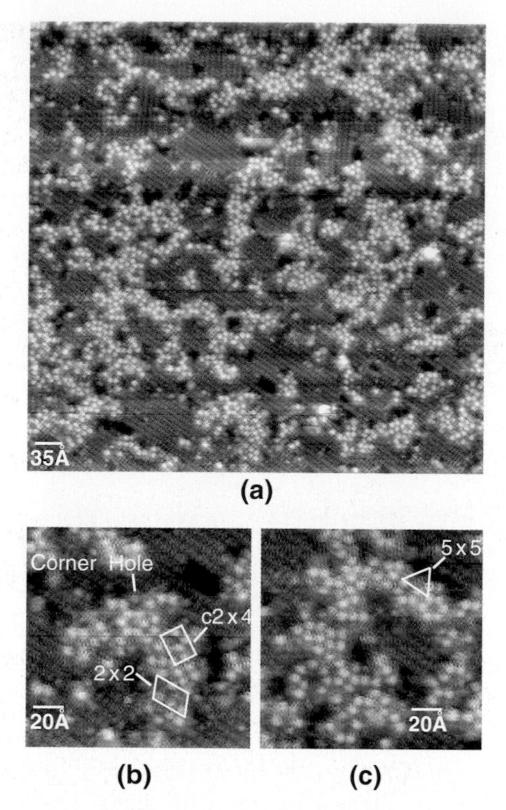

Fig. 2.13. The depassivated Si(111) surface after heating at 300°C. The π-bonded chains of the 2×1 reconstruction are seen in the dark areas, and the bright areas contain Si adatom structures. Some of the adatom structures are indicated in the higher magnification images in (**b**) and (**c**). (Reproduced from [45].)

passivated Si(111) surface. Since the depassivation occurred in air, an oxide grew on the depassivated region. The oxide lines were estimated to be one or two monolayers thick, had widths of 200 nm, and were fabricated in an array of parallel lines with neighbourinig lines separated by 300 nm.

The oxide pattern formed by STM can be etched in an aqueous HF solution [54] or it can be used as a mask for selective etching of the Si substrate in hydrazine or potassium hydroxide [55]. To selectively etch Si it is best to use a Si(100) substrate rather than a Si(111) substrate since the etching of Si(111) occurs very slowly in the chemical solutions which do not attack the oxide [56]. It has been demonstrated that hydrogen passivated regions of a surface can be etched to a depth of 120 nm with no change in the oxide pattern, and that lines as narrow as 25 nm can be fabricated on a 50 nm pitch grating [55]. These closely packed features were fabricated without taking proximity corrections into account, an advantage of STM lithography over conventional e-beam lithography.

The procedure was extended to the fabrication of a MOSFET device [57]. The processing steps included the growth of a gate oxide on a Si substrate, and the capping of the gate oxide with a 1100 Å thick layer of amorphous Si. The gate electrode was defined by selective etching of the amorphous Si. The oxide mask for the etching was fabricated by wet-chemical hydrogen passivation of the amorphous Si, followed by depassivation in air by an atomic force microscope tip that had been made conductive by coating with titanium. A line of oxide 33 Å high and 0.42 μm wide was formed by setting the tip bias voltage to 12 V and scanning the tip in a line across the amorphous Si. The oxide pattern was transferred into the amorphous Si by plasma etching resulting in a line of amorphous Si 0.21 μm wide. After ion implantation and further processing the effective gate length was 0.1 μm.

Metal Atom Deposition

The fabrication of metal lines by deposition of metal atoms onto the patterned surface utilizes the fact that metal atoms form clusters on hydrogen passivated Si and a more continuous film on bare Si. A metal atom evaporated onto the hydrogen passivated surface will diffuse until it nucleates a cluster, often at a Si dangling bond site, or until it meets and bonds to an existing cluster. There are two methods to fabricate metal nanostructures on the hydrogen passivated surface. In the first method, metal atoms are deposited onto the hydrogen passivated surface, and then the surface is depassivated with an STM tip. The second method consists first of depassivation to form a template, followed by deposition of metal atoms onto the surface.

An example of the first method is the fabrication of a Pb nanowire on Si(111) [45]. A Si(111) surface was cleaned in UHV and hydrogen passivated by atomic hydrogen exposure. Clusters of Pb atoms were formed on the hydrogen passivated surface after a room-temperature deposition of Pb. A nanowire and two contact pads connected to the nanowire were fabricated by desorption of hydrogen from selected areas of the surface after deposition of Pb. Figure 2.14(a) is an STM image of the nanowire and its two connecting contact pads. Each contact pad was formed by tracing the STM tip over a 330 nm $\times$ 330 nm area, and the nanowire between the two contact pads was formed by tracing the tip over four forward and reverse scan lines with a spacing of 3.3 Å between the traces. During fabrication, the bias voltage was -8 V, the current was set to 10 nA, and the feedback loop remained on. The nanowire was 20 nm wide and 770 nm long.

The depassivated region appears darker than the passivated region since the removal of the hydrogen atoms allowed the Pb atoms in the clusters to bond to the Si atoms on the depassivated region, decreasing the height of the clusters and spreading a film of Pb over the depassivated region. The structure of the Pb film within the depassivated region was seen in high magnification STM images. The Pb film was continuous, but it contained many vacancies.

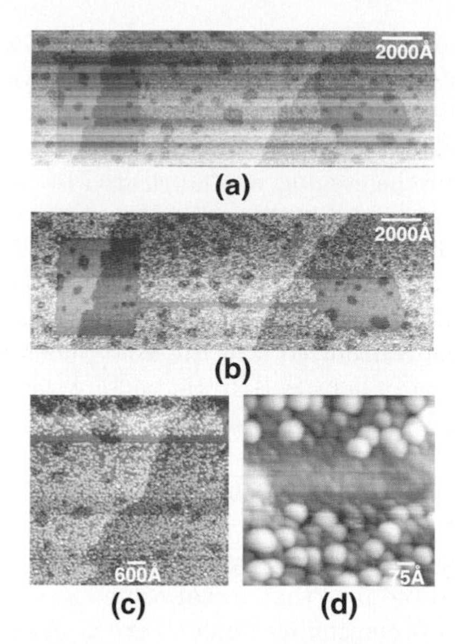

Fig. 2.14. STM images of a Pb nanostructure fabricated on a hydrogen passivated Si(111) surface. A nanowire and two connecting contact pads are shown (**a**) before and (**b**) after heating at 350°C. The nanowire is shown at higher magnification in (**c**) and (**d**) after the heat treatment. The hydrogen passivated region surrounding the nanostructure contains clusters of Pb atoms while a more continuous Pb film is observed within the nanowire and contact pads. The horizontal lines in (**a**) are noise, and the pits which are seen covering the surface were the result of surface etching by atomic hydrogen during hydrogen passivation. (Reproduced from [45].)

To reduce the vacancy defects, and to increase the size and spacing of the clusters on the hydrogen passivated region, the sample was heated to 350°C for 47 s. The STM images in Figs. 2.14(b)–(d), taken after heating, show that the dimensions of the contact pads and nanowire were unchanged by the heat treatment. On the hydrogen passivated region the heat treatment increased the average width of the Pb clusters from 32 Å to 89 Å. High magnification STM images showed that the heat treatment also reduced the number of vacancy defects on the depassivated region.

Template formation followed by metal deposition was the method used to fabricate Al [58], Ag [59], Ga [60] and Co [61] nanostructures on the hydrogen passivated Si(100) surface. Each of the elements formed clusters when they were deposited onto the hydrogen passivated surface. Single element metallic nanostructures were formed by Al, Ag and Ga deposition onto the bare-Si of the surface template, but Co reacted with Si to form a silicide. Lines of width as narrow as 8 nm were fabricated by Co deposition.

When Al atoms were deposited onto a surface containing parallel lines of depassivation, the Al atoms formed isolated clusters on the hydrogen passivated regions and more dense structures on the bare Si regions [58].

The density of Al was greater along the boundaries between the passivated and depassivated regions than in the interiors of the depassivated regions. This indicates that the growth of Al islands within the depasivated lines was mainly due to diffusion of Al atoms from the passivated regions. This is consistent with the observation of a lower density of clusters on the hydrogen passivated regions between the lines of depassivation than on the hydrogen passivated regions far from the lines of depassivation. The growth of Ag nanostructures by Ag evaporation onto a patterned hydrogen passivated surface exhibited a similar partial selectivity. The initial stage of Ag growth on a $20\,\mathrm{nm}\times 20\,\mathrm{nm}$ area of depassivation was along the edges of the depassivated region indicating growth by diffusion of Ag atoms from the passivated region [59].

Nanostructure fabrication by metal deposition onto a patterned hydrogen-passivated surface is best accomplished by increasing the selectivity of the metal deposition. This can be achieved by decreasing the line spacing or by increasing the rate of diffusion of metal atoms on the hydrogen passivated surface. The rate of diffusion can be increased by raising the temperature or by judicious selection of the metal species. An alternative method that gives better selectivity is the adsorption of molecules on the patterned surface.

Molecular Adsorption

A greater preference for adsorption on either the bare Si regions or the hydrogen passivated regions of the patterned surface can be achieved by dosing with molecules. A particular atom can be deposited onto the surface by adsorbing a molecule that contains the atom followed by heating, electron irradiation or photon irradiation to decompose the molecule. It has been demonstrated that molecules containing oxygen [62, 63], nitrogen [63], metal atoms [64, 65], dopant atoms [66], and C–C double bonds [65] can be selectively adsorbed on the patterned surface.

Oxygen and Ammonia

Oxygen and ammonia were among the first molecules to be dosed onto the patterned surface [62, 63]. The work was carried out on a Si(100) substrate in UHV. The goal of the study was to create thin oxides and nitrides that could be used as the active regions of a nanoscale electronic device or as masks for pattern transfer. Both molecules did not adsorb on the hydrogen passivated regions. Molecular oxygen adsorbed on the depassivated regions, fully covering the regions with oxygen. No change in the widths of patterned lines before and after dosing was observed. As observed in other studies of ammonia adsorption on Si(100) [67], the ammonia dissociated into H and NH_2 when it adsorbed on the depassivated regions. The two fragments bonded to Si dangling bonds, resulting in a partial repassivation of the surface. Although ammonia dissociation eliminates its usefulness as a precursor for a nitride layer, it can be used to dose a pattern of isolated dangling bond sites in order to create an array of isolated N atoms.

Molecules Containing Metals

Metal nanostructures can be fabricated by dosing the patterned surface with molecules that contain metal atoms. Selective deposition of Ti, Al and Fe was attained by dosing a patterned Si(100) surface in UHV with $TiCl_4$ [65], $H_2Al[(C_2H_5)NC_2H_4N(CH_3)_2]$ [65], and $Fe(CO_5)$ [64] respectively. The molecules adsorbed on the bare Si regions but not on the hydrogen passivated regions.

After dosing a patterned surface at room temperature with $TiCl_4$, the film growth within the depassivated areas was observed to be irregular and amorphous [65]. X-ray photoelectron spectroscopy clearly showed a large signal from Ti after dosing a bare Si surface, but only a small Ti signal after dosing a hydrogen-passivated surface. The trace amount of Ti on the hydrogen-passivated surface may have been due to adsorption on isolated Si dangling bond sites on the hydrogen-passivated surface.

In experiments with other molecules, the surface was heated during dosing in order to dissociate the molecule and deposit the metal atom onto the surface. The aluminum precursor, $H_2Al[(C_2H_5)NC_2H_4N(CH_3)_2]$, was specially designed for Al nanostructure growth on Si surfaces [68]. The substrate was heated to 200°C during exposure in order to decompose the molecule and deposit Al [65]. Scanning tunneling microscopy images showed that the molecule reacted with the Si dangling bond sites but not with the hydrogen passivated Si. After decomposition small areas of 2×2 reconstruction, similar to the surface structure that is formed in the initial stage of Al growth on Si(100) [68], were observed within the depassivated regions.

To deposit Fe onto the surface, $Fe(CO_5)$ was thermally decomposed by heating the surface to 275°C during dosing [64]. Decomposition occurred at Si dangling bond sites, but not at sites occupied by hydrogen atoms because of the increased activation barrier to decomposition at hydrogen passivated sites. About two monolayers of Fe were deposited within lines as narrow as 10 nm. The linewidth after dosing was no larger than that before dosing suggesting that even narrower lines can be achieved by decreasing the width of the line of depassivation.

The hydrogen passivation does not prevent the adsorption of all molecules. There are molecules which can adsorb on the hydrogen passivated Si surface but not on the Si oxide surface. To fabricate nanostructures with these molecules, an oxide mask is first fabricated by electron-induced desorption of hydrogen atoms from selected areas of the surface followed by oxide growth on the depassivated areas. This approach was followed to fabricate an epitaxial Al wire on a Si(111) substrate using dimethyl-aluminum hydride as the precursor gas [69]. The surface was patterned with an e-beam of 0.13 μm in diameter, and then an oxide was grown in the patterned area by exposing the surface to a clean room ambient. After oxide growth, the patterned substrate was dosed with dimethyl-aluminum hydride while it was heated to 270°C. The molecules did not adsorb on the oxide, but they dissociatively adsorbed on the hydrogen passivated regions. Linewidths as narrow as 500 nm were achieved. It should be possible to fabricate even narrower lines with the finer resolution of STM lithography.

Phosphine

Phosphine is another molecule that adsorbs on the depassivated regions of the patterned surface in preference to the hydrogen passivated regions. One proposal for a Si-based quantum computer uses phosphorous ^{31}P nuclei as two-level nuclear spins, called 'qubits', embedded in isotopically pure ^{28}Si [70]. The first stage in the fabrication of such a structure is the adsorption of phosphorous atoms at equally-spaced sites on a Si surface. It was demonstrated that the first stage in the fabrication can be accomplished by using the preference for phosphine adsorption on the depassivated regions of a hydrogen passivated Si surface [66]. Hydrogen atoms were desorbed from isolated areas of a hydrogen passivated Si(100)2 $\times$ 1 surface by applying -6 V, 10 ms pulses to an STM tip. Each isolated region of depassivation had a diameter less than 1 nm. The diameters were sufficiently small that only one phosphine molecule could adsorb on each region of depassivation. Phosphine has been shown to molecularly bond as PH_3 to one Si atom of a dimer and to then dissociate into adsorbed PH_2 and H, provided a neighbouring Si atom dangling bond is available for the hydrogen atom [71]. The P atoms can be incorporated into the crystal lattice as Si–P dimers by heating to 550°C for 5 minutes after adsorption [72]. This heat treatment also desorbs H from the surface.

Attachment by Cycloaddition Reactions

In the previous examples of molecular adsorption, the orientation of the molecules was not well controlled. Better control can be achieved by utilizing molecules containing C–C double bonds. It was noted earlier that the Si dimer of Si(100)2 $\times$ 1 is analogous to the alkene group of organic chemistry. The alkene group consists of two C atoms joined by a double bond that is comprised a σ bond and a π bond. The two Si atoms of the Si dimer are also joined by a σ bond and a π bond, although the π bond between Si atoms is weaker than that between C atoms. An organic molecule containing a C–C double bond can attach to a Si dimer in a process similar to the cyclodaddition reaction between the molecule and an alkene group. In the cycloaddition reaction the π bonds of the C–C double bonds are severed and the orbitals which participated in the π bonds form σ bonds between the C atoms in the alkene and the C atoms in the molecule. This results in the bonding together of the molecule and alkene group in a ring of C atoms.

Figure 2.15 illustrates the attachment of a 1,3–butadiene molecule to a Si dimer in a process analogous to the cycloaddition reaction. In the [2 + 2]

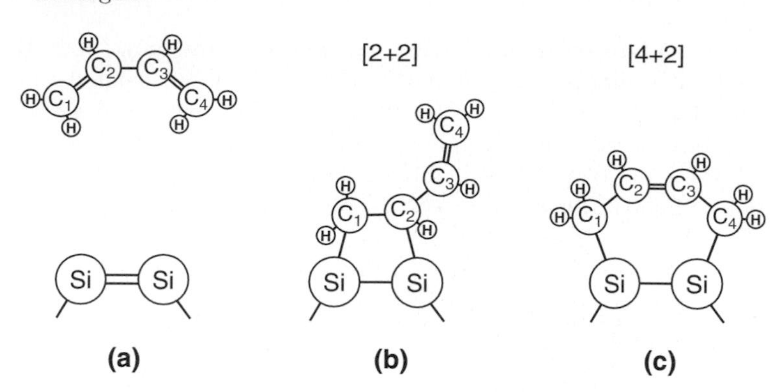

Fig. 2.15. Attachment of 1,3–butadiene to a Si dimer. The molecule in (**a**) attaches to the Si dimer in a [2 + 2] cycloaddition reaction (**b**) or a [4 + 2] cycloaddition reaction (**c**)

cycloaddition reaction of Fig. 2.15(b) the π bonds between the Si atoms of the dimer and between C atoms 1 and 2 of the molecule are severed. C atom 1 forms a σ bond with one of the Si atoms and C atom 2 forms a σ bond with the other Si atom. A four-membered ring of C atoms results. The notation [2 + 2] indicates that two π-electrons in the molecule and two π-electrons in the Si dimer are involved in the attachment reaction. In the [4 + 2] cycloaddition reaction of Fig. 2.15(c), the π bonds in the Si dimer and in the two C–C double bonds of the molecule are severed. Carbon atom 1 forms a σ bond with one Si atom, and C atom 4 forms a σ bond with the second Si atom. The remaining orbitals on C atoms 2 and 3 interact to form a double bond between the two C atoms. The final result is a six-membered ring of atoms. This reaction is given the notation [4 + 2] since it involves four π-electrons in the molecule and two π-electrons in the Si dimer. The two attachment geometries in Fig. 2.15 are consistent with STM images of the surface after molecular adsorption [73].

In addition to accounting for the attachment of 1,3–butadiene to Si(100) 2×1, the cycloaddition reaction accounts for the attachment of cyclopentene [74], benzene [75], chlorinated benzenes [76], Norbornadiene (bicyclo [2.2.1]hepta–2,5–diene) [65], and other molecules to the surface [77]. Norbornadiene contains two C–C double bonds separated by 2.43 Å. The two C–C double bonds can, respectively, bond in a [2 + 2] cycloaddition reaction to two adjacent Si dimers of a dimer row of Si(100)2×1. It was demonstrated that the molecule does not bond to hydrogen passivated Si(100)2×1, but that it does attach to areas of the surface that have been depassivated with an STM tip [65]. Figure 2.16 is an STM image of a patterned Si(100)2×1 surface after exposure to norbornadiene molecules showing the molecules selectively adsorbed within a line of depassivation [65].

The cycloaddition reactions force the molecules to bond to the surface at specific locations and in particular geometries, contributing to the long range

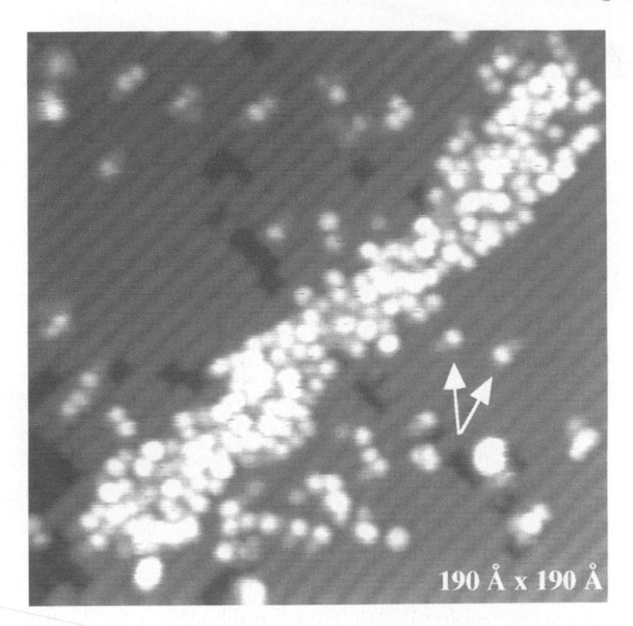

Fig. 2.16. Norbornadiene molecules adsorbed within a line of depassivation on the hydrogen passivated Si(100)2 × 1 surface. The dark lines are the hydrogen-passivated Si-dimer rows. The isolated bright spot are molecules that have bonded to defects on the hydrogen passivated surface. The arrows point to two different orientations of the adsorbed molecule. The line of bright spots are the molecules adsorbed within the line of depassivation. The distance between neighbouring dimer rows is 7.68 Å. (Reproduced from [65].)

order of the adsorbed layer. If the molecules are functionalized with atomic groups, the atomic groups will align with one another and point in distinct directions after adsorption. Because of the alignment of the atomic groups, any reactions induced after adsorption can deposit the atomic groups onto the surface in a pattern of repeating atomic sites, as discussed in the next section.

2.3.5 Adsorbate-Surface Reactions

A particular atom or group of atoms can be deposited onto a surface by adsorbing a molecule on the surface and dissociatively reacting the molecule with the surface by electron irradiation, photon irradiation or heating. It has been demonstrated for chlorinated benzene molecules adsorbed on Si(111) 7 × 7 that the reactions induced by electron and photon irradiation result in the severing of the C–Cl bond of the molecule and the deposition of the Cl atom onto a Si adatom adjacent to the site of adsorption of the molecule [5,78–80]. The reaction was postulated to be a concerted process in which the bond that was formed (Cl–Si) assisted in the rupture of the bond that was broken (C–Cl). Such a process can only occur if the new bond is adjacent to the old bond, i.e. if the reaction is localized. The localization of the reaction is expected to occur for other molecules adsorbed on surfaces with dangling bonds.

If the molecules are adsorbed in a pattern on the surface, the localization of the reactions can transfer the pattern of molecules into a similar pattern of adsorbates. As previously discussed, one way to form a pattern of molecules is to adsorb the molecules on a template created by STM-induced desorption of hydrogen from a hydrogen passivated Si surface. Instead of adsorbing the molecules on a template, a sufficiently high bias voltage can be applied to the STM tip while it is traced in a pattern over a surface covered by adsorbates. If the deposition reaction is localized then the decomposition products will be deposited onto the surface in the pattern traced by the STM tip. This procedure was followed for a Si(111)7$\times$7 surface with adsorbed chlorobenzene [78]. Chlorobenzene molecules were dissociated by applying -4 V, 10 ms pulses to an STM tip at 60 Å intervals along a straight line. The electron-induced reactions deposited Cl-atom clusters at 60 Å intervals along the straight line traced by the STM tip. The line of Cl clusters was 15 Å wide.

Alternatively, some molecules self assemble into a particular pattern on the surface. For instance, chlorobenzene self-assembles on the middle adatom sites of Si(111)7 $\times$ 7 in preference to the corner adatom sites [81]. When a molecule adsorbed on a middle adatom site dissociatively reacts with the surface, the Cl-atom product is depositied onto a neighbouring corner adatom site. Thus, electron irradiation of the surface with adsorbed chlorobenzene transfers the pattern of molecules on middle adatom sites into a similar pattern of Cl atoms on corner adatom sites [78].

The adsorbate molecule may contain two or more atomic groups that are distributed over the molecule in a particular geometry. The geometry of the molecule can be imprinted onto a surface by electron or photon irradiation leading to molecular dissociation and localized deposition of the atomic groups. The imprinting of molecular geometry was demonstrated first for 1,2–dichlorobenzene and 1,4–dichlorobenzene adsorbed on Si(111)7$\times$7 [79]. Irradiation with 193 nm photons reacted the adsorbed molecules with the surface resulting in the localized deposition of Cl atoms on Si adatoms. Scanning tunneling microscopy measurements showed that the average separation between neighbouring Cl atoms was 8 Å for Cl atoms from 1,2–dichlorobenzene and 14 Å for Cl atoms from 1,4–dichlorobenzene. As illustrated in Fig. 2.17, the different Cl-atom separations can be understood as being due to the different orientations of the C–Cl bonds in 1,2–dichlorobenzene and 1,4–dichlorobenzene. In 1,2–dichlorobenzene the two C–Cl bonds make a 60° angle with one another resulting in Cl deposition onto closely spaced Si adatoms, while for 1,4–dichlorobenzene the two C–Cl bonds make a 180° angle with one another resulting in Cl atom deposition onto Si adatoms which are farther apart.

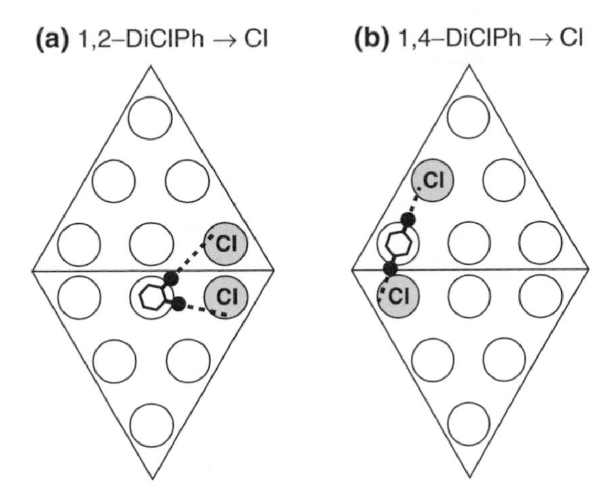

Fig. 2.17. Cl atom imprints from 1,2–dichorobenzene (**a**) and 1,4–dichlorobenzene (**b**) adsorbed on the 7×7 unit cell of Si(111). For 1,2–dichlorobenzene the Cl atoms are deposited onto neighbouring Si adatoms, while for 1,4–dichlorobenzene the Cl atoms are deposited onto Si adatoms separated by a greater distance

2.4 Conclusion

Nanostructure fabrication by surface template formation, adsorption, and reaction has been described in this chapter. Three kinds of templates were considered: (1) the pattern of bare Si atoms formed by electron-irradiation of a hydrogen atom resist on Si, (2) the pattern of molecules that self assemble on a surface, and (3) the pattern of atomic groups within a molecule. The pattern of the template is transferred into a similar pattern of atoms or molecular fragments by adsorption on the template followed, for molecular adsorbates, by surface-molecule reactions induced by heating, electron irradiation or photon irradiation. The localization of the reaction to the vicinity of the adsorbate ensures accurate pattern transfer.

The pattern of bare Si atoms on a hydrogen passivated Si surface acts as a template because atoms and molecules bond to the bare-Si regions in preference to the hydrogen passivated regions. Metal atoms that are deposited onto the surface form metallic clusters on the hydrogen-passivated regions and a continuous film on the bare Si regions. Nanostructures of Pb, Al, Ag and Co were fabricated by this approach [45, 58–61]. To increase the selectivity to the bare-Si regions, the surface can be exposed to molecules. By selecting or designing a molecule to have desired atomic groups at specific locations, the desired atoms or molecular fragments can be deposited onto the surface by inducing a reaction during or after adsorption. Examples include the thermal decomposition of $Fe(CO_5)$ to form Fe nanostructures [64] and the electron-induced decomposition of chlorobenzene to form Cl nanostructures [78]. Control of the position and orientation of the molecule on the surface controls the location of the deposited atom to a particular atomic site. There is a class of molecules that offers the required control: organic molecules containing C–C double bonds. They attach to the Si dimers of Si(100)2×1 by cycloaddition reactions, enabling excellent control of their position and orientation.

Many of the steps in nanostructure fabrication are facilitated by the scanning tunneling microscope. The properties of STM that make it useful for nanofabrication are the narrow tip radius which confines the emitted electrons, and thus surface modifications, to a narrow region of the surface and the capability of accurately positioning the tip to within a few Å. Two examples demonstrate these properties of STM. Firstly, the template that was fabricated by STM-induced desorption of hydrogen atoms from a hydrogen passivated Si surface had lines of 7 Å in width [7]. Secondly, the line of Cl atoms that was fabricated on a Si surface by STM-induced dissociation of adsorbed chlorobenzene molecules had a width of 15 Å [78].

Instead of forming the template by patterning with an STM, molecules can be self-assembled into a pattern, as seen for chlorobenzene on Si(111)7×7 [78,81]. Widespread irradiation with electrons or photons transfers the pattern of molecules into a pattern of atoms by decomposition of the molecules [78]. Another option is to eliminate the template on the surface in favour of a template on the molecule, as discussed in this chapter for dichlorobenzene [79]. This approach has the advantage that identical molecules are easily fabricated in great numbers. Widespread irradiation of the surface after adsorption transfers the pattern in the molecules into a similar pattern of adsorbed atoms or molecular fragments.

The methods described in this chapter can be applied to the fabrication of quantum effect and molecular electronic devices, in addition to conventional MOSFET devices. Those procedures requiring STM are limited by low throughput. They are not manufacturable procedures, but can be used to develop devices for the research community. A working micro-STM that includes integrated actuators and a doped Si tip has been fabricated by micromachining a Si wafer [82]. Advances in Si micromachining leading to arrays of these devices could lead to a more manufacturable STM-based nanolithography process.

References

1. X.L. Zhou, X.Y. Zhu, and J.M. White, Surf. Sci. Rep. **13**, 77 (1991)
2. D. Menzel and R. Gomer, J. Chem. Phys. **41**, 3311 (1964)
3. P.A. Redhead, Can. J. Phys. **42**, 886 (1964)
4. P.R. Antoniewicz, Phys. Rev. B **21**, 3811 (1980)
5. G. Jiang, J.C. Polanyi, and D. Rogers, Surf. Sci. **544**, 147 (2003)
6. R.A. Wolkow and D.J. Moffatt, J. Chem. Phys. **103**, 10696 (1995)
7. T.C. Shen, C. Wang, G.C. Abeln, J.R. Tucker, J.W. Lyding, P. Avouris, and R.E. Walkup, Science **268**, 1590 (1995)

8. P. Avouris, R.E. Walkup, A.R. Rossi, H.C. Akpati, P. Norlander, T.C. Shen, G.C. Abeln, and J.W. Lyding, Surf. Sci. **363**, 368 (1996)
9. J. Tersoff and D.R. Hamann, Phys. Rev. Lett. **50**, 1998 (1983); Phys. Rev. B **31**, 805 (1985)
10. N.D. Lang, Phys. Rev B **34**, 5947 (1986)
11. A. Selloni, P. Carnevali, E. Tosatti, and C.D. Chen, Phys. Rev. B **31**, 2602 (1985)
12. R.H. Fowler and L.W. Nordheim, Proc. R. Soc. London Ser A **119**, 173 (1928)
13. R. Martel, P. Avouris, and I.W. Lyo, Science **272**, 385 (1996)
14. B.C. Stipe, M.A. Rezaei, W. Ho, S. Gao, M. Persson, and B.I. Lundqvist, Phy. Rev. Lett. **78**, 4410 (1997)
15. P. Avouris, I.W. Lyo, and Y. Hasegawa, J. Vac. Sci. Technol. A **11**, 1725 (1993)
16. P. Avouris and I. W. Lyo Appl. Surf. Sci. **60/61**, 426 (1992)
17. T.T. Tsong, *Atom Probe Field Ion Microscopy* (Cambridge University Press, Cambridge, 1990)
18. I.W. Lyo and P. Avouris, Science **253**, 173 (1991)
19. H.M. Mamin, P.H. Guether, and D. Rugar, Phy. Rev. Lett. **65**, 2418 (1990)
20. R. Gomer and L.W. Swanson, J. Chem. Phys. **38**, 1613 (1963)
21. F. Thibaudau, J.R. Roche, and F. Salvan, Appl. Phys. Lett. **64**, 523 (1994)
22. A. Kobayashi, F. Grey, E. Snyder, and M. Aono, Phys. Rev. B **49**, 8067 (1994)
23. B.N.J. Persson and P. Avouris, Chem. Phys. Lett. **242**, 483 (1995)
24. J.J. Lander, G.W. Gobeli, and J. Morrison, J. Appl. Phys. **34**, 2298 (1963)
25. J.A. Stroscio, R.M. Feenstra, and A.P. Fein, J. Vac. Sci. Technol. A **4**, 1315 (1986)
26. K.C. Pandey, Phys. Rev. Lett. **47**, 1913 (1981); **49**, 223 (1982)
27. H. Iwasaki, S. Hasegawa, M. Akizuki, S. Li, S. Nakamura, and J. Kanamori, J. Phys. Soc. Jpn. **56**, 3425 (1987); J. Chevrier, L.T. Vinh, and A. Cruz, Surf. Sci. Lett. **268**, L261 (1992)
28. R.E. Schlier and H.E. Farnsworth, J. Chem. Phys. **30**, 917 (1959)
29. R.M. Feenstra and M.A. Lutz, Phys. Rev. B **42**, 5391 (1990)
30. K. Takayanagi, Y. Tanishiro, M. Takahashi and S. Takahashi, J. Vac. Sci. Technol A **3**, 1502 (1985)
31. B.D. Thoms and J.E. Butler, Surf. Sci. **328**, 291 (1995); T.W. Mercer and P.E. Pehrsson, Surf. Sci. **399**, L327 (1998)
32. R.A. Wolkow, Phys. Rev. Lett. **68**, 2636 (1992)
33. W. Kerns and D.A. Puotinen, RCA Rev. **31**, 187 (1970)
34. G.S. Higashi, Y.J. Chabal, G.W. Trucks, and K. Raghovachari, Appl. Phys. Lett. **56**, 656 (1990)
35. G.S. Higashi, R.S. Becker, Y.J. Chabal, and A.J. Becker, Appl. Phys. Lett. **58**, 1656 (1991)
36. S.R. Kasi, M. Liehr, P.A. Thiry, H. Dallapostok, and M. Offenberg, Appl. Phys. Lett. **59**, 108 (1991)
37. J.N. Smith and W.L. Fite, J. Chem. Phys. **37**, 898 (1962)
38. J.J. Boland, Surf. Sci. **261**, 17 (1992)
39. F. Owman and P. Mårtensson, Surf. Sci. Lett. **303**, L367 (1994); Surf. Sci. **324**, 211 (1995)
40. J.J. Boland, Phys. Rev. Lett. **65**, 3325 (1990)
41. J.A. Dagata, J. Schneir, H.H. Harary, C.J. Evans, M.T. Postek, and J. Bennett, Appl. Phys. Lett. **56**, 2001 (1990)

42. R.S. Becker, Y.J. Chabal, G.S. Higashi, and A.J. Becker, Phy. Rev. Lett. **65**, 1917 (1990)
43. J.J. Boland, Surf. Sci. **244**, 1 (1991)
44. M. Schwartzhopff, P. Radojkovic, M. Enachescu, E. Hartmann, and F. Koch, J. Vac. Sci. Technol. B **14**, 1336 (1996)
45. D. Rogers and H. Nejoh, J. Vac. Sci. Technol B **17**, 1323 (1999)
46. J. Wintterlin and P. Avouris, J. Chem. Phys. **100**, 687 (1994)
47. D.P. Adams, T.M. Mayer, and B.S. Swartzentruber, J. Vac. Sci. Technol B **14**, 1642 (1996)
48. J.W. Lyding, T.C. Shen, G.C. Abeln, C. Wang, and J.R. Tucker, Nanotechnology **7**, 128 (1996)
49. C. Syrykh, J.P. Nys, B. Legrand, and D. Stivenard, J. Appl. Phys. **85**, 3887 (1999)
50. M. Schlter and M.L. Cohen, Phys. Rev. B **17**, 716 (1978)
51. J.R. Tucker, C. Wang, and T.C. Shen, Nanotechnology **7**, 275 (1996)
52. G. Schulze and M. Henzler, Surf. Sci. **124**, 336 (1983); P. Gupta, V.L. Colvin, and S.M. George, Phys. Rev. B **37**, 8234 (1988)
53. D. Rogers and T. Tiedje, Phys. Rev. B **53**, R13227 (1996)
54. P. Avouris, R. Martel, T. Hertel, and R. Sandstrom, Appl. Phys. A **66**, S659 (1998)
55. E.S. Snow, P.M. Campbell, and P.J. McMarr, Appl. Phys.Lett. **63**, 749 (1993)
56. E.D. Palik, V.M. Bermudez, and O.J. Glembocki, J. Electrochem. Soc. **132**, 871 (1985)
57. S.C. Minne, H.T. Soh, P. Flueckiger, and C.F. Quate, Appl. Phys. Lett. **66**, 703 (1995)
58. T.C. Shen, C. Wang, and J.R. Tucker, Phy. Rev. Lett. **78**, 1271 (1997)
59. M. Sakurai, C. Thirstrup, and M. Aono, Phy. Rev. B **62**, 16167 (2000)
60. T. Hashizume, S. Heike, M.I. Lutwyche, S. Watanabe, and Y. Wada, Surf. Sci. **386**, 161 (1997)
61. G. Palasantzas, B. Ilge, J. De Nijs, and L.J. Geerligs, J. Appl. Phys. **85**, 1907 (1999)
62. T.C. Shen, C. Wang, J.W. Lyding, and J.R. Tucker, Appl. Phys. Lett. **66**, 976 (1995)
63. J.W. Lyding, T.C. Shen, G.C. Abeln, C. Wang, and J.R. Tucker, Nanotechnology **7**, 128 (1996)
64. D.P. Adams, T.M. Mayer, and B.S. Swartzentruber, Appl. Phys. Lett. **68**, 2210 (1996)
65. G.C. Abeln, M.C. Hersam, D.S. Thompson, S.T. Hwang, H. Choi, J.S. Moore, and J.W. Lyding, J. Vac. Sci. Technol. B **16** 3874 (1998)
66. J.L. O'Brien, S.R. Schofield, M.Y. Simmons, R.G. Clark, A.S. Dzurak, N.J. Curson, B.E. Kane, M.S. McAlpine, M.E. Hawley, and G.W. Brown, Phys. Rev. B **64**, 161401 (2001)
67. R.J. Hamers, P. Avouris, and F. Bozso, Phys. Rev. Lett. **59**, 2071 (1987)
68. H. Choi and S.T. Hwang, Chem. Mater. **10**, 2323 (1998)
69. K. Masu and K. Tsubouchi, J. Vac. Sci. Technol. B **12**, 3270 (1994)
70. B.E. Kane, Nature (London) **393**, 133 (1998)
71. D.S. Lin, T.S. Ku, and R.P. Chen, Phys. Rev. B **61**, 2799 (2000)
72. L. Oberbeck, N.J. Curson, M.Y. Simmons, R. Brenner, A.R. Hamilton, S.R. Schofield, and R.G. Clark, Appl. Phys. Lett. **81**, 3197 (2002)

73. J.S. Hovis, H. Lu, and R.J. Hamers, J. Phys. Chem. **102**, 6873 (1998)
74. R.J. Hamers, J.S. Hovis, S. Lee, H. Liu, and J. Shan, J. Phys. Chem. **101**, 1489 (1997)
75. R.A. Wolkow and D.J. Moffatt, J. Chem. Phys. **103**, 10696 (1995)
76. F.Y. Naumkin, J.C. Polanyi, D. Rogers, W. Hofer, and A. Fisher, Surf. Sci. **547**, 324 (2003)
77. R.A. Wolkow, Annu. Rev. Phys. Chem. **50**, 413 (1999)
78. P.H. Lu, J.C. Polanyi, and D. Rogers, J. Chem. Phys. **111**, 9905 (1999)
79. P.H. Lu, J.C. Polanyi, and D. Rogers, J. Chem. Phys. **112**, 11005 (2000)
80. I.D. Petsalakis, J.C. Polanyi, and G. Theodorakopoulos, Surf. Sci. **544**, 162 (2003)
81. X.H. Chen, Q. Kong, J.C. Polanyi, and D. Rogers, Surf. Sci **376**, 77 (1997)
82. Y. Xu, N.C. MacDonald, and S.A. Miller, Appl. Phys. Lett. **67**, 2305 (1995)

3 Adsorption Behavior of Single Molecules on Surfaces Formed by Molecular Assemblies Studied by Scanning Tunneling Microscopy

C. Wang, Q.D. Zeng, S.B. Lei, Y.L. Yang, D.X. Wu, and X.H. Kong

3.1 Introduction

Knowledge of molecular self-assemblies has greatly advanced in the past decade. The principles governing the assembling process of molecular building blocks, envisioned by pioneering researchers, have been manifested in a growing number of systems. Among these studies, scanning tunneling microscopy (STM) has added appreciably to the ability to resolve the fine details of molecular assemblies, presenting us with a wide range of assembled 2D structures. The involvement of weak intermolecular interactions, van der Waals and hydrogen bonding, is common to molecular assemblies [1]. The novel properties associated with the assembled molecular architectures have been the subject of many investigations. Among these studies, using the assembled molecular structures as a support medium for further building of heterogeneous structures is one of the important topics. Molecular assemblies in these circumstances behave very similarly to typical surface structures. The introduction of molecular assemblies as support structures leads to novel structural and electronic characteristics that could open up broad opportunities for designing functional surfaces.

One direct outcome of such a molecular surface could be the controlled growth of a molecular overlayer structure. This would be essential for understanding the interface structures and properties of molecular-based devices. As a matter of fact, such investigations have been actively pursued in designing liquid-crystal displays using molecularly decorated surfaces to direct the texture of overlayer liquid-crystal films [2, 3]. The molecular-level understanding of the adsorption and alignment effect of the molecular surface is deemed critical for mechanistic studies on the formation process of overlayer films.

It has been demonstrated by a number of studies that such molecular assemblies can possess properties analogous to those of pristine solid surfaces. Both functional groups of the assemblies and the cavities within the assembly structures can be explored as the adsorption sites. Adsorption behavior, or the response of the molecular assemblies to the guest species introduced, can be studied with typical methodology for pristine surfaces. The guest species or adsorbate molecules can be introduced through either codeposition from solutions or evaporation under vacuum conditions. The nature of the adsorbate–substrate interaction in the resulting complex structure is identical for both preparation approaches. The uniqueness of such molecularly decorated surfaces is that the adsorption and diffusion barrier can be readily designed and constructed by utilizing a rich variety of functional groups that can be incorporated into the molecular assemblies.

As the basis of studying adsorption properties of molecular assemblies, it should be noted that self-assembling of nanometer-sized building blocks into designed molecular architectures represents one of the major goals of supramolecular chemistry and material science [1]. Covalent and noncovalent interactions such as metal–ligand binding, hydrogen bonding and electrostatic interactions have been successfully used for the design of supramolecular architectures. Hydrogen bonds have the advantage of selectivity and directionality, which are especially important in building biological nanostructures. For other interactions, such as van der Waals and hydrophobic interaction, the lack of directional selectivity makes them generally less desirable in constructing molecular structures. The extensive studies of self-assembled molecular structures have greatly enhanced our knowledge of the optical, electrical, frictional and other properties of molecular materials. The expanding interest and potential applications of the vast variety of molecular structures have stimulated many studies in this general field.

The construction of 2D molecular nanostructures could be achieved at clean metallic and semiconductor surfaces, or at passivated surfaces that are covered with inert spacer molecules. The assembling behavior on those surfaces could be very different as a result of surface chemical compositions. For example, the existence of an alkane layer is deemed an approach to electrically isolate the adsorbate molecules from the conductive substrate, while still allowing electrons to tunnel through. The effect can also be reflected through the changes in the magnitude of the adsorption and diffusion barriers. With the presence of various functional groups periodically distributed in the buffer layer, one could encounter anisotropic diffusion barriers which are dependent on the functional groups in the molecular assembly.

We will illustrate in this work that the molecular surfaces can respond to the molecular adsorbates. Selective adsorption and inclusive adsorption are two major types of responses identified by using STM. These effects are associated with the weak interactions between the adsorbates and molecular assemblies as the supporting surface without altering the general characteristics of the molecular surface.

3.2 Selective Adsorption Behavior of Molecular Surfaces

The stability of adsorbed organic species on surfaces results from the interplay of adsorbate–substrate interactions, where the substrate is formed by molecular assemblies. In the case of the assembled adsorbate structure, adsorbate–adsorbate interaction will also contribute to the adsorption stability. Common to these intermolecular interactions are the typically electrostatic and van der Waals forces. Understanding and controlling these interactions are essential in preparing a designed organic molecular architecture [4].

The driving mechanisms for molecular adsorption atop molecular assemblies can be derived from the typical adsorption principles on solid surfaces. Considerable overlap of common concerns are shared by these studies. The stability of adsorbed molecules within the basal plane may be affected by thermal motion of molecules or the disturbances of the scanning tunneling microscope tip in the scanning process. A significant amount of effort has been put into improving substrate and preparation conditions to achieve optimal adsorbate stability. Another aspect of concern is the interaction between adsorbate and substrate. It should be cautioned that strong interactions may have an appreciable impact on the molecules of concern, such as in systems of organic molecules adsorbed on metal and semiconducting surfaces. Studies on organic layers grown on top of an alkanethiol layer also revealed an interesting modulation effect [5].

3.2.1 Physisorption of Single Molecules on Molecular Surfaces

Single-molecule adsorption on molecular modified surfaces has been observed in several systems mainly using alkane derivatives as the support surface. It is well known that linear alkane derivatives form parallel lamella patterns on graphite surfaces in ambient conditions. When precovered on the support surface, the molecular lamella structures introduce inherently the heterogeneous adsorption sites and anisotropic diffusion barriers in association with the functional groups. The presence of heterogeneous adsorption sites could result in selective adsorption of single molecules. The adsorbed species would also experience the anisotropic diffusion barrier and organize in a restricted manner.

Site-selective adsorption of copper phthalocyanine (CuPc) has been observed atop organic surfaces of monolayers of various alkane derivatives (stearic acid, 1-octadecanol and 1-iodooctaecane) adsorbed on the surface of highly oriented pyrolytic graphite (HOPG). Isolated and paired CuPc molecules were detected on top of the stearic acid monolayer as seen in Fig. 3.1 [6]. As observed in the experiments, these CuPc molecules all exclusively located on top of the alkane part of stearic acid lamellae; CuPc molecules adsorbed on the regions that were ascribed to the location of carboxyl groups were never detected.

Two representative adsorption sites were considered in the molecular mechanics simulations shown schematically in Fig. 3.2, i.e., site II, in which CuPc adsorbs on the top of a trough linked by head-to-head functional groups, and site I, in which CuPc adsorbs on the alkyl moiety.

The calculated results indicated that the system potentials on site I are higher than those on site II by above 21 kJ/mol for three alkane derivative systems (Fig. 3.3). This may be caused by two factors. First, the trough linked

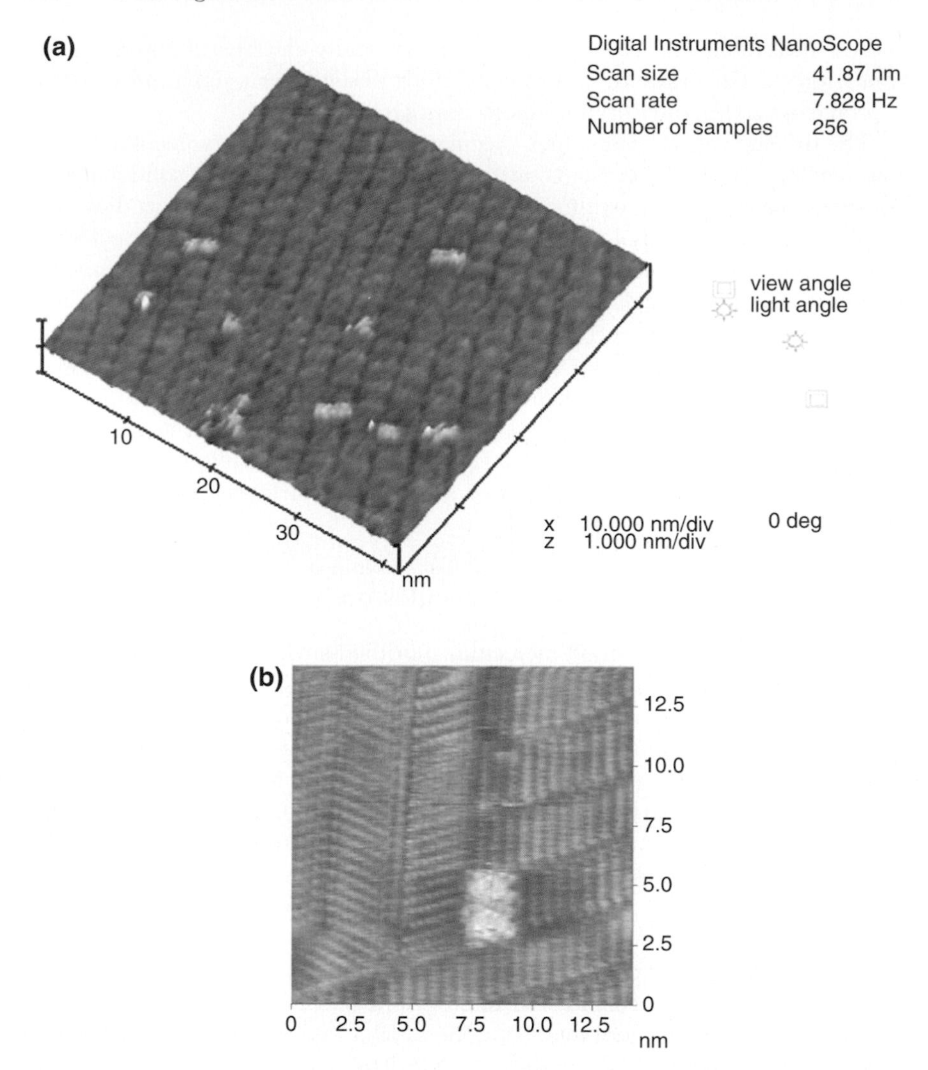

Fig. 3.1. *Top*: Isolated copper phthalocyanine (CuPc) selectively adsorbed on the hydrocarbon-chain portion of stearic acid. *Bottom*: CuPc adsorbed on a monolayer of 1-octadecanol [6]

by head-to-head functional groups is about 3 Å in width and the number of atoms in the trough is less than for other sites of the organic monolayers. Consequently the van der Waals interaction between adsorbed CuPc and underlying organic substrate decreases when CuPc adsorbs on the top of the trough. Moreover, because the trough is linked by atoms and groups with large polarity, the electrostatic repulsion between the π ring of CuPc and the functional groups will be stronger than that between CuPc and the alkyl

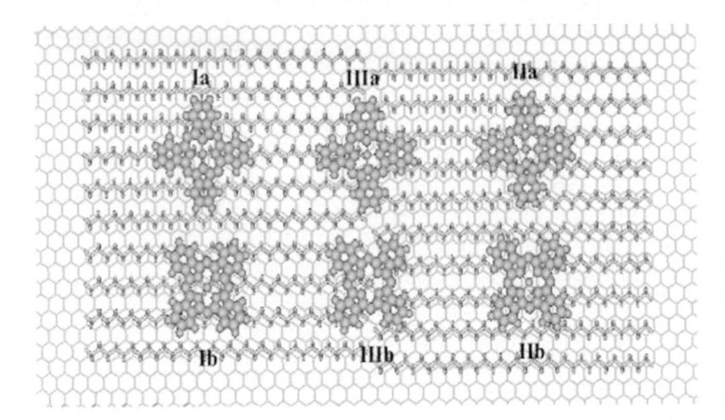

Fig. 3.2. Molecular mechanics simulation of different adsorption sites of a CuPc molecule atop alkane lamellae

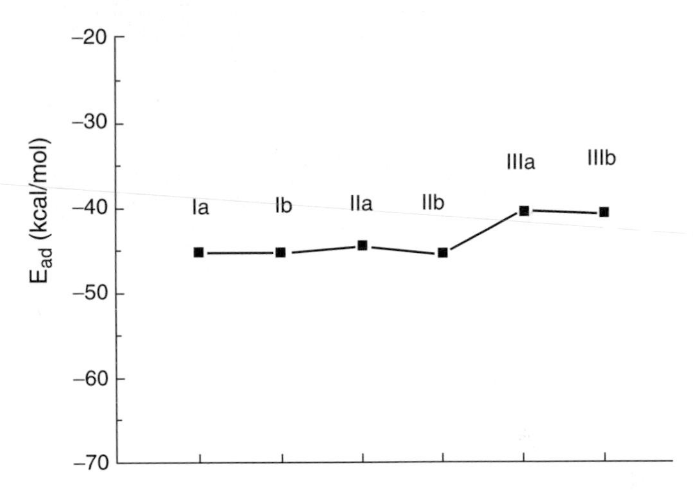

Fig. 3.3. Comparison of adsorption energy at the different adsorption sites shown in Fig. 3.2 revealing the existence of a diffusion barrier

moiety. As a consequence, selective adsorption of CuPc on the alkyl moiety could achieve a stable adsorption state. The selectivity is dependent on the relevant functional groups and could vary among alkane derivatives.

In another study, isolated CuPc molecules and clusters could be stabilized at the alkane part of the tridodecyl amine (TDA) lamellae (Fig. 3.4) [7]. When coadsorbed with TDA at a low CuPc to TDA ratio, CuPc dimers are most commonly observed, with a smaller population of quadrumers and hexamers. From the large-scale view in Fig. 3.5, one can observe two CuPc molecules located on both sides of the amine group of the TDA lamellae. The CuPc dimers appear to adsorb on top of the alkane part of the TDA lamellae.

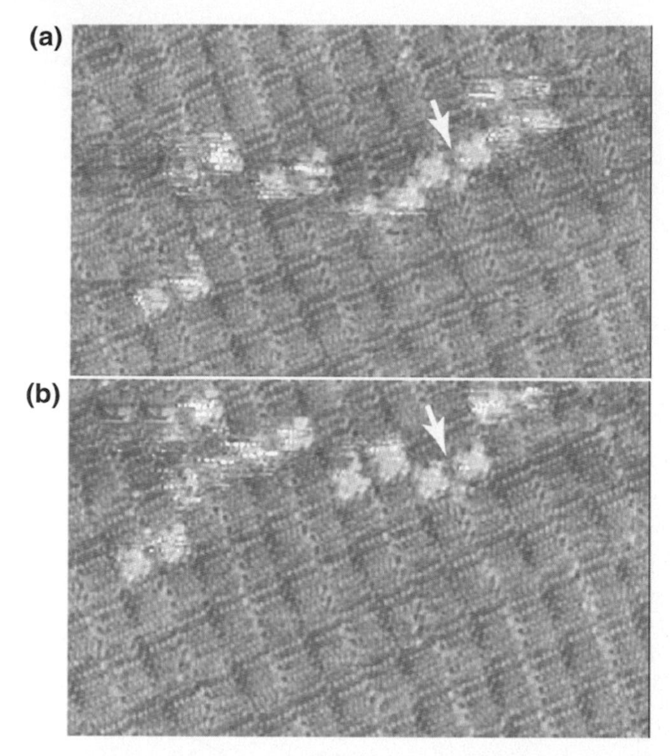

Fig. 3.4. Isolated CuPc molecules observed on the lamellae of tridodecyl amine (*TDA*). *Arrows* indicate the migration of the molecules in consecutive scans [7]

The lateral diffusion of the single CuPc molecules as well as clusters of adsorbed CuPc molecules was found exclusively along the direction of the TDA lamellae. Such highly directional diffusion behavior can be direct evidence for the 1D template effect of TDA lamellae. Such effects have never been observed on lamellae of simple alkanes, possibly owing to the lack of functional groups that could establish sufficient diffusion barriers for the overlayer adsorbates. This concept could be generalized to the construction of molecular templates for novel molecular nanostructures.

In addition to the previously described isolated adsorbates, well-ordered double-rowed CuPc molecular bands were developed at higher CuPc coverage as the result of a molecular template of TDA lamellae. A bimolecular CuPc band can be readily observed in Fig. 3.5 [7]. These bimolecular bands could extend to hundreds of nanometers without appreciable distortion. As measured from the STM images, the repeat period of this assembling structure is consistent with the 3.37 ± 0.05 nm width of the TDA lamellae, which suggests the ordering of CuPc molecules is induced by the TDA lamellae. High-resolution STM observation reveals that the arrangement of CuPc in these double molecular bands is the same as that in the 2D crystals: joggled

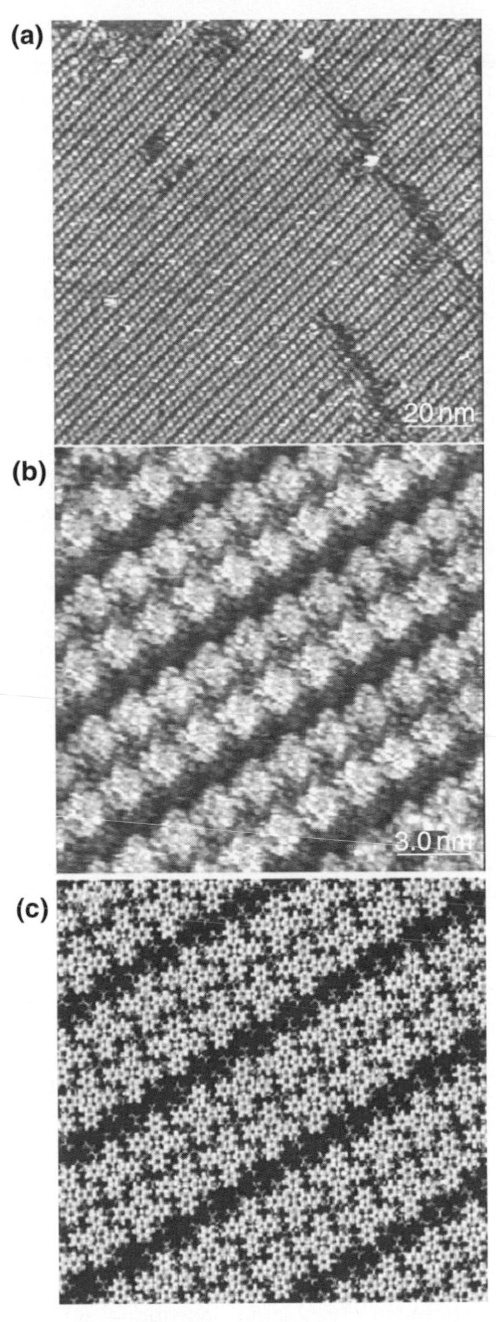

Fig. 3.5. Large-scale view (**a**) and high-resolution image (**b**) of CuPc double molecular bands. (**c**) The molecular model corresponding to the structure shown in (**b**) [7]

together with each other, showing an intermolecular distance of 1.45 nm. There is an apparent dark trough between these bimolecular molecular bands. The width of this trough is about 0.7 nm, larger than the 0.3–0.4-nm trough which separates the TDA lamellae. A molecular model of this band structure has been proposed as shown in Fig. 3.5c.

The assembled individual molecular domains can be identified atop the support molecular structures. As an example, using solutions of binary mixtures of CuPc and tetratriaconta ($C_{34}H_{70}$) as a sample system, nanometer-sized domains of CuPc molecules (both unsubstituted and octa-alkylated ones) can form with high stability on top of the alkane overlayer adsorbed on a graphite surface [8]. The CuPc molecules are close-packed within the domains.

It can be seen from images that the packing symmetry of the Pc domain is a quasi-fourfold one that can be attributed to the symmetry of the CuPc molecules, and is independent of that of the alkane buffer layer. Moreover, no preferential orientation of the CuPc domain is observed relative to the underneath alkane lamellae. This could serve as an indication of reduced adsorbate – substrate interaction. In general, the adsorbed layer could be influenced by substrate structures, causing distortions of the bulk molecular structure. A weak molecule – substrate interaction would lead to a molecular architecture with less distortion compared with its bulk phase, as in the case for Pc on graphite and MoS_2 [9], as well as 3,4,9,10-perylenetetracarboxylic dianhydride (PTCDA) on a alkanethiol layer [5]. For a strong molecule – substrate interaction, the substrate will have significant impact on the structure of the molecular overlayer, causing distortions of the molecular lattice. It should be mentioned here that the experimental conditions described in this study are very different from those of standard epitaxial methods, in which the growth rate and temperature can be precisely controlled. The domain structures observed in this work may also be affected by the kinetic factors that are prevalent in the process of solvent evaporation.

It may be worth mentioning that the existence of the alkane buffer layer rules out the electrostatic interactions between CuPc layers and substrate. It is well known that electrostatic interaction is important in stabilizing metal Pc molecules on metal and semiconductor surfaces. This is also the origin of the effect of the substrate lattice (symmetry and spacing) on the molecular overlayer [4]. However, the enhanced electrostatic interaction between adsorbate and substrate could hamper the formation of large and uniform-sized domains, caused by the enhanced intermolecular electrostatic repulsion. This is an important factor in forming molecular layers. Efforts have been seen to reduce this effect by passivated Si surfaces [10].

We now switch to another molecule—coronene, which can be considered the smallest possible flake of a graphite sheet saturated by hydrogen atoms with D_{6h} symmetry [11]. A 2D assembled structure of coronene was obtained using the lamellar structure of 4-octadecylnitrobenzene (ONB) molecules as a buffer layer. Atop the ONB lamellae, 2D islands of coronene can be readily observed, in which coronene molecules are arranged sixfold symmetry. The coronene monolayer with a 2D crystalline structure can extend to several hundred nanometers and is stable enough to endure repeated scanning. High-resolution images can be obtained for this assembly as shown in Figs. 3.6 and 3.7. Owing to the highly symmetric structure of coronene molecules and the weak interactions between coronene and the buffer layer molecules, it is energetically favorable to form a close-packed 2D monolayer.

It is worthwhile noticing that the measured coronene lattice constants ($a = b = 10.9 \pm 0.2$ Å, and the angle α in the unit cell, $60 \pm 2°$) are in good agreement with the those of a closely packed coronene monolayer adsorbed on Ag(111) (measured lattice constant is 1.15 nm) [12] and on HOPG (measured lattice constant is 1.1 nm) [13]. The high-resolution image in Fig. 3.7, top, reveals coronene molecules as a ring-type structure. This ring structure is in qualitative agreement with the calculated density of state of the coronene molecule calculated at a 2-Å plane above the surface using density functional theory computation (shown in Fig. 3.7, bottom). The specific positions with higher density cannot be distinguished clearly in the ring structure of the STM image.

3.2.2 Hydrogen-Bond Assisted Selective Adsorption

In addition to the selective adsorption behavior on alkane lamellae, hydrogen-bond interaction can also be employed to facilitate the adsorption process. The directionality and selectivity of the hydrogen-bond characteristics are of special importance in construction of chemically decorated surfaces.

Site-selective adsorption of urea molecules was observed on the lamella templates of a double-alkyl amino acid [14]. The unprotected amino acid groups were found to be the preferential adsorption sites for urea molecules as shown in Fig. 3.8. In contrast, the lamella structure of a single-alkyl amino acid did not show any adsorption effect as the result of dimer formation that saturates the functional groups.

The hydrogen-bond-assisted adsorption effect has also been demonstrated with the lamella of TDA molecules. The conformation of the nitrogen atom in the amine group of TDA is tetrahedral, in which the nitrogen atom sits on one peak of this tetrahedron (Fig. 3.9). Since the C–N bond is dipolar, amine molecules are also dipolar, in which the nitrogen is partially negatively charged. Thus, when amine molecules adsorb onto an inert surface of graphite, there exists a net dipole moment pointing nearly perpendicular to the surface. As an example, benzoic acid was found exclusively adsorbed onto the TDA assembly at the sites of amino groups [15]. This shows the possibility of using alkane derivative lamellae as templates to direct site-specific adsorption, diffusion and assembly structures of other organic molecules.

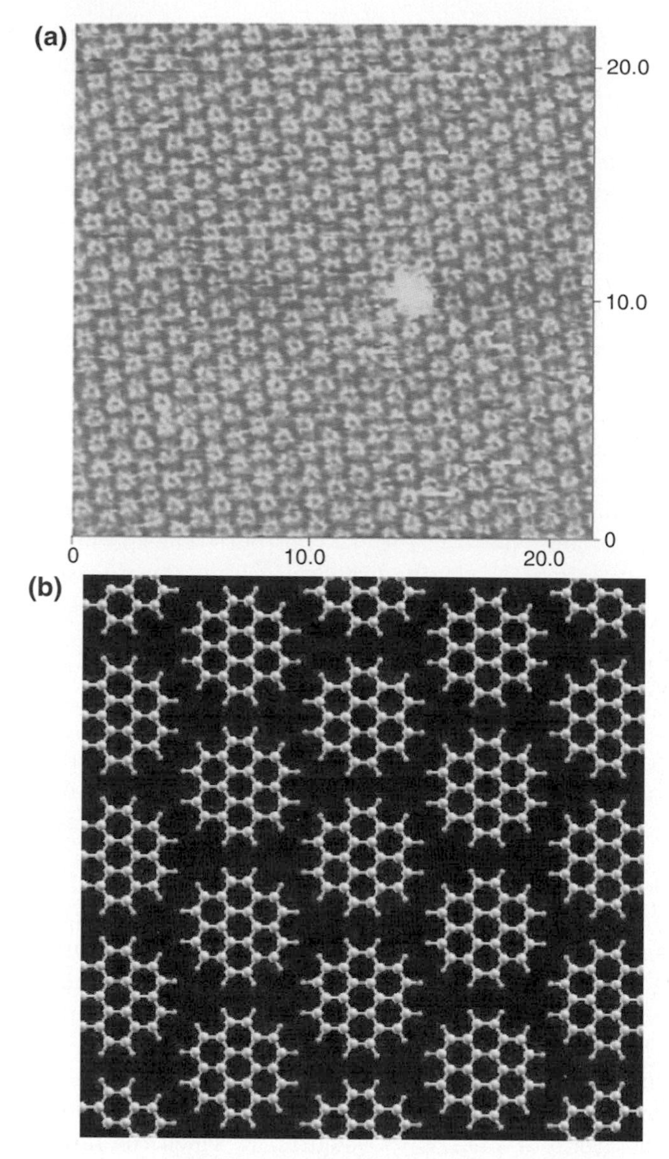

Fig. 3.6. *Top*: Scanning tunneling microscopy (*STM*) image (21.8 nm × 21.8 nm) of a 2D assembly structure of coronene using 4-octadecylnitrobenzene lamellae as a buffer layer. The imaging conditions are 338 pA and 930 mV. *Bottom*: The proposed assembly structure of coronene domains [11]

3.3 Inclusive Adsorption Behavior of Molecular Surfaces

In addition to the results already presented concerning the chemical functional groups of the molecular assemblies being utilized as the direct sites

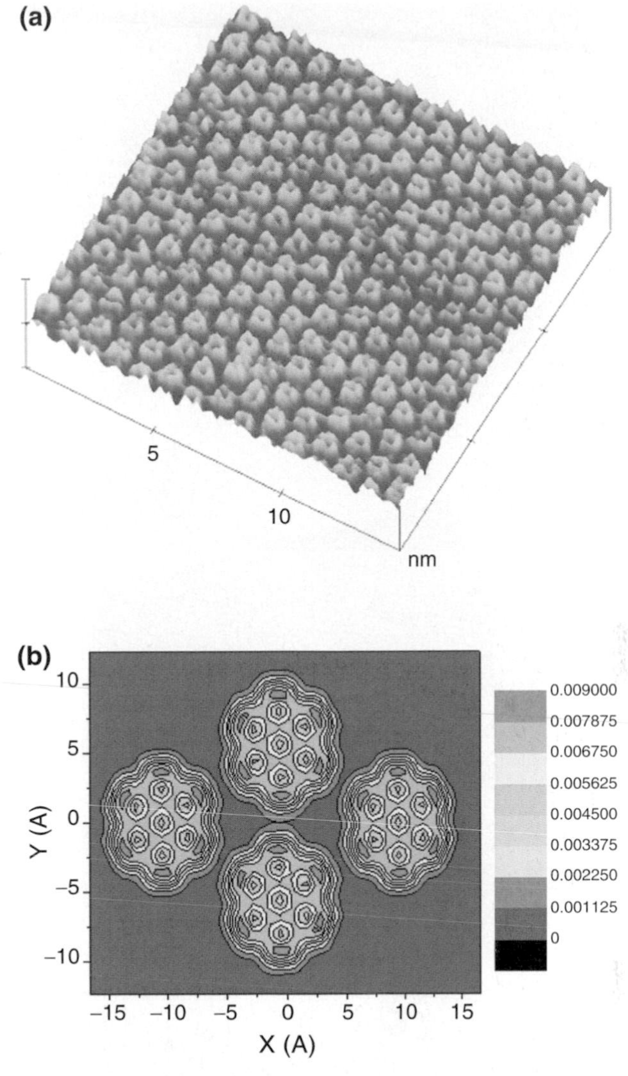

Fig. 3.7. *Top*: Surface view of a high-resolution STM image (14.6 nm × 14.6 nm) of coronene crystalline structure. The imaging conditions are the same as for Fig. 3.6. *Bottom*: The calculated density of state of coronene molecules 2 Å above the surface [11]

for adsorption, cavities with various symmetries and dimensions within the assembly structures can also be used as adsorption sites. Instead of staying atop the surfaces formed by molecular assemblies, the adsorbates are incorporated into the molecular assemblies. The molecules introduced interact with the original molecular assemblies and form various guest–host complex structures.

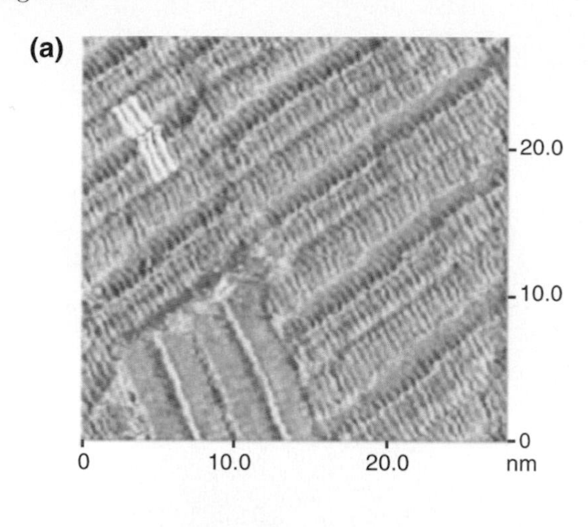

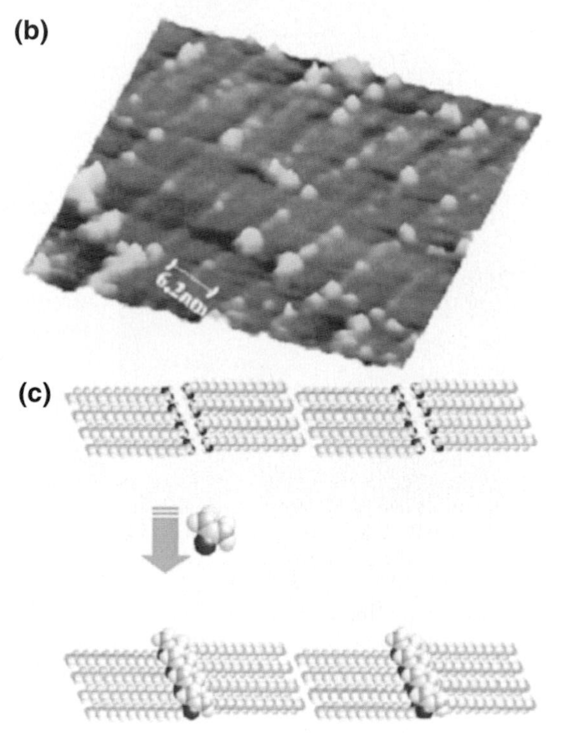

Fig. 3.8. (**a**), (**b**) Selective adsorption of urea molecules at the sites of amino groups. (**c**) Site-selective adsorption of guest molecules atop molecular templates [14]

Upon introducing the guest molecule (or adsorbate molecule) into the molecular assemblies, it will inevitably interact with the host structure. Such interaction may lead to certain structural effects on the host lattices,

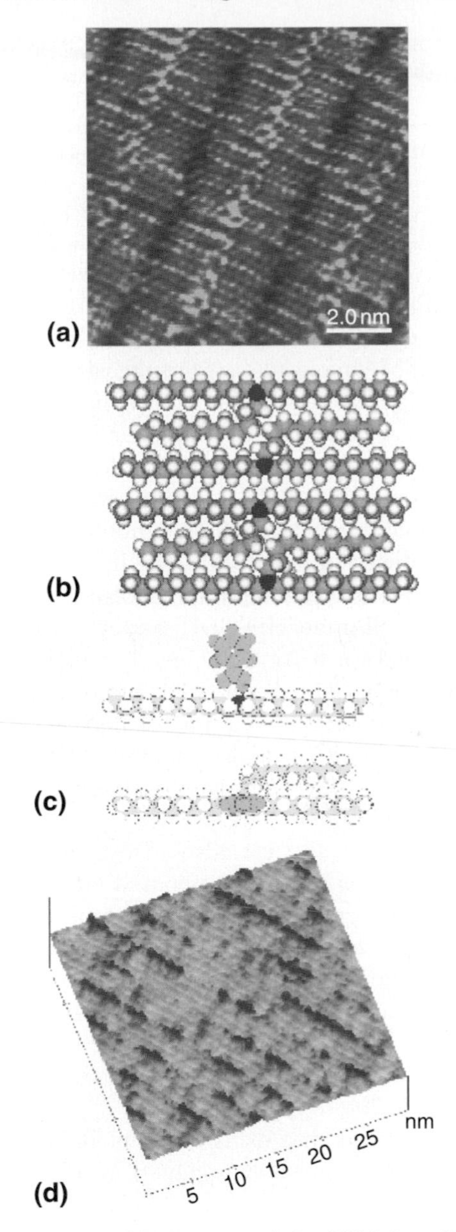

Fig. 3.9. (**a**) High-resolution STM image of the TDA lamellae structure and (**b**) proposed packing model of the lamellae structure. (**c**) Structural schematics illustrating the side and top views of the binding of a benzoic acid molecule to the TDA template. (**d**) STM image of benzoic acid adsorbed on TDA lamellae [14]

depending on the magnitude of the interaction. The category of molecular assemblies discussed in the following section has shown no appreciable structural deformation upon inclusion of the guest species, while the subsequent section presents examples of more flexible molecular lattices that display measurable structural deformations due to inclusion of guest molecules.

3.3.1 Rigid Supramolecular Networks for Inclusive Adsorptions

Two types of 1,3,5-Benzenetricarboxylic (Trimesic) acid (TMA) molecular networks have been identified on graphite surface at 25 K [16]. In addition to the hexagonal honeycomb lattice, the "flower" structure is also stabilized by hydrogen bonding with two different sized cavities [16]. Single TMA molecules could be observed as guest species in the cavities of the existing networks. The entrapped guest TMA molecules were suggested to be stabilized by two hydrogen bonds to the host network as shown in Fig. 3.10. Little effect on the host lattice geometry from the guest molecule can be observed. Recent studies showed that coronene and C_{60} molecules can also be immobilized within the TMA networks [17,18]. In addition, appreciable migration of C_{60} molecules can be induced at relatively high tunneling current (approximately 150 pA), providing evidence of mobility at the liquid–solid interface [17].

A large area of a hydrogen-bonded molecular network formed by perylenetetracarboxylic diimide (PTCDI) and 1,3,5-triazine-2,4,6-triamine (melamine) was obtained on a Ag/Si(111)-$\sqrt{3} \times \sqrt{3}$R30° surface under ultrahigh vacuum conditions [19]. Such a supramolecular network adopts a hexagonal pattern with a lattice constant of 34.6 Å. The open pores can be packed with up to seven C_{60} molecules (Fig. 3.11). The adsorption registry of C_{60} heptomers is clearly resolved and differs from that of C_{60} on Ag/Si(111)-$\sqrt{3} \times \sqrt{3}$R30°. This is a reflection of the effect due to the host lattice structure. In addition, C_{60} can also adsorb directly on top of PTCDI and melamine molecules, leading to a replicated lattice structure.

In a different approach, metal–organic coordination networks (MOCNs) have attracted much attention in recent years because of their potential applications in molecular recognition, gas storage and separation [20, 21]. A rich variety of nanostructures with well-defined shapes and geometries have been achieved using organic ligands interconnected by metallic nodes [22–24], among which the assembly of transition metal carboxylates at surfaces has received much effort [25–29].

It has been demonstrated that using different linker molecules of benzoic acid derivatives, one can obtain ladder-shaped MOCNs on Cu(100) surface. The cavities within the networks can effectively immobilize individual C_{60} molecules [30].

3.3.2 Flexible Supramolecular Networks

Measurable deformations of molecular lattices resulting from inclusive adsorption have been observed in several studies. In comparison with the robust host lattices connected by coordination or covalent bonds, flexible host

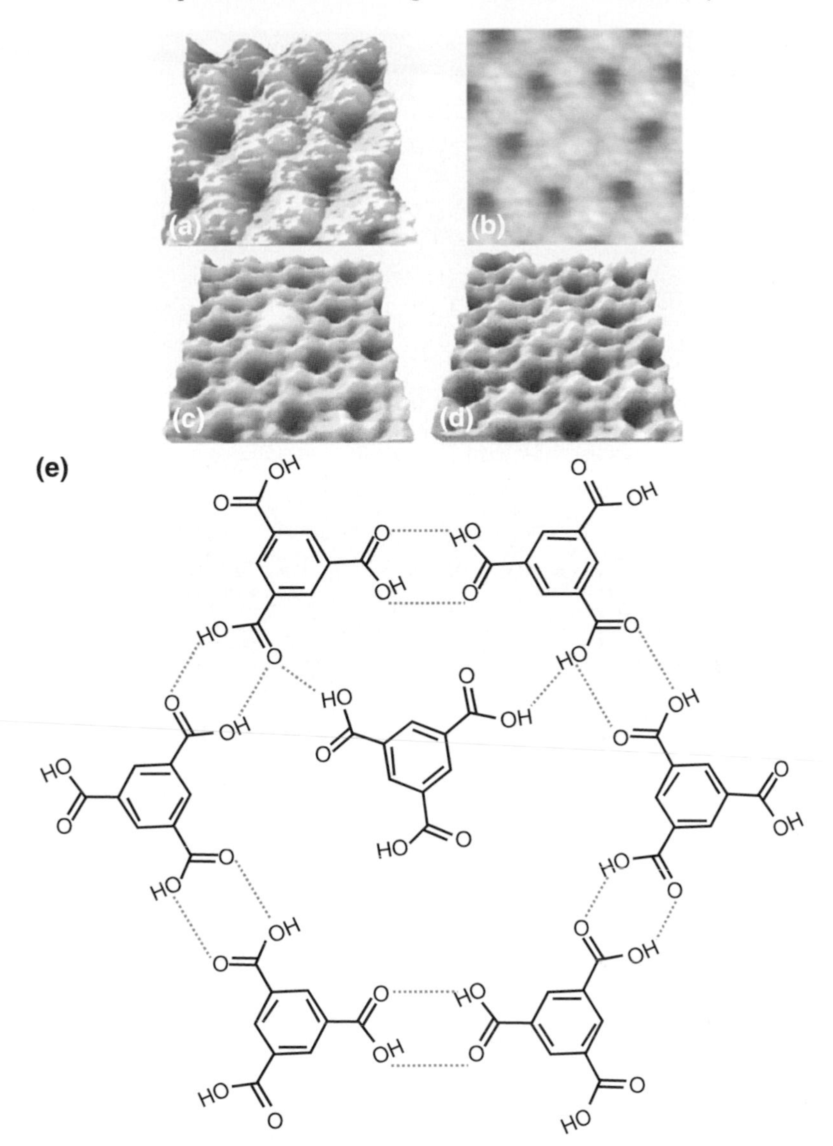

Fig. 3.10. (**a**)–(**d**) Inclusion of single TMA molecules in the TMA networks. (**e**) Structural model of a guest molecule trapped inside the host network [16]

structures can be obtained by using either rigid building blocks with weak interaction or flexible building blocks with strong interaction [31–33]. Such flexible host frameworks are important for molecular recognition such as in drug–receptor complexes [34]. In such studies, molecules with different shape and size are encapsulated within the same receptor cavity. The hydrogen

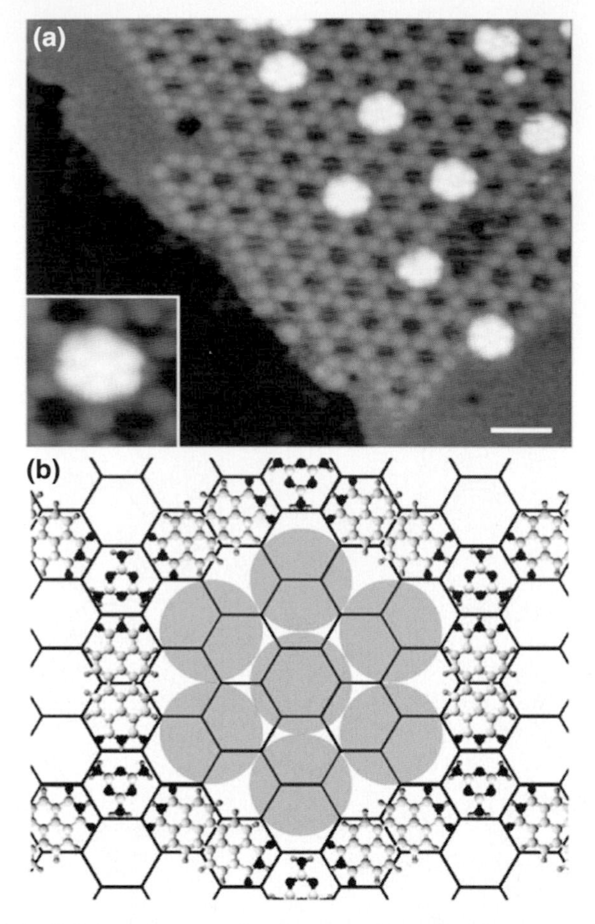

Fig. 3.11. (**a**) STM image of the hexagonal molecular network formed by perylenetetracarboxylic diimide and melamine adsorbed on a Ag/Si(111)-$\sqrt{3} \times \sqrt{3}$R30° surface. The *bright clusters* correspond to seven C_{60} molecules within a network cavity as illustrated in (**b**). The *scale bar* represent 5 nm [19]

bond and hydrophobic interactions between drug molecules and receptor enzymes could cause significant conformational changes of the host enzyme structures.

In an early study, quasi-quadratic molecular lattices was shown to be constructed by interdigitated alkylated CuPc molecules [35]. The trapping effect of 2D assembly of octa-alkoxyl-substituted Pc (PcOC8) for individual molecules of phthalocyanine, porphyrins, and calix[8]arene has been observed as shown in Fig. 3.12 [36].

It was observed that single molecules are trapped in quadratic lattices rather than hexagonal lattices, and domain boundaries are the preferential trapping sites as compared with the sites within the domains as shown in

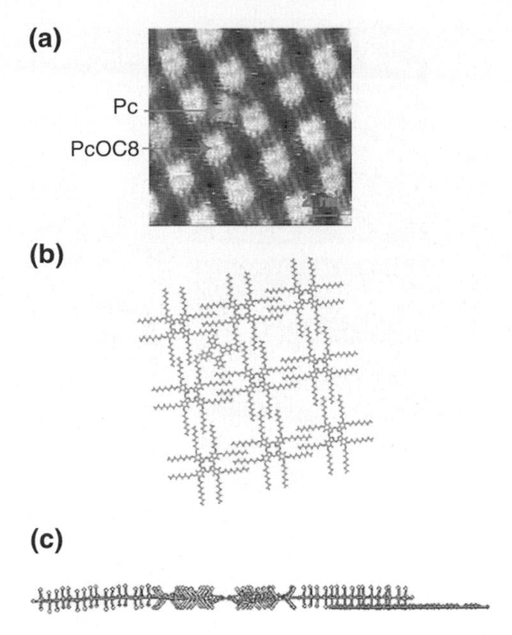

Fig. 3.12. (**a**) Immobilization of single molecules inside the quadratic lattices of alkylated phthalocyanines. No distortion of the host lattice geometry is observed. (**b**) Structural model of the guest molecule entrapment. (**c**) Side view of the assembly with the guest molecule [36]

Fig. 3.13 [36]. The observed trapping behaviors had analogues in the impurity segregation phenomena of point defects, dislocations and grain boundaries in solid-state materials.

The spacing increase between host molecular rows induced by molecular insertion is presented in Fig. 3.13c. It is evident that, with the increase of the intercalating molecular size, the distances of neighboring PcOC8 rows also increase correspondingly. This could be an indication of the flexibility of the PcOC8 lattice. The insertion of molecules in the initially packed lattices led to enhanced repulsive interaction between molecules. On the other hand, the overlapped alkyl parts could readjust to accommodate additional molecules and compensate for the associated increase in repulsions. The molecular networks have thus shown certain degrees of flexibility in trapping single molecules of different size and shape.

A monolayer of 1,3,5-tris(carboxymethoxy)benzene (TCMB) shows 2D hexagonal networks formed by hydrogen bonding, while a monolayer of 1,3,5-tris(10-carboxydecyloxy)benzene (TCDB) shows 2D tetragonal networks on HOPG [37]. The inclusion effect of hydrogen-bonded 2D networks of TCDB was demonstrated on the surface of HOPG in ambient conditions [32]. With the TCDB network as the host structure and CuPc, coronene, decacyclene and pentacene as guest molecules, the host–guest architectures were

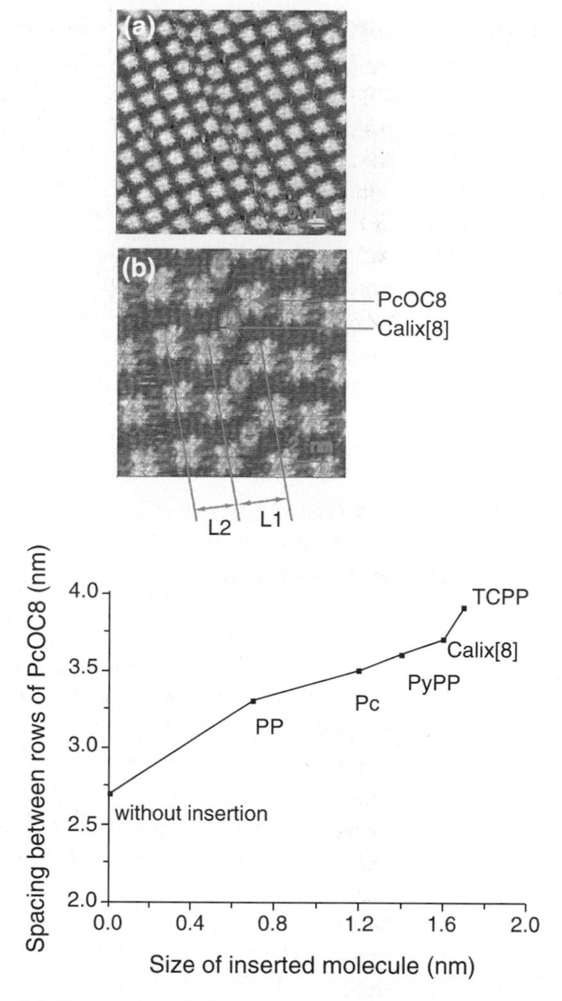

Fig. 3.13. (**a**), (**b**) Examples of the entrapment of single calix[8]arene molecules at different sites of the assembled host lattices. (**c**) Effect of the guest molecule insertion on the measured lattice spacing [36]

achieved (Fig. 3.14). An appreciable variation of the lattice dimension was observed as the result of the guest–host interaction. It has also been observed that either one or two Pc or coronene molecules can be found within the TCDB network, again suggesting that the network possesses a certain degree of flexibility.

In a separate work, it was reported that a supramolecular lamella-type host structure using terpyridine derivative [bis(2,2':6',2"-terpyridine)-4'-oxyhexadecane (BT-O-C16)] can accommodate short linear molecules (Fig. 3.15) [38]. It was identified that the inclusion of the guest species at

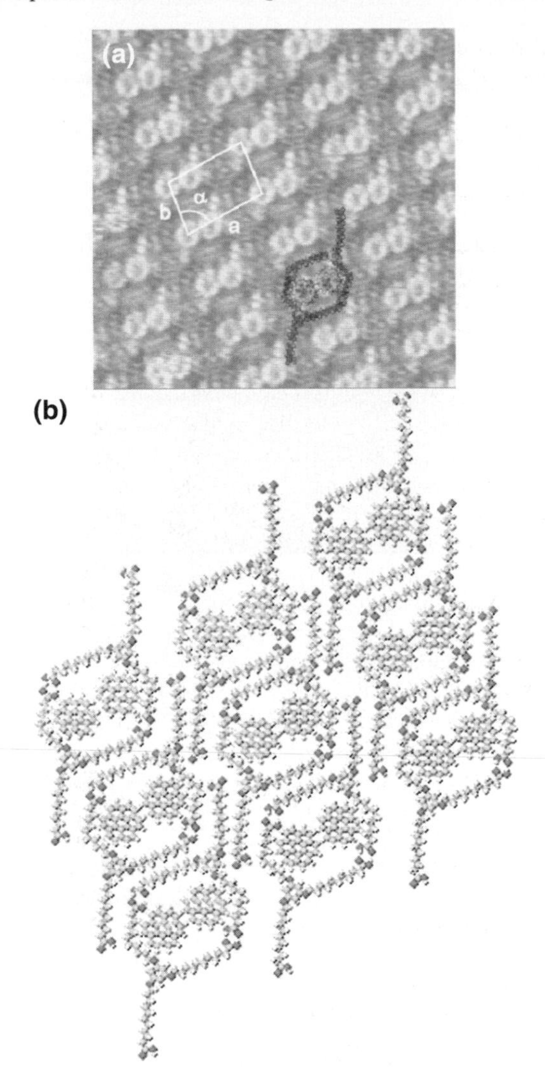

Fig. 3.14. (**a**) Guest–host structure formed by a 1,3,5-tris(10-carboxydecyloxy) benzene (TCDB) host lattice and coronene guest molecules. The scan size is 11.2 nm × 11.2 nm. (**b**) Molecular model for the TCDB network which is connected by hydrogen bonds [37]

ambient conditions is highly dependent on hydrogen bonds and molecular length, i.e., only molecules with appropriate length and terminal hydrogen bonds can be immobilized under ambient conditions.

From a combination of experimental results with theoretical analysis, it was concluded that the guest molecules of terephthalic acid (TPA) and azobenzene-4,4'-dicarboxylic acid are connected to the host lattice via

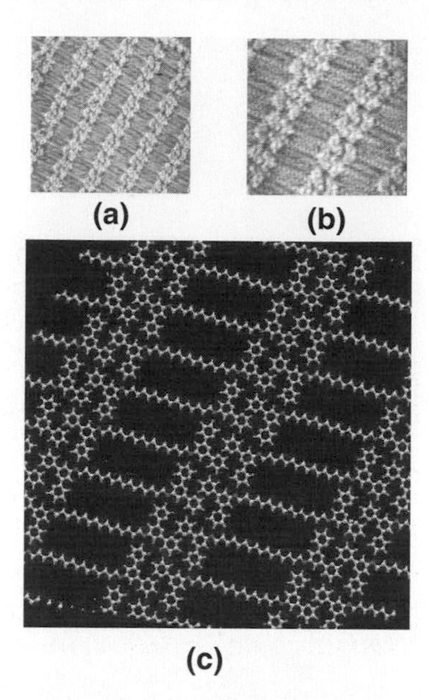

Fig. 3.15. (**a**) STM image (19 nm × 19 nm) of a self-assembled monolayer of bis(2,2':6',2"-terpyridine)-4'-oxyhexadecane (BT-O-C16) on highly oriented pyrolytic graphite (HOPG). The tunneling conditions are $I = 580\,\text{pA}$ and $V = 750\,\text{mV}$. (**b**) High-resolution STM image (10.3 nm × 10.3 nm) of BT-O-C16 self-assembled on HOPG. The tunneling conditions are $I = 580\,\text{pA}$ and $V = 750\,\text{mV}$. (**c**) Molecular model for the 2D packing of BT-O-C16 molecules [38]

hydrogen bonds. Furthermore, the guest species can induce varied conformational change of the host lattice, manifested as altered cavity dimensions. Such a study demonstrates that the STM method could provide useful insight into the flexibility of 2D host structures under the influence of guest molecule inclusions.

The inclusion properties of BT-O-C16 host networks can be readily observed by adding a small linear molecule such as TPA and azobenzene-4,4'-dicarboxyl acid. The general lamella characteristics of the host almost persist in the homogenous assembly of BT-O-C16, and the guest molecules were seen to insert themselves into the cavity of the original BT-O-C16 network, and stabilize within the moieties of the alkane ladders (Fig. 3.16). The TPA molecules appear in dimers via hydrogen bonding between carboxyl groups, while only one azobenzene-4,4'-dicarboxyl acid molecule can be inserted into the alkane ladder moiety.

Theoretical study shows that both types of the guest TPA dimer and azobenzene-4,4'-dicarboxyl acid strongly interact with the host BT-O-C16 framework by forming an O-H···N hydrogen bond through the end carboxyl group and the N atom in the terpyridine head group of BT-O-C16. The calculated the distance of O-H···N hydrogen bond is about 2.6 Å and the binding

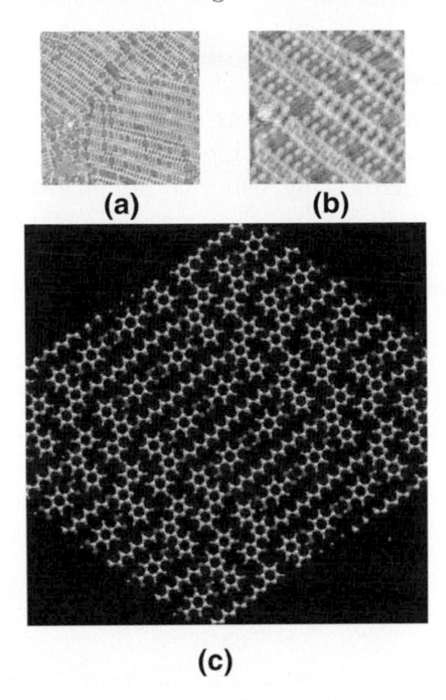

Fig. 3.16. (**a**) STM image (50 nm × 50 nm) of a self-assembled monolayer of terephthalic acid (TPA)/BT-O-C16 on HOPG. The tunneling conditions are $I = 600$ pA and $V = 600$ mV. (**b**) High-resolution STM image (15 nm × 15 nm) of TPA/BT-O-C16 coadsorbed on HOPG. The tunneling conditions are $I = 600$ pA and $V = 600$ mV. (**c**) Molecular model representing the assembly structure of TPA/BT-O-C16 molecules [38]

energy of the interaction is about 59.5 kcal/mol for TPA, and 2.7 Å and about 31.3 kcal/mol for azobenzene-4,4'-dicarboxyl acid.

Further analysis suggests that since the length of the azobenzene-4,4'-dicarboxyl acid molecule (about 1.5 nm) does not match the length of the alkane ladder moiety of BT-O-C16 (about 2.1 nm), the original O-H···O hydrogen bonds between the side-by-side BT-O-C16 dimer were broken, and the chamber length shrinks to 1.9 nm. However, the lamella network remains stable because the inserted azobenzene-4,4'-dicarboxyl acid molecule forms a strong O-H···N hydrogen bond with the BT-O-C16 dimer molecules.

3.4 Future Perspectives

It has been illustrated in this work that surfaces precovered with molecular assemblies with functional groups or cavities can develop unique adsorption characteristics such as high selectivity. These molecular assemblies can develop unique properties as a support surface. Such molecular surfaces could serve as a model system for studying bulk molecular materials that interact with various environments. The relevant properties of such molecular surfaces, such as site selectivity and diffusion barrier, can be readily studied just as those of metal or semiconductor surfaces.

The advances in molecular assembly architectures have added a new dimension to surface engineering using combinations of functional groups. Such capability will surely expand the scope of surface studies from pristine surfaces to chemically designed surfaces. Molecular-level studies of adsorption behavior at heterogeneous molecular interfaces could facilitate the mechanistic understanding of functional molecular structures which are central to various device applications, such as displays and sensing.

An example of the importance of molecular surface studies could be seen in template-assisted processing of materials, which has been a ubiquitous practice using lithographically prepared templates or self-assembled monolayer templates. The progress could help gain better control of the functional selectivity of the self-assembled molecular structures. The rich variety of organic functional groups may lead to a wide selection of possible templates.

References

1. Jean-Marie Lehn: *Supramolecular Chemistry: Concepts and Perspectives*, (VCH, Weinheim, Germany 1995)
2. for example, J. Hoogboom, P.M.L. Garcia, M.B.J. Otten, J.A.A.W. Elemans, J. Sly, S.V. Lazarenko, T. Rasing, A.E. Rowan, R.J.M. Nolte, J. Am. Chem. Soc. **127**, 11047 (2005)
3. X. Tong, G. Wang, A. Yavrian, T. Galstian, Y. Zhao, Adv. Mater. **17**, 370 (2005)
4. S.R. Forrest, Chem. Rev. **97**, 1793 (1997)
5. M.C. Gerstenberg, F. Schreiber, T.Y.B. Leung, G. Bracco, S.R. Forrest, G. Scoles, Phys. Rev. B **61**, 7678 (2000)
6. S.B. Lei, S.X. Yin, C. Wang, L.J. Wan, C.L. Bai, J. Phys. Chem. B **104**, 224 (2004)
7. S.B. Lei, C. Wang, L.J. Wan, C.L. Bai, J. Phys. Chem. B **108**, 1173 (2004)
8. B. Xu, S. Yin, C. Wang, X. Qui, Q. Zeng, C. Bai, J. Phys. Chem. B **104**, 10502 (2000)
9. (a) R. Strohmaier, C. Ludwig, J. Petersen, B. Gompf, W.J. Eisenmenger, Vac. Sci. Technol. B **14**, 1079–1082 (1996). (b) C. Ludwig, R. Strohmaier, J. Petersen, B. Gompf, W.J. Eisenmenger, Vac. Sci. Technol. B **12**, 1963–1966 (1994)
10. M. Nakamura, H. Tokumoto, Surf. Sci. **398**, 143–153 (1998)
11. Y.L. Yang, K. Deng, Q.D. Zeng, C. Wang, Surf. Interf. Anal. **38**, 1039 (2006)
12. M. Lackinger, S. Griessl, W.M. Hechael, M. Hietschold: J. Phys. Chem. B **106**, 4482 (2002)
13. K. Walzer, M. Sternberg, M. Hietschold: Surf. Sci. **415**, 376 (1998)
14. S. Hoeppener, J. Wonnemann, L.F. Chi, G. Erker, H. Fuchs, ChemPhysChem **4**, 490 (2003)
15. S.B. Lei, C. Wang, L.J. Wan, C.L. Bai, Langmuir **19**, 9759 (2003)

16. S.J.H. Griessl, M. Lackinger, M. Edelwirth, M. Hietschold, W.M. Heckl: Single Mol. **3**, 25 (2002)
17. S.J.H. Griessl, M. Lackinger, F. Jamitzky, T. Markert, M. Hietschold, W.M. Heckl, Langmuir **20**, 9403 (2004)
18. S.J.H. Griessl, M. Lackinger, F. Jamitzky, T. Markert, M. Hietschold, W.M. Heckl, J Phys Chem B **108**, 11556 (2004)
19. J.A. Theobald, N.S. Oxtoby, M.A. Phillips, N.R. Champness, P.H. Beton, Nature **424**, 1029 (2003)
20. M.E. Davis, Nature **417**, 813–821 (2002)
21. A. Stein, Adv. Mater. **15**, 763–775 (2003)
22. S. Leininger. B. Olenyuk, P. J. Stang, Chem. Rev. **100**, 853 (2000)
23. G. F. Swieger, T. J. Malefets, Chem. Rev. **100**, 3483 (2000)
24. B. J. Holliday, C. A. Mairkin, Angew. Chem. **113**, 2076 (2001); Angew. Chem. Int. Ed. **40**, 2022 (2001)
25. A. Dmitriev, H. Spillmann, N. Lin, J.V. Barth, K. Kern, Angew. Chem. Int. Ed. **42**, 2670–2673 (2003)
26. M. Lingenfelder, H. Spillmann, A. Dmitriev, S. Stepanoww, N. Lin, J.V. Barth, K. Kern, Chem. Eur. J. **10**, 1913–1919 (2004)
27. D.G. Kurth, N. Sevein, J.P. Rabe, Angew. Chem. Int. Ed. **41**, 3681 (2002)
28. N. Lin, A. Dmitriev, J. Weckesser, J.V. Barth, K. Kern, Angew. Chem. Int. Ed., 4973 (2002); Angew. Chem. Int. Ed. **41**, 4779 (2002)
29. P. Messina, A. Dmitriev, N. Lin, H. Spillmann, M. Abel, J.V. Barth, K. Kern, J. Am. Chem. Soc. **124**, 14000 (2002)
30. S. Stepanow, M. Lingenfelder, A. Dmitriev, H. Spillmann, E. Delvigne, N. Lin, X. Deng, C. Cai, J.V. Barth, K. Kern, Nature Mater. **3**, 229–233 (2004)
31. R. Thaimattam, F. Xue, J.A.R.P. Sarma, T.C.W. Mak, G.R. Desiraju, J. Am. Chem. Soc. **123**, 4432 (2001)
32. R. Matsuda, R. Kitaura, S. Kitagawa, Y. Kubota, T.C. Kobayashi, S. Horike, M. Takata, J. Am. Chem. Soc. **126**, 14063 (2004)
33. J. Lu, S.B. Lei, Q.D. Zeng, S.Z. Kang, C. Wang, L.J. Wan, C.L. Bai J. Phys. Chem. B **108**, 5161 (2004)
34. A.M. Davis, S.J. Teague, Angew. Chem. Int. Ed. **38**, 736 (1999)
35. X.H. Qiu, C. Wang, Q.D. Zeng, B. Xu,, S.X. Yin, H.N. Wang, S.D. Xu, C.L. Bai, J. Am. Chem. Soc. **122**, 5550 (2000)
36. Y.H. Liu, S.B. Lei, S.X. Yin, S.L. Xu, Q.Y Zheng, Q.D. Zeng, C. Wang, L.J. Wan, C.L. Bai, J. Phys. Chem. B **106**, 12569 (2002)
37. J. Lu, Q.D. Zeng, C. Wang, Q.Y. Zheng, L.J. Wan and C.L. Bai: J. Mater. Chem. **12**, 2856 (2002)
38. D.X. Wu, K. Deng, Q.D. Zeng, C. Wang, J. Phys. Chem. B **109**, 22296 (2005)

4 Fabricating Nanostructures via Organic Molecular Templates

Yunshen Zhou and Bing Wang

4.1 Introduction

In the rapid growth of nanoscience and technology, organic compounds occupy a prominent position and play comprehensive roles as stabilizers [1–5], protective masks [6], templates [7–9], surface modifiers [10], position indicators [11], functional units and building blocks [12–17], and molecular "ink" [18–21], etc. Further more, organic compounds are key components in the design and fabrication of nanomachines and nanodevices. Utilization of organic molecules has allowed prototypes of nanomachines and nanodevices, such as molecular motors [22–26], conductors [27–30], logic gates [31], memories [32], rectifiers [33–36], negative differential resistance devices [37], single electron tunneling devices [38], and gears [39–41], to be developed [42].

Using organic compounds as templates to transfer the desirable and controllable patterned structures into mechanically and chemically stable, durable, and processable nanostructures has received tremendous attention for decades [43–47]. In general, template materials require some characteristics [48]:

1. Have featured structures for the final materials
2. Be stiff and durable enough to maintain their structures during the templating processes
3. Be easily removed without disrupting the final materials, if the final materials do not require the templates

Templating nanostructures includes several steps: firstly, organic templates with featured structures are fabricated by advanced techniques; secondly, the organic templates are brought into contact with precursors or preformed building units (the procedure generally occurs in solution, but can also occur in the gas phase), then the template-directed formation of nanostructures results in hybrid structural materials; finally, the organic templates may be removed if necessary by heat treatment [49], or washed with organic solvents [50] to produce the isolated structural materials. Easy removal of organic templates is one of the advantages of using organic templates instead of inorganic templates [51, 52].

Organic templates include any templates consisting of organic species [53]. Organic molecular templates, polymer templates, carbon nanotubes, and biological templates are a broad subject for discussion. Organic templates as polymer templates [54–64], carbon nanotubes [65–70], and biological templates [71–80] will not be included in this chapter. This chapter focuses on the templating of nanostructures based on organic molecular templates, which we refer to as organic templates consisting of organic molecules. According to the fabricating methods, organic molecular templates can be classified into several groups as surfactant mesophases, Langmuir–Blodgett (LB) monolayers, self-assembled monolayers (SAMs), and self-organized structures, which will be introduced in the following sections: Sect. 4.2, 0-D nanostructures; Sect. 4.3, 1-D nanostructures; Sect. 4.4, 2-D nanostructures.

4.2 0-D Nanostructures

0-D nanostructures are important for their intrinsic properties, as ideal models for fundamental researches, and as building units for fabricating more complicated nanostructures [81–85]. Particle size, composition, and configuration of nanoparticles are of significant importance for researches and applications, and are of eternal pursuit for nano scientists. However, bare metal and semiconductor nanoparticles are unstable owing to their huge surface-to-volume ratio. They intend to agglomerate in order to reduce the surface-to-volume ratio and reach an energetically stable situation. They are oxidized more easily when exposed to air, and are more reactive compared with their macro counterparts. The instability of metal and semiconductor 0-D nanostructures is a major obstacle for researches and applications. Organic molecules have been successfully applied to stabilize nanoparticles and prevent oxidation [1, 86]. Meanwhile, organic molecules also provide versatile templates for size and shape control of nanoparticles [87–91]. Coatings of organic molecules can significantly influence the properties of the nanoparticles, such as hydrophilicity, hydrophobicity, photoactive behavior, and electroactive behavior [53].

Surfactant mesophases, such as microemulsion systems, reverse micelles, and lamellar bilayers [81], have been proved to be applicable and versatile soft templates [81,87,88,90,92]. Reverse micelles and microemulsion systems with spherical, cylindrical, or bicontinuous aqueous phases have been used for synthesis of nanoparticles and nanowires [88]. Owing to the hydrophobic and hydrophilic head groups of surfactant molecules, surfactant molecules can readily form vesicular structures encapsulating an aqueous core, as displayed in Fig. 4.1. More complicated multilamellar vesicles composed of several concentric surfactant bilayers surrounding an aqueous core can also be elaborately prepared [88], as displayed in Fig. 4.1. The as-prepared vesicles are excellent templates for preparing organic and inorganic nanoparticles, as displayed in Fig. 4.2. One prominent application is fabricating 0-D polymer nanostructures via vesicle templating [88]. By introducing a hydrophobic monomer, which swells the surfactant bilayer, the surfactant vesicle is used as

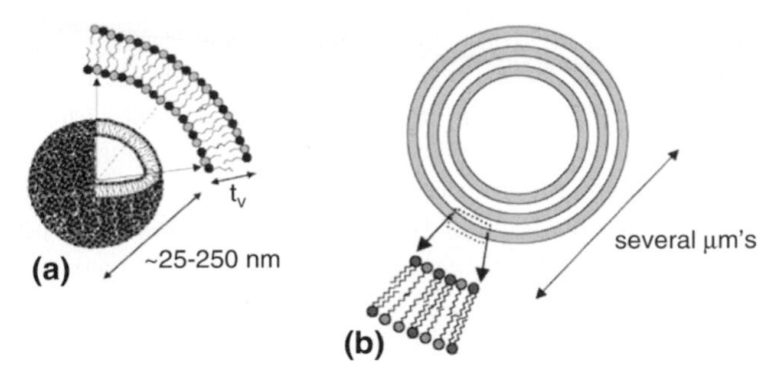

Fig. 4.1. Vesicle architectures: (**a**) unilamellar vesicle, typical radii range from 25 to 250 nm, bilayer thicknesses are about 2–4 nm; (**b**) multilamellar vesicle, composed of several concentric surfactant bilayers surrounding an aqueous core [88]

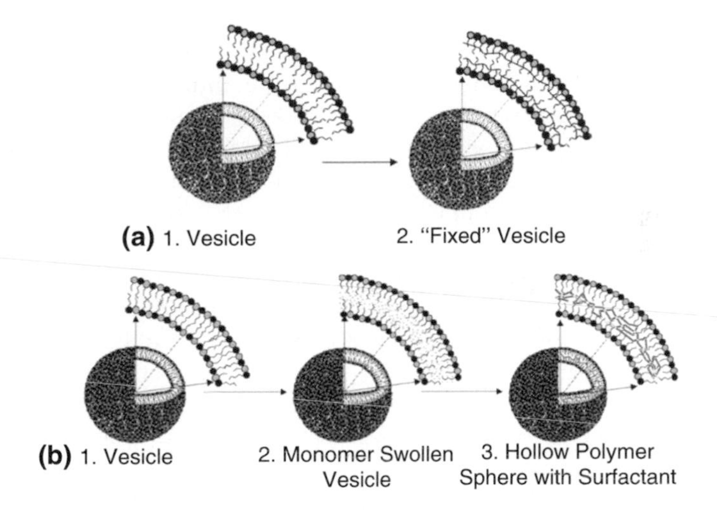

Fig. 4.2. Templating vesicles: (**a**) vesicle-forming polymerizable surfactants are fixed into place by polymerization; (**b**) a hydrophobic monomer swells the interior of a vesicle bilayer and is subsequently polymerized to form a hollow polymer shell [88]

a template directing the polymerization process and produces a hollow polymer shell, and surfactant molecules themselves are not incorporated in the polymerization. The polymer shells can be easily isolated from the templating surfactant vesicles for further applications. By choosing suitable monomers, polymer shells with different properties can be easily obtained. Other structures, such as parachutes, matrioshka structures, and necklaces, have also been reported by using surfactant vesicle templates in polymerization [88]. Templated polymer synthesis within vesicular solutions and microemulsions provides ways to fabricate hollow nanospheres and nanoparticles.

Fabricating inorganic nanoparticles is another important application of vesicle templates, in which surfactant molecules provide nucleating sites, shape control, and stabilization of particles. The strong affinity between the metal salts and the polar head groups of amphiphilic molecules and the anisotropic structure of the microemulsion systems plays important roles in the anisotropic growth. Schmid et al. [93] reported the synthesis of ligand-stabilized gold nanoparticles, $Au_{55}(PPh_3)_{12}Cl_6$, in 1981. Then, in 1994, Brust et al. [1] reported the synthesis of thiol-stabilized gold nanoparticles, the method reached its golden era. In general, surfactant vesicles in microemulsion reactions can be used to produce spherical nanoparticles, such as Fe [94, 95], Pt [96], Cd [97], Pd [98, 99], Ag [100], Au [101], Cu [102], ZnS [103, 104], $BaSO_4$ [105–107] particles, Au/Pd [108, 109], Pd/Pt [110], Au/Pt [111], Au/Ag [112–114] bimetallic particles, and Au/CdSe core–shell structures [115]. Nonspherical inorganic nanoparticles can be produced using the similar methods. Pinna et al. [116] reported the synthesis of triangular CdS nanocrystals in a $Cd(AOT)_2$/isooctane/water system, where AOT is sodium bis(2-ethylhexyl) sulfosuccinate. The group of Pileni [117–121] studied the $Cu(AOT)_2$/isooctane/water system and obtained spherical, rod-shaped, and irregularly shaped Cu particles by adjusting system compositions. Johnson et al. [122] and Jana et al. [123] reported the transformation from gold nanoparticles to nanorods with controllable aspect ratios by adding cetyltrimethylammonium bromide (CTAB) to the system. Summers et al. [107] used polymerizable surfactants to modify the interfacial curvature of reverse micelles, and obtained spherical $BaSO_4$ nanoparticles in the unpolymerized case as well as cylindrical $BaSO_4$ nanoparticles when the polymerization changed the micellar curvature. Hubert et al. [124, 125] described using unilamellar vesicles as templates for transcription into silica. Vesicle-templated structures of more complex morphologies could be obtained by other methods. Silica spheres containing pillarlike structures have been synthesized by controlling the acidification rate of a solution of sodium silicate, butanol, and myristyltrimethylammonium bromide [126]. In order to improve the nonlinear optical response and catalytic properties of materials, nanometer-sized composite particles containing a core–shell structure, including inorganic–inorganic, inorganic–organic, organic–organic, organic–inorganic, and inorganic–biomolecule nanoparticles, were designed and synthesized in reverse-micelle systems templated by surfactant molecules, and have been systematically reviewed by Adair et al., [81] and Kickelbick and Liz-Marzan [127].

Surfactant monolayers have been well established as templates for nucleating and growing nanoparticles, in which geometrical and stereochemical matching of the monolayers and crystal lattices can result in epitaxial crystal growth [53, 92, 128, 129]. Among various surfactant monolayers, LB monolayers have been demonstrated to be excellent templates for growing nanocrystallites with morphology and orientation control. In 1985, Landau et al. [130] first introduced growing crystals based on the LB films of amphiphilic molecules as artificial templates for the promotion of crystal nucleation. Since the packing of the polar head groups can determine the nucleation rate and the degree of orientation of the attached growing crystals, adjustment of polar terminating groups of the amphiphilic molecules or dilution of the monolayers with other amphiphiles allows the the packing arrangements of the head groups and the domain sizes of the films to be adjusted and restricted, thus influencing the crystallization process in a controlled manner. The lattice-matching between the functional head groups in the monolayer and the nanocrystals will result in epitaxial growth of corresponding nanocrystallites on a monolayer [7, 131–138]. Yang et al. [129, 139, 140] have carried out fruitful work on this subject, and demonstrated the close relationship between the nucleation faces and the monolayer structures. By floating arachidic acid (AA) monolayers as templates on aqueous lead nitrate and cadmium chloride solutions, Fendler et al. obtained epitaxially grown lead sulfide (PbS), lead selenide (PbSe), and cadmium sulfide (CdS) semiconductor nanocrystallites. In order to investigate the monolayer-directed epitaxy control of nanocrystallites, Fendler et al. performed experiments on growing PbS nanocrystallites under mixed monolayers of AA and octadecylamine (ODA). By adjusting the AA-to-ODA ratio, PbS nanocrystallites with different symmetry and orientations were obtained. The experimental results clearly exhibited excellent monolayer-directed orientational control of nano-crystallite growth. Li et al. [141] carried our a similar experiment on growing CdS nanoparticles using AA and ODA mixed LB films as templates. An ordered array of CdS particles with a particle separation of 4.2 nm was obtained in the absence of ODA. Mixing ODA disturbed the crystalline structure of the AA film, and resulted in randomly distributed particles with average particle size reduced from 4 to 1 nm when the amount of ODA was increased. No CdS particles were produced when AA was absent. The experimental results demonstrated the close relation between the template structure and the growth of nanoparticles. Kmetko et al. [142] reported work on in situ synchrotron X-ray diffraction studies of the nucleation of barium fluoride at fatty acid monolayer templates, and revealed the commensurate relationship between the interfacial lattices of the organic molecules and the inorganic atoms.

Although Langmuir monolayers are excellent templates for synthesizing nanoparticles, most surface structures of Langmuir monolayers are difficult to manipulate and characterize, which causes instability in particles size and shape control. Highly ordered structures of SAMs [12, 143, 144], especially alkanethiolates on a gold/silver surface [145], have been extensively studied and abundant structural information has been accumulated, such as surface lattice spacing and the tilting angle of the molecules. So, compared with Langmuir monolayers, SAMs would be more favorable for use as templates for controllable nucleation and growth of nanoparticles. From another point of view, SAMs are more controllable and could be easily patterned with featured structures, which promise SAMs as versatile and controllable templates for fabricating nanostructures.

The synthetic growth of nanocrystallites can be guided by molecular recognition at an interface using patterned SAMs as templates [7, 146, 147]. By the microcontact printing (CP) technique, Whitesides and coworkers patterned metal substrates with SAMs having areas of different terminating groups (such as arrays of acid-terminated regions separated by methyl-terminated regions), which exhibited different nucleating activity, and immersed the patterned substrate in salt solution. After being taken out from the solution, the liquid droplets were retained in the polar region on the substrate, where the rate of nucleation was fastest but the growth of crystals was confined by the droplets and a suitable choice of array spacing, as displayed in Fig. 4.3. By controlling pattern density and feature sizes of the patterned substrates, terminal functional groups of the SAMs, and the concentration of crystallizing solutions, they could control key parameters of the crystallization process, including the location and density of nucleating regions on the substrate, the density of crystallites nucleated within each regions, and the crystallographic orientation of the crystals. Successful examples were demonstrated by growing oriented calcite nanoparticles by using patterned SAMs of ω-terminated alkanethiols [$HS(CH_2)_nX$}, in which X is CO_2^-, SO_3^-, PO_3^-, OH, $N(CH_3)^+$, and CH_3, on metal substrates as templates [146]. Compared with bare metal substrates, SAMs terminated with CO_2^-, SO_3^-, PO_3^-, and OH groups were more active in inducing nucleation, whereas SAMs terminated with $N(CH_3)^+$ and CH_3 groups inhibited nucleation. The orientation of the crystallization of the calcite particles on each substrate was unique and highly homogeneous, but distinct on different substrates, such as SAMs of CO_2^-/Au, which resulted in face-selective nucleation of calcite from the (015) crystallographic planes, and CO_2^-/Ag (012), OH/Au (104), OH/Ag (103), SO_3^-/Au (1012), and SO_3^-/Ag (107), respectively. Whitesides and coworkers demonstrated a promising approach for the templated growth of nanoparticles in a controllable manner. By patterning SAMs with different structural features, similar processes produce 1-D and 2-D nanostructures, which will be introduced in the next sections.

The group of Sagiv [8, 148–151] developed an all-chemical strategy for the bottom-up fabrication and spatial fixation of nanoparticles and nanowires electrically connected to patterned electrodes. The approach included two processes, surface self-assembly and "constructive nanolithography", a chemical patterning process utilizing electrical pulses delivered by a conductive atomic force microscope (AFM) tip for the nondestructive nanometer-scale inscription of chemical information on the top surfaces of certain highly ordered long-tail organosilane SAMs on silicon, as displayed in Fig. 4.4. One of the most attractive applications of the method is to produce patterned organic thin films as templates for the precisely controlled assembly of metal and semiconductor nanoparticles. Several successful examples were demonstrated. By tip-induced oxidation of surface-exposed vinyl ($-CH{=}CH_2$) and methyl (CH_3) groups to carboxyl (–COOH) and mercapto (–SH) groups, which would act as efficient binding sites for cations, they produced

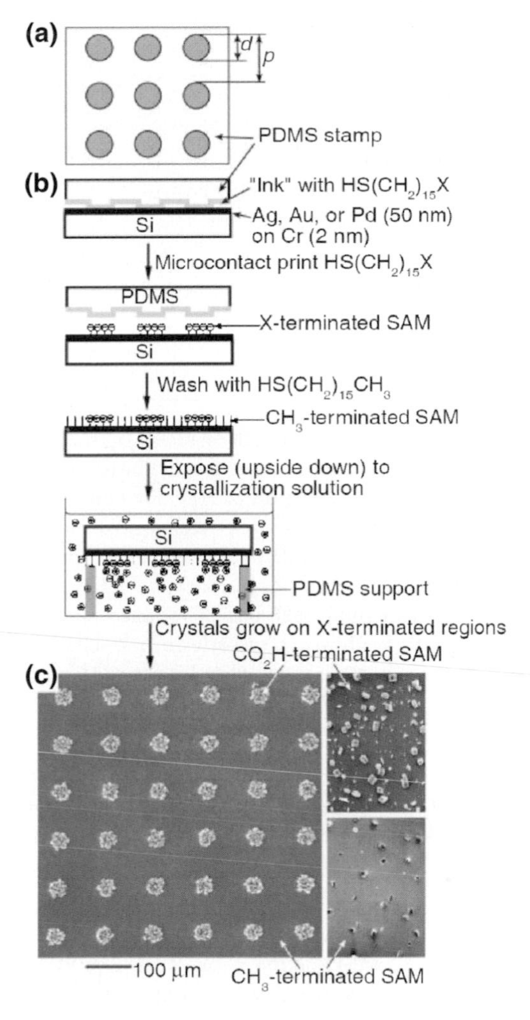

Fig. 4.3. Experimental design of crystallization on patterned self-assembled monolayers (*SAMs*). (**a**) Relief structure of the patterned poly(dimethylsiloxane) (*PDMS*) stamps used for microcontact printing. (**b**) The experimental steps. (**c**) Scanning electron micrograph of the sample patterned surface-printed circles of $HS(CH_2)_{15}CO_2H$ in a background of $HS(CH_2)_{15}CH_3$ supported on Ag(111) overgrown with calcite crystals. The images on the *right* illustrate the wide distribution of the sizes of the crystals formed on the nonpatterned SAMs. To prepare substrates, silicon wafers (test grade, n or p type, Silicon Sense, Nashua, NH, USA) were coated with 2.5 nm of Cr, to promote adhesion, and then coated with metal (Ag, Au, Pd; typically, 50 nm) using an electron-beam evaporator (base pressure 10^{-7} Torr). The stamps were prepared by casting and curing (PDMS) against rigid masters bearing a photoresist pattern formed using conventional lithographic techniques [7]

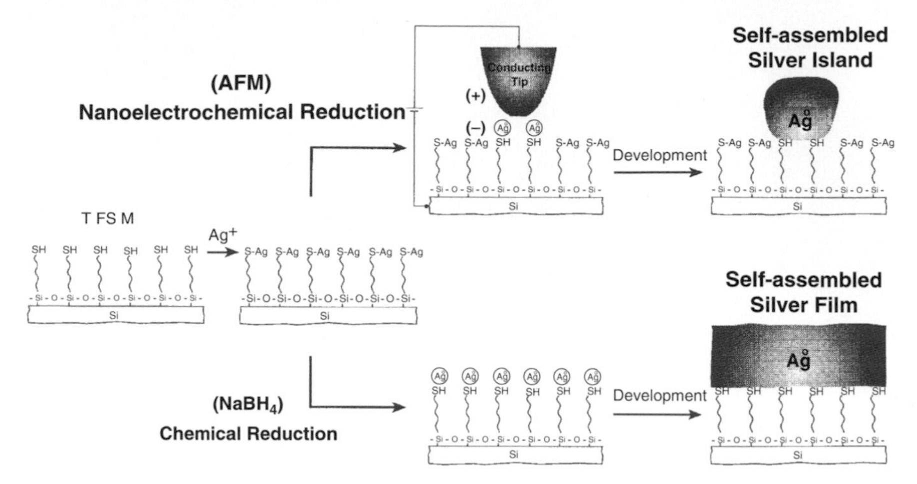

Fig. 4.4. The site-defined self-assembly of silver metal on a thiol-top-functionalized silane monolayer (*TFSM*) preassembled on silicon. The silver thiolate (*S-Ag*) template surface obtained by the chemisorption of Ag^+ ions on the TFSM surface (*left*) is nondestructively patterned using either a wet chemical reduction process (*lower path*) or a nanoelectrochemical process (*upper path*) involving the application of a DC voltage to a conducting atomic force microscope (*AFM*) tip. Further development of the macropatterns and nanopatterns of reduced silver particles imprinted on the Ag^+–TFSM template is shown to result in a thicker self-assembled silver film (*lower path*), or self-assembled silver islands selectively grown at tip-defined sites (*upper path*) [151]

site-selectively patterned substrates with precise spatial control. After treatment with a solution containing Cd^+ ions, the substrates were exposed to gaseous H_2S and generated template-controlled arrays of CdS nanoparticles [148]. Similar processes could be applied in fabricating template-controlled arrays of other metal and semiconductor nanoparticles [8,149–151]. The procedure can also be applied in fabricating more complex structures, which will be introduced in the following part of this chapter.

Piner et al. [18] presented another important technique for patterning SAMs, named "dip-pen" nanolithography (DPN). By using an AFM tip as a "pen" and organic molecules (especially alkanethiols) as "ink", they transferred molecules to a solid substrate of interest via capillary transport in an ambient atmosphere, as displayed in Fig. 4.5. The general process began with fabricating a pattern on a solid substrate consisting of arrays of "molecular ink" dots, which would act as binding and nucleating sites for nanoparticle growth. The patterned substrates were then treated with suitable solutions and consequent processes to obtain nanoparticles, 1-D wires, and 2-D arrays, such as growing functionalized gold nanoparticles on a gold substrate patterned with 16-mercaptohexadecanoic acid (MHA).

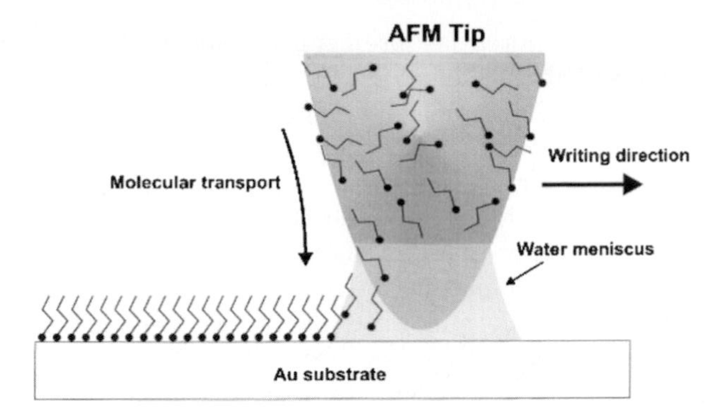

Fig. 4.5. Dip-pen nanolithography (*DPN*). A water meniscus forms between the AFM tip coated with *n*-octadecanethiol (ODT) and the Au substrate. The size of the meniscus, which is controlled by relative humidity, affects the ODT transport rate, the effective tip–substrate contact area, and DPN resolution [18]

Chen and Lin [5] fabricated cubic cadmium sulfide (CdS) nanoparticles using patterned SAMs as templates. By using CP, they fabricated patterned SAMs of alkanethiolates on gold. The patterned SAMs were then immersed sequentially into cadmium and sulfide ion solutions; each drop of the solution remaining on the hydrophilic regions was regarded as tiny vessel. The CdS particles growth was restricted in each droplet. So the crystal sizes might be controlled by the amount and concentration of the ion solutions. Then, a narrow size distribution and uniform morphology were obtained. The average particles sizes could be adjusted by using different solvents in the synthesis. Adjusting patterned SAMs with different terminating groups resulted in the formation of nanoparticles with different crystallographic directions. Woo et al. [152] fabricated tellurium nanoclusters at Au(111) electrodes modified with thiolated β-cyclodextrin molecular templates. Besides the aforementioned techniques for patterning SAMs, Hayes and Shannon [153] reported growth of polyaniline nanostructures on gold electrodes using a two-component SAM consisting 4-aminothiophenol (4-ATP) and *n*-octadecanethiol (ODT) as a template, in which 4-ATP served as a nucleating site.

4.3 1-D Nanostructures

1-D nanostructures have become the focus of intensive research interest owing to their unique application in mesoscopic physics and fabrication of nanodevices [57, 154–162]. 1-D nanostructures provide an ideal system for investigating the dependence of electrical and thermal transport, as well as mechanical properties on dimensionality and quantum confinement [154].

1-D nanostructures are potential interconnects and building units in fabricating nanodevices. Fabrication of 1-D nanostructures can be realized by several advanced nanolithographic techniques, such as template-directed synthesis, chain polymerization, focused ion beam lithography, electron-beam lithography, photolithography applying X-ray and extreme UV, scanning-probe lithography, vapor–liquid–solid method, solution–liquid–solid method, solvothermal method, vapor-phase method, and self-assembly [6,13–15]. One of the most successful examples of 1-D nanostructures is carbon nanotubes [163]. Synthesis of inorganic nanowires/nanotubes/nanobelts/nanorods has been extensively investigated and reviewed [164–176]. Organic species also play important roles in fabricating 1-D nanostructures. In this section we will introduce organic molecular templated 1-D nanostructure fabrication.

By adsorbing a $C_{90}H_{98}$ molecule, named as the Lander, on a clean Cu(110) surface, Rosei et al. [177] found that a single Lander molecule could act as a template inducing the formation of metallic nanostructures at step edges two Cu atoms wide and eight Cu atoms long, namely, 0.75-nm wide and 1.85-nm long, which were adapted to the dimension of the Lander molecule, as displayed in Fig. 4.6.

Assembling behaviors of bis(N-α-amido-glycylglycine)-1,7-heptane dicarboxylate, one of the bolaamphiphiles, exhibit pH sensitivity [178]. In an acidic solution, the heptane bolaamphiphile forms a crystalline tubule with average size of 700 nm in diameter and 10 m in length [179]. The non-hydrogen-bonded amide groups in the assembled heptane bolaamphiphile nanotubes show strong affinity to metal ions such as Pt, Pd, Cu, and Ni, which can act as organic–inorganic junctions. So the heptane bolaamphiphile nanotubes have the potential to be excellent templates for inorganic and metallic nanowires. Matsui et al. [179] reported work on fabrication of Ni and Cu nanowires by electroless metallization of the heptane bolaamphiphile nanotubes in Ni and Cu baths with reducing agents, in which metal ions were coordinated between non-hydrogen-bonded amide groups of neighboring bolaamphiphile molecules. The results indicate a possible way to attach metallic nanowires onto nanoelectrodes by self-assembling the heptane bolaamphiphile nanotubes onto the electrodes by carboxylic acid–thiol self-assembly monolayers.

Vesicular structures formed by surfactant molecules also provide versatile templates for fabricating 1-D nanostructures. Simmons et al. [180] reported the synthesis of high-aspect-ratio CdS nanorods in reverse micelles of AOT and phosphatidylcholine. Hopwood and Mann [181] reported the fabrication of highly elongated $BaSO_4$ crystalline filaments in BaNaAOT microemulsions with length up to 100 m and aspect ratio of 1000. Chen et al. [182] reported the synthesis of vanadium oxide nanotubes by hydrothermal self-asembly from NH_4VO_3 and using organic molecules as structure-directing templates. Organic molecules including primary amines ($C_2H_{2n-1}NH_2$), α,ω-diamines [$H_2N(CH_2)_nNH_2$], and quaternary ammonium salt (CTAB) were demonstrated as suitable structure-directing templates for the formation

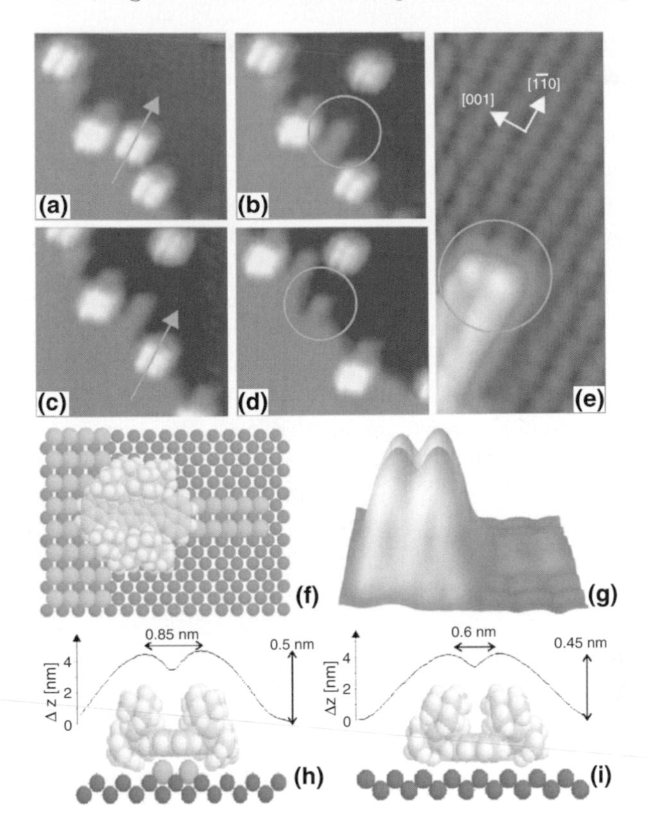

Fig. 4.6. (**a**)–(**d**) Manipulation sequence of the Lander molecules from a step edge on Cu(110). The *arrows* show which molecule is being pushed aside; the *circles* mark the toothlike structures that are visible on the step where the molecule was docked. All image dimensions are 13 nm by 13 nm. Tunneling parameters for the imaging are $I_t = -0.47$ nA; $V_t = -1.77$ V; tunneling parameters for manipulation are $I_t = -1.05$ nA; $V_t = -55$ mV. (**e**) Zoom-in smooth-filtered scanning tunneling microscope (*STM*) image showing the characteristic two-row width of the toothlike structure (*right corner*) after removal of a single Lander molecule from the step edge. The Cu rows are also visible. The arrows show the directions on the surface. $I_t = -0.75$ nA; $V_t = -1.77$ V. Image dimensions are 5.5 nm by 2.5 nm. (**f**)–(**i**) Details of the conformation of the Lander molecule on the tooth. (**f**) Molecular structure, extracted from a comparison between experimental and calculated STM scans, showing that the board is parallel to the tooth. (**g**) Calculated constant-current 3-D STM image. (**h**) Cross section of the tooth. (**i**) Cross section on a terrace. Tunneling parameters are $I_t = -0.47nA$; $V_t = -1.77$ V [177]

of nanotubes. They also suggested a rolling mechanism for the formation of nanotubes from lamellar structures. Niederberger et al. [183] reported synthesis of vanadium oxide nanotubes by novel non-alkoxide routes by using either vanadium(V) oxytrichloride or vanadkum(V) pentoxide as a vanadium source and primary amines as templates, respectively. Qi et al. [184] synthesized $BaCO_3$ crystal nanowires with lengths up to 100 m and diameters of 10–30 nm in a barium ions/carbonate ions/tetraethylene glycol monododecyl ether)/cyclohexane reverse-micelle system.

CTAB is a cationic surfactant which can form a CH_3–CH_3–CH_3–N structure and induce sphere–rod micelle structures in aqueous solution when some salts such as NaCl and Na_2SO_4 are added [185]; therefore, CTAB can be used as a soft template for synthesizing materials with special morphologies [123, 186]. Using Ag seeds for the growth of silver in solution, Jana et al. [187] obtained silver nanorods and nanowires from $AgNO_3$/ascorbic acid/CTAB/NaOH solution with a rodlike micelle as a template. Silver nanorods of controllable aspect ratio of 2.5–15 (10–15-nm short axes) and nanowires of 1–4-m length and 12–18-nm short axes were produced and effectively separated. Liu et al. [188] synthesized single crystalline SnS nanowires with average diameter of 30 nm and length of hundreds of microns in a SnC_{12}/Na_2S/CTAB/oxalic acid system with CTAB as a soft template. Yao et al. [189] reported the synthesis of a hydroxyapatite nanostructure using CTAB as a template. Perez-Juste et al. [190] reported the synthesis of Au nanorods with controllable aspect ratio in aqueous surfactant solutions in the presence of CTAB, to which gold ions were attached. From the experiments, it was demonstrated that the size and the aspect ratio of the gold nanorods could be controlled through the use of different sized seed particles with length tuned from 25 to 170 nm, while the width remained almost constant at 22–25 nm. The presence of CTAB and lower temperature are favored. The addition of chloride ions or the use of dodecyltrimethylammonium bromide caused shorter-aspect rods. Synthesis of Cu and Cu_2O nanotubes were reported by reducing $Cu(OH)_4^{2-}$ with hydrazine hydrate and glucose in the presence of CTAB as a structure-directing template [191]. Yu et al. [192] reported the synthesis of porous Cu_2O nanowhiskers with shape control by using CTAB as a template. The porous nanowhiskers exhibit a well-crystallized 1-D structure of more than 200 nm in length and a diameter of 15–30 nm, growing mainly along the <111> direction. The 1-D structures cannot be obtained when poly(ethylene glycol), glucose and sodium dodecyl benzenesulfonate are used as templates. From transmission electron microscopy studies, it was found that the role of CTAB was to interact with tiny $Cu(OH)_2$, which could bind OH^- and be negatively charged, to disperse the tiny $Cu(OH)_2$ solid and induce the growth of Cu_2O along the 1-D direction. When Cu^{2+} was used as a precursor, no nanowhiskers were obtained, which demonstrated the importance of the ion character of the precursor [$Cu(OH)_2 \cdot OH^-$ or Cu^{2+}] in the formation of 1-D nanostructures.

Huang et al. [193, 194] used lyotropic reverse hexagonal liquid-crystalline phases as templates, which contained 1-D aqueous channels for fabricating nanowires such as those of Cu_2O and Ag. Cu_2O nanowires growing from anionic surfactant AOT reverse hexagonal liquid-crystalline phases could reach tens of micrometers in length and 25–100 nm in diameter. High-aspect-ratio crystalline silver nanowires with a uniform wire diameter of 20–30 nm were obtained from a similar system. Huang et al. [193–195] and Luo et al. [196, 197] demonstrated the potential of lyotropic liquid-crystalline phases used as soft templates for fabricating inorganic nanostructures.

Li et al. [198] reported direct formation of 1-D and more complex structures by combining nanoparticle synthesis with self-assembly. The experiments were carried out by adding $Ba(AOT)_2$ reverse micelles into Na_2CrO_4 containing NaAOT microemulsion droplets. By adjusting the reactant molar ratio and the water content, they obtained different structures, such as linear chains, rectangular superlattices, and long filaments. The difference was ascribed to the interfacial activity of reverse micelles and microemulsions arising from the interdigitation of surfactant molecules attached to specific nanoparticle crystal faces.

SAMs are another kind of important technique for templated fabrication of 1-D nanostructures. Cheung et al. [199] demonstrated the fabrication of virus arrays consisting of 1-D virus nanowires by a multistep approach combining scanning probe nanolithography with chemoselective deposition of a virus onto functionalized SAMs. Choi and Park [200] developed an approach using SAMs as molecular templates to grow nano-sized conducting polymer wires and rings tens of nanometer thick and a few micrometers long.

The surface patterning technique, combining surface self-assembly with "constructive nanolithography," developed by Sagiv and coworkers [8, 148, 201], as displayed in Fig. 4.4, can also be used in fabricating 1-D nanostructures by inscribing 1-D patterns on corresponding SAMs using conductive AFM tips. 1-D gold wires inscribed on smooth silicon surfaces were produced by selectively assembling gold clusters on a purpose-designed organic template via a hierarchical layer-by-layer self-assembly strategy with lateral confinement produced by constructive nanolithography. The all-chemical technique provides attractive options for the nanofabrication in different dimensionalities, and might have real impact on future nanofabricating applications.

The DPN technique developed by Piner et al. [18] also provides a promising approach for future nanofabrication. By combining DPN and wet chemical etching, Zhang et al. [202] reported the synthesis of a class of open-ended cylindrical Au–Ag alloy nanostructrues on a Si/SiO_x substrate. Ag dots with average diameter of 585 ± 60 nm were firstly prepared by etching MHA-patterned Ag/Si film. The average height of the Ag dots was 48 ± 3 nm, which is consistent with the thickness of the Ag film. Then, the as-prepared Ag nanodot pattern was transferred into aqueous $HAuCl_4$ (a known redox etchant for Ag) for 5 min, and the solid Ag nanodots were transformed into open-ended hollow cylindrical nanostructures with average diameter of 570 ± 80 nm and height of 83 ± 6 nm.

Annealing treatments of alkanethiol SAMs have been extensively studied [203–206]. By carefully controlling the annealing treatment, condensed alkanethiol SAMs could be converted to periodic concave–convex striped structures. The striped structures provide excellent spatial confinement in nanoscale, and could be applied as templates for fabricating 1-D nanostructures. Zeng et al. [9] have successfully utilized annealed alkanethiol SAMs for growing 1-D C_{60} molecular chains. By annealing decanethiol SAMs in vacuum, they achieved alkanethiol templates with a mixed striped phase, as displayed in Fig. 4.7. Three kinds of typical grooves were observed, with widths of about 3.3, 1.9, and 1.0 nm, respectively. The grooves provided lateral confinement for growing 1-D C_{60} molecular chains, since their widths were about integer multiples of the C_{60} bulk spacing. By depositing C_{60} onto the annealed templates, they observed only bimolecular C_{60} chains with maximal length up to 20 nm grown along grooves of 1.9 nm in width, as displayed in Fig. 4.8. C_{60} molecules were preferentially located at the S terminus sites of the flat-lying decanethiols, and only the grooves of 1.9 nm in width had S terminuses in the middle of the groove, so only C_{60} molecular chains were observed in this kind of groove. During the process of annealing treatment, it was noted that C–S bonds were easily broken in the SAMs of alkanethiols on Au(111), with alkane chains desorbing from the surfaces and S atoms aggregating and being confined by the striped structures to form 1-D atomic chains under well-controlled annealing condition, as displayed in Fig. 4.9. After the annealing treatment, flat-lying striped structures were observed similar to the well-defined $(11.5 \times \sqrt{3})$ phase [204]. Bright chains between adjacent stripe pairs were observed as indicated by arrows with a spacing of about 5 Å along the stripes and were attributed to S atom chains.

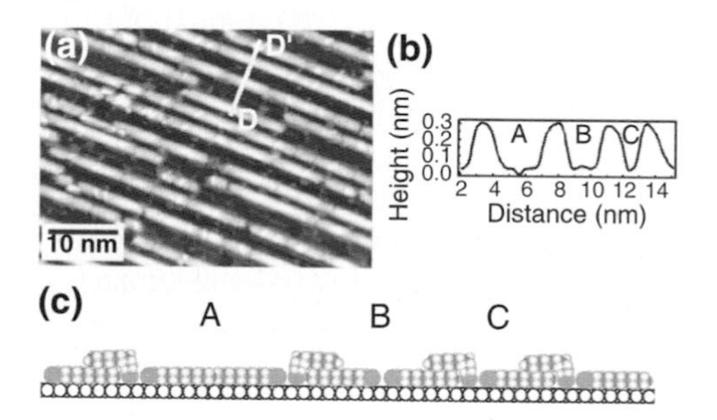

Fig. 4.7. (**a**) STM image of the striped template with furrowlike structures obtained by annealing decanethiol SAMs. (**b**) Height profile along line DD′ in (**a**). Three furrows with different widths are marked as *A*, *B*, and *C*, respectively. (**c**) Corresponding structural model for the furrow structures along line DD′ in (**a**). *Open circles* represent Au atoms, *shaded gray circles* C atoms, *shaded white circles* H atoms, and *light gray circles* S atoms [9]

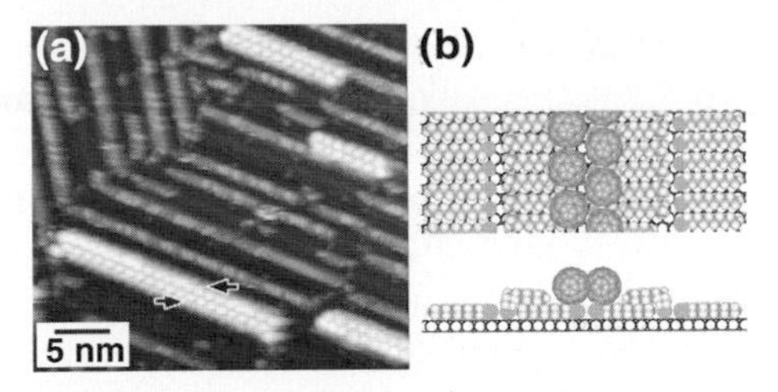

Fig. 4.8. (**a**) STM image after deposition of 0.1 monolayers of C_{60} onto the striped thiol template. C_{60} bimolecular chains are observed in furrow B. *Arrows* indicate the flanges of furrow B. (**b**) Proposed structural model for a C_{60} chain in furrow B in top view (*upper*) and side view (*lower*), respectively. C_{60} molecules are located at the sites of S terminuses of the flat-lying decanethiols [9]

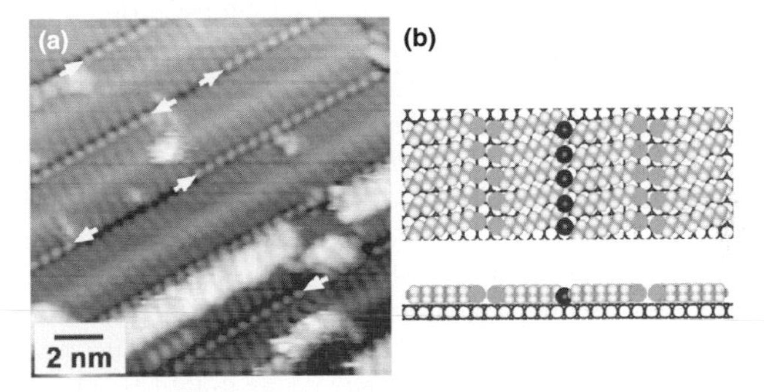

Fig. 4.9. (**a**) STM image of the S atomic chains (indicated by the *arrows*) formed between adjacent flat-lying decanethiol stripe pairs on the Au(111) surface. (**b**) Corresponding structural model for a S atomic chain in top view (*upper*) and side view (*lower*), respectively. *Shaded black circles* represent individual S atoms. The S atoms in the atomic chains are located at the threefold Au(111) hollow sites [9]

They ascribed the existence of spatial confinement of the decanethiol stripes as the key element in the formation of S atomic chains.

Self-organized molecular assemblies provide a direction for templated fabrication of nanostructures. Baral and Schoen [207] reported the synthesis of hollow submicron-diameter silica cylinders by using a self-organized molecular assembly, an aqueous dispersion of phospholipid tubules, as a template for silica film deposition by the sol–gel method. The silica film coated phospholipid tubules were then treated with freeze-drying following heating at 600°C and resulted in hollow silica cylinders. Hoeppener et al. [201] reported the formation of linearly arranged Au_{55} clusters on a molecular template at the solid–liquid interface. The strategy utilized the incorporation of chemically modified Au_{55} clusters into highly oriented molecular templates by substitution of individual molecules of the template. The organic templates required were formed spontaneously at the solid–liquid interface of highly oriented pyrolytic graphite (HOPG) by self-organization of a fatty acid or a linear alkane. By adjusting the length of the linear alkyl molecule, they could control the distance between gold cluster rows.

4.4 2-D Nanostructures

2-D nanostructures, including monolayers and arrays, exhibit unique topological ordering and surface-bound characteristics, which promise the importance of 2-D nanostructures in fundamental research and technological application [208]. 2-D nanostructures also provide an ideal platform for study of surface chemistry, friction and lubrication, molecular biological behavior, wettability, corrosion resistance, nanobioelectronics, and functional devices, and will help us extend our investigations to nano and quantum worlds [27, 42, 128, 143, 168, 209]. 0-D and 1-D nanostructures are more likely to act as potential functional devices, building units, and interconnectors, but 2-D nanostructures are intended to be integrated and more applicable systems [42, 168]. Intensive exploration has been carried out aimed at achieving 2-D nanostructures with controllable structures and desired properties, which have several requirements: stability, orderliness, orientation of adsorbates, composition, morphology, symmetry, and location of active functional units. Among the numerous methods explored, SAMs play important roles in fabricating and patterning 2-D nanostructures [42, 128, 143, 168, 209–211]. Structures, stability, and characteristics of SAMs have been widely studied and reviewed [12, 144, 145], and will not be discussed here. The main theme of this section focuses on the formation of 2-D nanostructures using SAMs as templates.

Using functionalized SAMs as templates for the controlled formation of thin films from aqueous solutions has been demonstrated as an effective technique, such as for growing lead lanthanum zirconate titanate thin films using silane/quartz SAMs as templates [212], TiO_2 films growing on octadecyltrichlorosilane (OTS)/SiO_2 SAMs using the sol–gel method [213], γ-FeO(OH) films growing on alkylthiols/Au(111) SAMs using a soft-chemical method [214], and magnetic CoNiFe films growing on thiol/Cu SAMs using electrodeposition [215]. K. Koumoto et al. [216] demonstrated the formation of ordered 2-D structures consisting -CN terminated silica spheres using a patterned silane SAMs as template, in which SiO_2 spheres were selectively attached to silanol surfaces via ester bonds. Cao et al. reported the fabrication of $SrTiO_3$ thin films on patterned silane SAMs by liquid-phase deposition, in which $SrTiO_3$ films were selectively deposited in the silanol regions [217]. By annealing at 500°C for 2 h in air, the amorphous $SrTiO_3$ films obtained were crystallized.

The DPN technique developed by the group of Mirkin [18, 218–222] provides a powerful tool for patterning 2-D monolayers and surfaces. Using MHA as molecular ink, the AFM tip as a pen, positively charged protonated amine- or amidine-modified polystyrene (PS) spheres as building units, they used a gold surface patterned with negatively charged carboxyl terminating groups as a template for growing PS particle arrays [222]. Different 2-D nanostructures, such as a DNA pattern [221] and an Fe_3O_4 nanoparticle pattern [220], could be obtained by adjusting building units while using similar templates. By using a binary ink, consisting of 11-mercaptoundecyl-penta(ethylene glycol) disulfide and a mixed disulfide substituted with one maleimide group, they patterned a gold surface with nanoscale features presenting functional terminal groups for the chemospecific immobilization of biomolecular particles [219]. Ordered 2-D cysteine mutant cowpea mosaic virus capsid particle arrays with well-defined spatial restriction were prepared via the procedure.

Functionalized organic monolayers with patterning features are promissing candidates for preparing templates with selectable surface chemical properties. Qin et al. [147] introduced a combinatorial method including CP of SAMs, surface-templated self-assembly, and confined crystallization or precipitation for fabricating ordered 2-D arrays of microparticles or nanoparticles on solid substrates. The gold surface was patterned by CP into grids of hydrophobic (CH_3-terminated) and hydrophilic (COOH-terminated) SAMs of alkanethiolates. The patterned substrate was then transferred into aqueous solution for site-selective deposition. The size and shape of the pattern and the contact angle of the liquid determined the volume of the liquid droplet absorbed on the surface. The particle arrays were obtained by evaporating solvents. The distribution of particles in hydrophilic regions could be adjusted by withdrawing the substrate from the solution with different orientations of the pattern relative to the direction of shear on the liquid film. The size of the particles could be controlled by several factors, such as the concentration and compositions of the solution, the shape and area of the hydrophilic region, and the parameters influencing the volume of the droplets deposited on the hydrophilic region. Tien et al. [223] produced 2-D nanoparticle arrays by electrostatic self-assembly on gold substrates with patterned surface charge. The whole process included preparation of a charged surface, fabrication of charged particles, and assembly of the charged particles onto the charged surface. The charged surface was prepared by modifying a gold surface with different terminal groups, $-NH_3Cl^-$, $N(CH_3)_2$, $N(CH_3)_3Br^-$, and $C(NH_2)_2Cl^-$ yielded a positively charged surface, while –COOH and PO_3H_2 yielded a negatively charged surface. Gold particles covered with ionized SAMs were prepared by immersing SAM-covered gold particles in corresponding solutions; the process also prevented the particle aggregation. The ionized particles were then assembled onto charged substrates through electrostatic self-assembly and formed 2-D structures. Sagiv et al. reported fabrication of spatially defined self-assembly of nanostructures through a

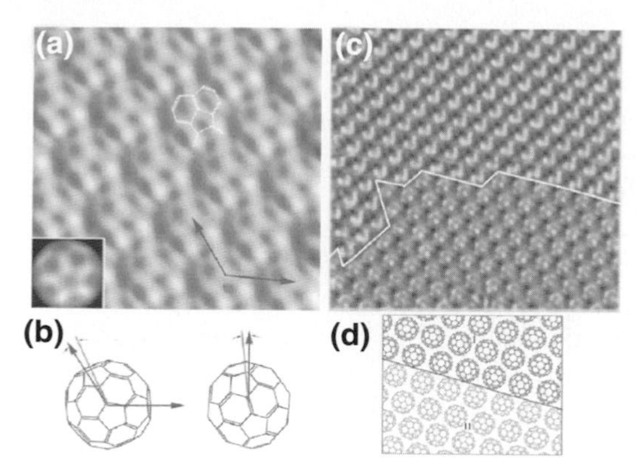

Fig. 4.10. (**a**) The native cage structure of C_{60} molecules. STM image (35 Å×35 Å) of a C_{60} lattice taken at −268°C with −2.0 V sample bias. Detailed internal features of the C_{60} molecule are evident that closely resemble the C_{60} cage structure and match the theoretical simulation shown in the *inset*. (**b**) Top view (*left*) and side view (*right*) of a stick model outlining the C_{60} orientations obtained by simulation. The *c*-axis points out of the image. (**c**) Domain boundary of a 2-D C_{60} array. STM image (100 Å×100 Å) of two molecular orientational domains and the domain boundaries. Both the positional order and the bond-orientational order are fully preserved and no defect exists along the domain boundaries. (**d**) The two molecular orientations derived by comparison with the theoretically simulated images [226]

nanoelectrochemically patterned monolayer based on nondestructive patterning techniques.

Being extensively studied and well understood, alkanethiol SAMs are promising candidate for use as molecular templates [53, 92, 128, 143, 144, 168, 224]. The use of alkanethiol SAMs as templates has demonstrated that 2-D C_{60} monolayers/domains on SAMs of alkanethiol can exhibit molecular orientational domains at low temperature [225, 226], as displayed in Fig. 4.10. Unlike strong binding between C_{60} and metal/semiconductor substrates, which freeze molecular rotation of C_{60} at room temperature, C_{60} molecules adsorbed on the SAMs of alkanethiol rotate freely and display smooth hemispherical protuberance. C_{60} molecules can detach readily and diffuse easily on the templates at room temperature. At −196°C, C_{60} molecules display a hemisphere consistent with a rotating pattern around a fixed axis. When the sample was cooled down to −268°C, well-known C_{60} cage structures were observed experimentally with novel orientationally ordered domains. An abrupt boundary separating two distinguishable domains with different orientations is clearly observed. No positional defect at the domain boundary is observed and centers of C_{60} maintain perfect translation symmetry. The topological orders observed in 2-D C_{60} are significantly different from those in 3-D cases and it is concluded that this is an intrinsic property of a 2-D system. The corresponding theoretical explanation points out that the reduced dimensionality allows the molecules a greater degree of freedom in adjusting mutual orientations. Although the 2-D orientations have lower symmetry than those of the 3-D counterparts, the presence of the SAMs adequately minimizes the system energy and the domain boundary energies and leads to a novel uniorientational molecular order for 2-D C_{60} and a new topological order for the orientational domains.

Phase-separated ultrathin organic films can serve as surface templates for the selective and patterned deposition of functional units on 2-D nanostructures [58, 153, 227–230]. The selective deposition is governed by the chemical differences in the domains or domain boundaries generated by phase separation. Fang and Knobler [231] reported the generation of phase-separated organosilane monolayers using OTS and 1*H*, 1*H*, 2*H*, 2*H*-perfluro-decyltrichlorosilane (FTS) as building units with well-defined distribution of two terminal groups by combining LB deposition with self-assembly. The as-prepared phase-separated OTS/FTS monolayers can be easily adjusted by controlling the deposition condition, and used effectively as templates for selective deposition of proteins. Hayes and Shannon [153] reported template-directed growth of polyaniline nanostructrues on patterned surfaces of two-componentSAMs consisting of 4-ATP and ODT, in which 4-ATP served as a nucleation site for the deposition of polyaniline. The coverage of the polyaniline nanostructrues is closely related to the monolayer composition. Moraille and Badia [232] reported the application of a chemically homogeneous surface with coexisting solid/fluid phase as templates for spatially directed adsorption. The system studied consisted of nanoscale parallel strips generated by the LB deposition composing of L-α-dipalmitoylphosphatidylcholine (DPPC) and L-α-dilauroylphosphatidylcholine (DLPC) in different phases. Generations of novel protein and Au nanoparticle/protein patterns from the selective deposition of human serum albumin and human γ-globulin to the DPPC/DLPC monolayers were observed.

More complex structures, such as superlattices of metal–metal or metal–semiconductor quantum dots can be readily prepared through layer-by-layer deposition of nanoparticle arrays on prepared self-assembled templates. Brust et al. [233] reported growth of multilayer thin films with alternating layers of 6-nm Au nanoparticles and α,ω-dithiols on –SH functionalized glass slides. Sarathy et al. [234] also reported similar experiments of growing a Au–Pt superlattice consisting of alternating layers of Au and Pt nanoparticles, and Pt–CdS heterostructures consisting of alternating layers of Pt and CdS nanoparticles using dithiol-modified Au substrates as templates.

Noncovalently connected self-organized structures provide another group of templates for fabricating nanostructures. Lei et al. [235] reported assembly of phthalocyanine (Pc) single molecular arrays using 2-D alkane lamellar structures on HOPG as templates, as displayed in Figs. 4.11 and 4.12. Linear alkanes adsorbed from nonpolar solutions could form close-packed monolayers through 2-D crystallization of alkane parts on a graphite surface with polar groups paired together, which were used as molecular templates for assembling ordered Pc arrays. Linear $C_{18}X$ (X is Cl, Br, I, CN, and SH) molecules

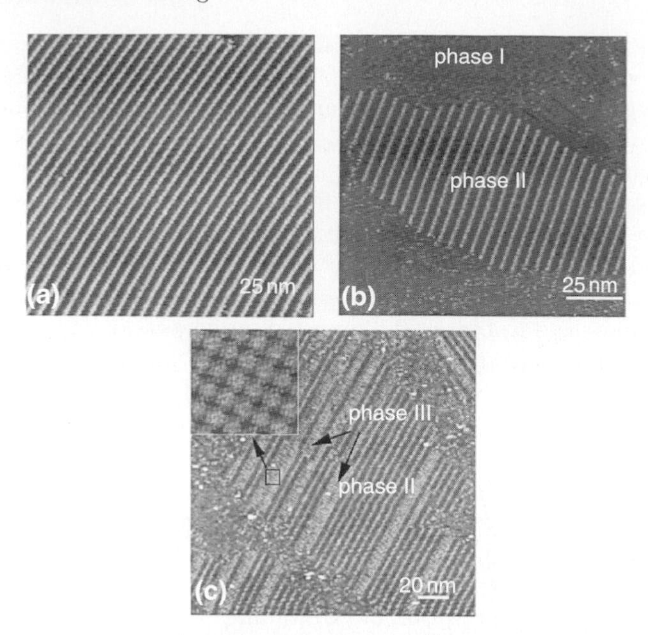

Fig. 4.11. (**a**) A large-scale image of the uniform assembly of phthalocyanine (*Pc*) with $C_{18}SH$ when the molar ratio was been adapted to 1:3 (667 mV, 1.019 nA). (**b**) Coexistence of phase I (domain of pure thiol) and phase II (uniform assembly) when the molar ratio was below 1:3 (−431 mV, 677 pA). (**c**) Coexistence of phase II and phase III (pure Pc domain) when the molar ratio was above 1:3 (715 mV, 1.136 nA). The *insert* shows a high-resolution image obtained of the Pc domain [235]

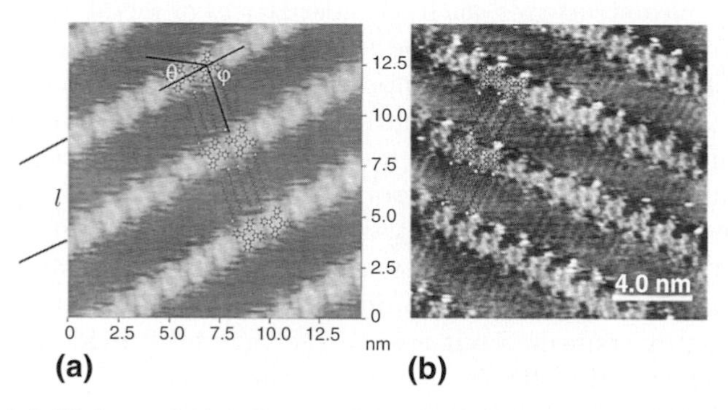

Fig. 4.12. (**a**) High-resolution image of the uniform assembly of Pc with $C_{18}SH$ (560 mV, 1.019 nA). (**b**) High-resolution image of the uniform assembly of Pc with $C_{18}I$ (−828 mV, 1.168 nA) [235]

were chosen for fabricating molecular templates. By adjusting the molar ratio of Pc to C18X to about 1:3, they obtained uniform assemblies of Pc molecules on $C_{18}X$ templates. The formation of the as-observed structures was ascribed to the interaction between Pc and the end group of the alkane derivative as well as the interaction between functional groups of the alkane derivative. Byrd et al. [236, 237] demonstrated that an organized organic monolayer could be used as a template for assembling an inorganic lattice by combining LB and self-assembly techniques. A LB monolayer of octadecylphosphonic acid formed on an octadecyltrichlorosilane-covered substrate was used as a template for growing Zr^{2+} ions from solution. A capping octadecylphosphonic acid LB monolayer was added to complete the thin film structure.

4.5 Summary

In this chapter, a comprehensive review on examples of organic molecular templates and methods for the fabrication of nanostructrues in different dimensionalities has been presented. Organic molecular templates can be divided into several groups, reverse-micelle and microemulsion system, LB monolayers, SAMs (including versatile patterned SAMs), and self-organized structures, in which SAMs occupy a prominent and unparalleled position owing to their excellent stability, orderliness, and versatile structures. SAMs provide not only techniques for fabricating nanostructures, but also methods of integrating systems for future applications. The fabrication and the applications of organic molecular templates have been investigated for decades and obtained fruitful results. It is believed that in the era of nanoscience and technology, organic molecular templates (especially SAMs) will play important roles for exploration and application of the amazing nano and quantum world.

References

1. M. Brust, M. Walker, D. Bethell, D.J. Schiffrin, R. Whyman: J. Chem. Soc. Chem. Commun., 801 (1994)
2. G. Carotenuto, L. Pasquini, E. Milella, M. Pentimalli, R. Lamanna, L. Nicolais: Eur. Phys. J. B. **31**, 545 (2003)
3. M. Aslam, G. Gopakumar, T.L. Shoba, I.S. Mulla, K. Vijayamohanan, S.K. Kulkarni, J. Urban, W. Vogel: J. Colloid Interface Sci. **255**, 79 (2002)
4. I. Quiros, M. Yamada, K. Kubo, J. Mizutani, M. Kurihara, H. Nishihara: Langmuir **18**, 1413 (2002)
5. C.C. Chen, J.J. Lin: Adv. Mater. **13**, 136 (2001)
6. Y.N. Xia, J.A. Rogers, K.E. Paul, G.M. Whitesides: Chem. Rev. **99**, 1823 (1999)
7. J. Aizenberg, A.J. Black, G.M. Whitesides: Nature **398**, 495 (1999)
8. S.T. Liu, R. Maoz, G. Schmid, J. Sagiv: Nano Lett. **2**, 1055 (2002)
9. C.G. Zeng, B. Wang, B. Li, H.Q. Wang, J.G. Hou: Appl. Phys. Lett. **79**, 1685 (2001)

10. B. Xu, S.X. Yin, C. Wang, X.H. Qiu, Q.D. Zeng, C.L. Bai: J. Phys. Chem. B. **104**, 10502 (2000)
11. S.L. Dawson, J. Elman, D.E. Margevich, W. McKenna, D.A. Tirrell, A. Ulman: In: *Hydrogels and Biodegradable Polymers for Bioapplications (ACS Symposium Series 627)*, ed by R. Ottenbrite, S. Hwang, K. Park, (American Chemical Society, Washington, DC, 1996), pp 187–196
12. A. Ulman: Chem. Rev. **96**, 1533 (1996)
13. Y. Okawa, M. Aono: J. Chem. Phys. **115**, 2317 (2001)
14. Y. Okawa, M. Aono: Nature **409**, 683 (2001)
15. A. Miura, S. De Feyter, M.M.S. Abdel-Mottaleb, A. Gesquiere, P.C.M. Grim, G. Moessner, M. Sieffert, M. Klapper, K. Mullen, F.C. De Schryver: Langmuir **19**, 6474 (2003)
16. J. Zak, H.P. Yuan, M.Ho, L.K. Woo, M.D. Porter: Langmuir **9**, 2772 (1993)
17. T.A. Postlethwaite, J.E. Hutchison, K.W. Hathcock, R.W. Murray: Langmuir **11**, 4109 (1995)
18. R.D. Piner, J. Zhu, F. Xu, S.H. Hong, C.A. Mirkin: Science **283**, 661 (1999)
19. Y.N. Xia, J.J. McClelland, R. Gupta, D. Qin, X.M. Zhao, L.L. Sohn, R.J. Celotta, G.M. Whitesides: Adv. Mater. **9**, 147 (1997)
20. X.M. Zhao, Y.N. Xia, G.M. Whitesides: J. Mater. Chem. **7**, 1069 (1997)
21. X.M. Zhao, Y.N. Xia, D. Qin, G.M. Whitesides: Adv. Mater. **9**, 251 (1997)
22. T.R. Kelly: Acc. Chem. Res. **34**, 514 (2001)
23. T.R. Kelly, H. De Silva, R.A. Silva: Nature **401**, 150 (1999)
24. T.R. Kelly, J.P. Sestelo: In: *Molecular Machines and Motors (Structure and Bonding), Series 99*, ed by J.-P. Sauvage, (Springer, Berlin Heidelberg New York, 2001), pp 19–53
25. N. Koumura, R.W.J. Zijlstra, R.A. van Delden, N. Harada, B.L. Feringa: Nature **401**, 152 (1999)
26. R.K. Soong, G.D. Bachand, H.P. Neves, A.G. Olkhovets, H.G. Craighead, C.D. Montemagno: Science **290**, 1555 (2000)
27. L.G. Zhu, S.M. Peng: Chin. J. Inorg. Chem. **18**, 117 (2002)
28. R.H. Mitchell, Y.S. Chen: Tetrahedron Lett. **37**, 6665 (1996)
29. X. Chi, M.E. Itkis, B.O. Patrick, T.M. Barclay, R.W. Reed, R.T. Oakley, A.W. Cordes, R.C. Haddon: J. Am. Chem. Soc. **121**, 10395 (1999)
30. T.M. Barclay, A.W. Cordes, R.C. Haddon, M.E. Itkis, R.T. Oakley, R.W. Reed, H. Zhang: J. Am. Chem. Soc. **121**, 969 (1999)
31. C.P. Collier, E.W. Wong, M. Belohradsky, F.M. Raymo, J.F. Stoddart, P.J. Kuekes, R.S. Williams, J.R. Heath: Science **285**, 391 (1999)
32. Z.Y. Hua, G.R. Chen, W. Xu, D.Y. Chen: Appl. Surf. Sci. **169**, 447 (2001)
33. R.M. Metzger, J.W. Baldwin, W.J. Shumate, I.R. Peterson, P. Mani, G.J. Mankey, T. Morris, G. Szulczewski, S. Bosi, M. Prato, A. Comito, Y. Rubin: J. Phys. Chem. B. **107**, 1021 (2003)
34. R.M. Metzger: J. Solid State Chem. **168**, 696 (2002)
35. R.M. Metzger, B. Chen, U. Hopfner, M.V. Lakshmikantham, D. Vuillaume, T. Kawai, X.L. Wu, H. Tachibana, T.V. Hughes, H. Sakurai, J.W. Baldwin, C. Hosch, M.P. Cava, L. Brehmer, G.J. Ashwell: J. Am. Chem. Soc. **119**, 10455 (1997)
36. B. Chen, R.M. Metzger: J. Phys. Chem. B. **103**, 4447 (1999)
37. C.G. Zeng, H.Q. Wang, B. Wang, J.L. Yang, J.G. Hou: Appl. Phys. Lett. **77**, 3595 (2000)

38. S.H.M. Persson, L. Olofsson, L. Gunnarsson, E. Olsson: Nanostruct. Mater. 12, 821 (1999)
39. A.M. Stevens, C.J. Richards: Tetrahedron Lett. **38**, 7805 (1997)
40. J. Han, A. Globus, R. Jaffe, G. Deardorff: Nanotechnology **8**, 95 (1997)
41. J. Clayden, J.H. Pink: Angew. Chem. Int. Ed. Engl. **37**, 1937 (1998)
42. D.A. Bonnell: J. Vac. Sci. Technol. A **21**, S194 (2003)
43. R.A. Caruso, M. Antonietti: Chem. Mater. **13**, 3272 (2001)
44. S.A. Davis, M. Breulmann, K.H. Rhodes, B. Zhang, S. Mann: Chem. Mater. **13**, 3218 (2001)
45. L.A. Estrof, A.D. Hamilton: Chem. Mater. **13**, 3227 (2001)
46. S. Mann, S.L. Burkett, S.A. Davis, C.E. Fowler, N.H. Mendelson, S.D. Sims, D. Walsh, N.T. Whilton: Chem. Mater. **9**, 2300 (1997)
47. N.K. Raman, M.T. Anderson, C.J. Brinker: Chem. Mater. **8**, 1682 (1996)
48. R.A. Caruso: Top. Curr. Chem. **226**, 91 (2003)
49. N. Kawahashi, E. Matijevic: J. Colloid Interface Sci. **143**, 103 (1991)
50. Y. Lu, Y.D. Yin, Y.N. Xia: Adv. Mater. **13**, 271 (2001)
51. C. Sanchez, G.J.D.A. Soler-Illia, F. Ribot, T. Lalot, C.R. Mayer, V. Cabuil: Chem. Mater. **13**, 3061 (2001)
52. P. Innocenzi, G. Brusatin, Chem. Mater. **13**, 3126 (2001)
53. K.J.C. van Bommel, A. Friggeri, S. Shinkai: Angew. Chem. Int. Ed. Engl. **42**, 980 (2003)
54. T.P. Lodge: Macromol. Chem. Phys. **204**, 265 (2003)
55. K.M. Ryan, N.R.B. Coleman, D.M. Lyons, J.P. Hanrahan, T.R. Spalding, M.A. Morris, D.C. Steytler, R.K. Heenan, J.D. Holmes: Langmuir **18**, 4996 (2002)
56. D.B. Zhang, L.M. Qi, J.M. Ma, H.M. Cheng: Chem. Mater. **13**, 2753 (2001)
57. C.R. Martin: Chem. Mater. **8**, 1739 (1996)
58. M. Goren and R.B. Lennox: Nano Lett. **1**, 735 (2001)
59. J.Y. Cheng, A.M. Mayes, C.A. Ross: Nat. Mater. **3**, 823 (2004)
60. I.W. Hamley: Nanotechnology **14**, R39 (2003)
61. W. Meier: Curr. Opin. Colloid Interface Sci. **4**, 6 (1999)
62. H.W. Li, W.T.S. Huck: Curr. Opin. Solid State Mater. Sci. **6**, 3 (2002)
63. M.F. Zhang, M. Drechsler, A.H.E. Muller: Chem. Mater. **16**, 537 (2004)
64. I.A. Ansari, I.W. Hamley: J. Mater. Chem. **13**, 2412 (2003)
65. S. Fullam, D. Cottell, H. Rensmo, D. Fitzmaurice: Adv. Mater. **12**, 1430 (2000)
66. Y. Zhang, H.J. Dai: Appl. Phys. Lett. **77**, 3015 (2000)
67. L. Hu, Y.X. Li, X.X. Ding, C. Tang, S.R. Qi: Chem. Phys. Lett. **397**, 271 (2004)
68. A. Govindaraj, B.C. Satishkumar, M. Nath, C.N.R. Rao: Chem. Mater. **12**, 202 (2000)
69. C.N.R. Rao, B.C. Satishkumar, A. Govindaraj, M. Nath: Chem. Phys. Chem. **2**, 78 (2001)
70. B.C. Satishkumar, A. Govindaraj, M. Nath, C.N.R. Rao: J. Mater. Chem. **10**, 2115 (2000)
71. S.A. Davis, S.L. Burkett, N.H. Mendelson, S. Mann: Nature **385**, 420 (1997)
72. E. Braun, Y. Eichen, U. Sivan, G. Ben-Yoseph: Nature **391**, 775 (1998)
73. C.J.F. Dupraz, P. Nickels, U. Beierlein, W.U. Huynh, F.C. Simmel: Superlattices Microstruct. **33**, 369 (2003)
74. Y. Eichen, E. Braun, U. Sivan, G. Ben-Yoseph: Acta Polym. **49**, 663 (1998)

75. M. Field, C.J. Smith, D.D. Awschalom, N.H. Mendelson, E.L. Mayes, S.A. Davis, S. Mann, Appl. Phys. Lett. **73**, 1739 (1998)
76. C.J. Loweth, W.B. Caldwell, X.G. Peng, A.P. Alivisatos, P.G. Schultz: Angew. Chem. Int. Ed. Engl. **38**, 1808 (1999)
77. J.E. Meegan, A. Aggeli, N. Boden, R. Brydson, A.P. Brown, L. Carrick, A.R. Brough, A. Hussain, R.J. Ansell: Adv. Funct. Mater. **14**, 31 (2004)
78. N.C. Seeman, Trends. Biotechnol. **17**, 437 (1999)
79. W. Shenton, D. Pum, U.B. Sleytr, S. Mann: Nature **389**, 585 (1997)
80. T. Shimizu: Macromol. Rapid Commun. **23**, 311 (2002)
81. J.H. Adair, T. Li, T. Kido, K. Havey, J. Moon, J. Mecholsky, A. Morrone, D.R. Talham, M.H. Ludwig, L. Wang: Mater. Sci. Eng. R **23**, 139 (1998)
82. P.M. Mendes, Y. Chen, R.E. Palmer, K. Nikitin, D. Fitzmaurice, J.A. Preece: J. Phys. Condens. Mater. **15**, S3047 (2003)
83. A.N. Shipway, E. Katz, I. Willner: Chem. Phys. Chem. **1**, 18 (2000)
84. Z.T. Liu, C. Lee, V. Narayanan, G. Pei, E.C. Kan: IEEE Trans. Electron Devices **49**, 1606 (2002)
85. C.N.R. Rao, G.U. Kulkarni, P.J. Thomas, P.P. Edwards: Chem. Soc. Rev. **29**, 27 (2000)
86. Y.S. Shon, H. Choo: Cr. Chim. **6**, 1009 (2003)
87. I. Capek: Adv. Colloid Interface Sci. **110**, 49 (2004)
88. H.P. Hentze, C.C. Co, C.A. McKelvey, E.W. Kaler: Top. Curr. Chem. **226**, 197 (2003)
89. H.P. Hentze, E.W. Kaler: Curr. Opin. Colloid Interface Sci. **8**, 164 (2003)
90. D. Horn, J. Rieger: Angew. Chem. Int. Ed. Engl. **40**, 4331 (2001)
91. T.B. Liu, C. Burger, B. Chu: Prog. Polym. Sci. **28**, 5 (2003)
92. V.T. John, B. Simmons, G.L. McPherson, A. Bose: Curr. Opin. Colloid Interface Sci. **7**, 288 (2002)
93. G. Schmid, R. Boese, R. Pfeil, F. Bandermann, S. Meyer, G. Calis, J. van der Velden: Chem. Ber. **114**, 3634 (1981)
94. M.A. Lopezquintela, J. Rivas: J. Colloid Interf. Sci. **158**, 446 (1993)
95. N. Lufimpadio, J.B. Nagy, E.G. Derouane: *Surfactants in Solution* (Plenum, New York, 1984), p 1483
96. H.H. Ingelsten, R. Bagwe, A. Palmqvist, M. Skoglundh, C. Svanberg, K. Holmberg, D.O. Shah: J. Colloid Interface Sci. **241**, 104 (2001)
97. A. Henglein: Chem. Rev. **89**, 1861 (1989)
98. M. Iida, S. Ohkawa, H. Er, N. Asaoka, H. Yoshikawa: Chem. Lett., 1050 (2002)
99. W. Lu, B. Wang, K.D. Wang, X.P. Wang and J.G. Hou, Langmuir **19**, 5887 (2003)
100. A. Manna, B.D. Kulkarni, K. Bandyopadhyay, K. Vijayamohanan: Chem. Mater. **9**, 3032 (1997)
101. A.B.R. Mayer, J.E. Mark: Eur. Polym. J. **34**, 103 (1998)
102. E.M. Egorova, A.A. Revina: Colloid Surf. A **168**, 87 (2000)
103. G.C. De, A.M. Roy, S. Saba, S. Aditya: J. Indian Chem. Soc. **80**, 551 (2003)
104. H. Ohde, M. Ohde, F. Bailey, H. Kim, C.M. Wai: Nano Lett. **2**, 721 (2002)
105. D. Adityawarman, A. Voigt, P. Veit, K. Sundmacher: Chem. Eng. Sci. **60**, 3373 (2005)
106. L.M. Qi, J.M. Ma, H.M. Cheng, Z.G. Zhao: Colloids Surf. A **108**, 117 (1996)
107. M. Summers, J. Eastoe, S. Davis: Langmuir **18**, 5023 (2002)
108. G. Cardenas, R. Segura: Mater. Res. Bull. **35**, 1369 (2000)

109. M.L. Wu, D.H. Chen, T.C. Huang: Langmuir **17**, 3877 (2001)
110. M.L. Wu, D.H. Chen, T.C. Huang: J. Colloid Interface Sci. **243**, 102 (2001)
111. M.L. Wu, D.H. Chen, T.C. Huang: Chem. Mater. **13**, 599 (2001)
112. L. Guczi: Catal. Today **101**, 53 (2005)
113. O.M. Wilson, R.W.J. Scott, J.C. Garcia-Martinez, R.M. Crooks: J. Am. Chem. Soc. **127**, 1015 (2005)
114. D.H. Chen, C.J. Chen: J. Mater. Chem. **12**, 1557 (2002)
115. W. Lu, B. Wang, J. Zeng, X. Wang, S. Zhang, J.G. Hou: Langmuir **21**, 3684 (2005)
116. N. Pinna, K. Weiss, H. Sack-Kongehl, W. Vogel, J. Urban, M.P. Pileni: Langmuir **17**, 7982 (2001)
117. A. Filankembo, S. Giorgio, I. Lisiecki, M.P. Pileni: J. Phys. Chem. B **107**, 7492 (2003)
118. M.P. Pileni, J. Tanori, A. Filankembo: Colloids Surf. A **123**, 561 (1997)
119. M.P. Pileni: Langmuir **17**, 7476 (2001)
120. I. Lisiecki, M.P. Pileni: J. Am. Chem. Soc. **115**, 3887 (1993)
121. J. Tanori, M.P. Pileni: Langmuir **13**, 639 (1997)
122. C.J. Johnson, E. Dujardin, S.A. Davis, C.J. Murphy, S. Mann: J. Mater. Chem. **12**, 1765 (2002)
123. N.R. Jana, L. Gearheart, C.J. Murphy: J. Phys. Chem. B. **105**, 4065 (2001)
124. D.H.W. Hubert, M. Jung, P.M. Frederik, P.H.H. Bomans, J. Meuldijk, A.L. German: Adv. Mater. **12**, 1286 (2000)
125. D.H.W. Hubert, M. Jung, A.L. German: Adv. Mater. **12**, 1291 (2000)
126. H.P. Lin, Y.R. Cheng, C.Y. Mou: Chem. Mater. **10**, 3772 (1998)
127. G. Kickelbick, L.M. Liz-Marzan: In: *Encyclopedia of Nanoscience and Nanotechnology, Series 10*, ed by H.S. Nalwa, (American Scientific, Stevenson Ranch, 2003), pp. 1–22
128. I. Soten, G.A. Ozin: Curr. Opin. Colloid Interface Sci. **4**, 325 (1999)
129. J.P. Yang, F.C. Meldrum, J.H. Fendler: J. Phys. Chem. **99**, 5500 (1995)
130. E.M. Landau, M. Levanon, L. Leiserowita, M. Lahav, J. Sagiv: Nature **318**, 353 (1985)
131. S. Mann, B.R. Heywook, S. Rajam, J.D. Birchall: Nature **334**, 692 (1988)
132. E.M. Landau, S.G. Wolf, M. Levanon, L. Leiserowita, M. Lahav, J. Sagiv: J. Am. Chem. Soc. **111**, 1436 (1989)
133. J.H. Fendler, F.C. Meldrum: Adv. Mater. **7**, 607 (1995)
134. J.M. Didymus, S. Mann, W.J. Benton, I.R. Collins: Langmuir **11**, 3130 (1995)
135. B.R. Heywood, S. Mann: J. Am. Chem. Soc. **114**, 4681 (1992)
136. B.R. Heywood, S. Mann: Chem. Mater. **6**, 311 (1994)
137. S. Mann, B.R. Heywood, S. Rajam, J.B.A. Walker: J. Phys. D Appl. Phys. **24**, 154 (1991)
138. S. Mann, G.A. Ozin: Nature **382**, 313 (1996)
139. J.P. Yang, J.H. Fendler: J. Phys. Chem. **99**, 5505 (1995)
140. X.K. Zhao, J. Yang, L.D. Mccormick, J.H. Fendler: J. Phys. Chem. **96**, 9933 (1992)
141. H.H. Li, G.Z. Mao, K.Y.S. Ng: Thin Solid Films **358**, 62 (2000)
142. J. Kmetko, C.J. Yu, G. Evmenenko, S. Kewalramani, P. Dutta: Phys. Rev. Lett. **89**, 186102 (2002)
143. J.J. Gooding, F. Mearns, W.R. Yang, J.Q. Liu: Electroanalysis **15**, 81 (2003)
144. F. Schreiber: Prog. Surf. Sci. **65**, 151 (2000)

145. G.E. Poirier: Chem. Rev. **97**, 1117 (1997)
146. J. Aizenberg, A.J. Black,G.M. Whitesides: J. Am. Chem. Soc. **121**, 4500 (1999)
147. D. Qin, Y.N. Xia, B. Xu, H. Yang, C. Zhu, G.M. Whitesides: Adv. Mater. **11**, 1433 (1999)
148. S. Hoeppener, R. Maoz, S.R. Cohen, L.F. Chi, H. Fuchs, J. Sagiv: Adv. Mater. **14**, 1036 (2002)
149. R. Maoz, S.R. Cohen, J. Sagiv: Adv. Mater. **11**, 55 (1999)
150. R. Maoz, E. Frydman, S.R. Cohen, J. Sagiv: Adv. Mater. **12**, 725 (2000)
151. R. Maoz, E. Frydman, S.R. Cohen, J. Sagiv: Adv. Mater. **12**, 424 (2000)
152. D.H. Woo, S.J. Choi, D.H. Han, H. Kang, S.M. Park: Phys. Chem. Chem. Phys. **3**, 3382 (2001)
153. W.A. Hayes, C. Shannon: Langmuir **14**, 1099 (1998)
154. N.I. Kovtyukhova, T.E. Mallouk: Chem-Eur. J. **8**, 4355 (2002)
155. Z.Y. Tang, N.A. Kotov: Adv. Mater. **17**, 951 (2005)
156. S.M. Prokes, K.L. Wang: MRS Bull. **24**, 13 (1999)
157. X.F. Duan, Y. Huang, Y. Cui, J.F. Wang, C.M. Lieber: Nature **409**, 66 (2001)
158. Y. Huang, X.F. Duan, Y. Cui, L.J. Lauhon, K.H. Kim, C.M. Lieber: Science **294**, 1313 (2001)
159. Y. Huang, X.F. Duan, Q.Q. Wei, C.M. Lieber: Science **291**, 630 (2001)
160. C.A. Huber, T.E. Huber, M. Sadoqi, J.A. Lubin, S. Manalis, C.B. Prater: Science **263**, 800 (1994)
161. T.M. Whitney, J.S. Jiang, P.C. Searson, C.L. Chien: Science **261**, 1316 (1993)
162. C.G. Wu, T. Bein: Science **264**, 1757 (1994)
163. S. Iijima: Nature **354**, 56 (1991)
164. B.T. Mayers, K. Liu, D. Sunderland, Y.N. Xia: Chem. Mater. **15**, 3852 (2003)
165. B. Mayers, Y.N. Xia: J. Mater. Chem. **12**, 1875 (2002)
166. Z.L. Wang: Annu. Rev. Phys. Chem. **55**, 159 (2004)
167. X.D. Wang, Y. Ding, C.J. Summers, Z.L. Wang: J. Phys. Chem. B. **108**, 8773 (2004)
168. B.A. Parviz, D. Ryan, G.M. Whitesides: IEEE Trans. Adv. Packaging **26**, 233 (2003)
169. P.D. Yang, F. Kim: Abstr. Pap. Am. Chem. Soc. **225**, U136 (2003)
170. C.N.R. Rao, G. Gundiah, F.L. Deepak, A. Govindaraj, A.K. Cheetham: J. Mater. Chem. **14**, 440 (2004)
171. C.N.R. Rao, F.L. Deepak, G. Gundiah, A. Govindaraj: Prog. Solid State Chem. **31**, 5 (2003)
172. C. Qian, F. Kim, L. Ma, F. Tsui, P.D. Yang, J. Liu: J. Am. Chem. Soc. **126**, 1195 (2004)
173. S.K. Lee, H.J. Choi, P. Pauzauskie, P.D. Yang, N.K. Cho, H.D. Park, E.K. Suh, K.Y. Lim, H.J. Lee: Phys. Status Solidi B. **241**, 2775 (2004)
174. M. Law, J. Goldberger, P.D. Yang: Annu. Rev. Mater Res. **34**, 83 (2004)
175. J.R. LaRoche, Y.W. Heo, B.S. Kang, L.C. Tien, Y. Kwon, D.P. Norton, B.P. Gila, F. Ren, S.J. Pearton: J. Electron. Mater. **34**, 404 (2005)
176. Y.W. Heo, D.P. Norton, L.C. Tien, Y. Kwon, B.S. Kang, F. Ren, S.J. Pearton, J.R. LaRoche: Mater. Sci. Eng. R **47**, 1 (2004)
177. F. Rosei, M. Schunack, P. Jiang, A. Gourdon, E. Laegsgaard, I. Stensgaard, C. Joachim, F. Besenbacher: Science **296**, 328 (2002)
178. H. Matsui, B. Gologan: J. Phys. Chem. B. **104**, 3383 (2000)

179. H. Matsui, S. Pan, B. Gologan, S.H. Jonas: J. Phys. Chem. B. **104**, 9576 (2000)
180. B.A. Simmons, S.C. Li, V.T. John, G.L. McPherson, A. Bose, W.L. Zhou, J.B. He: Nano Lett. **2**, 263 (2002)
181. J.D. Hopwood, S. Mann: Chem. Mater. **9**, 1819 (1997)
182. X. Chen, X.M. Sun, Y.D. Li: Inorg. Chem. **41**, 4524 (2002)
183. M. Niederberger, H.J. Muhr, F. Krumeich, F. Bieri, D. Gunther, R. Nesper: Chem. Mater. **12**, 1995 (2000)
184. L.M. Qi, J.M. Ma, H.M. Cheng, Z.G. Zhao: J. Phys. Chem. B. **101**, 3460 (1997)
185. L. Yan, Y.D. Li, Z.X. Deng, J. Zhuang, X.M. Sun: Int. J. Inorg. Mater. **3**, 633 (2001)
186. N.R. Jana, L. Gearheart, C.J. Murphy: Adv. Mater. **13**, 1389 (2001)
187. N.R. Jana, L. Gearheart, C.J. Murphy: Chem. Commun. 617 (2001)
188. Y.K. Liu, D.D. Hou, G.H. Wang: Chem. Phys. Lett. **379**, 67 (2003)
189. J. Yao, W. Tjandra, Y.Z. Chen, K.C. Tam, J. Ma, B. Soh: J. Mater. Chem. **13**, 3053 (2003)
190. J. Perez-Juste, L.M. Liz-Marzan, S. Carnie, D.Y.C. Chan: Adv. Funct. Mater. **14**, 571 (2004)
191. M.H. Cao, C.W. Hu, Y.H. Wang, Y.H. Guo, C.X. Guo, E.B. Wang: Chem. Commun. 1884 (2003)
192. Y. Yu, F.P. Du, J.C. Yu, Y.Y. Zhuang, P.K. Wong: J. Solid State Chem. **177**, 4640 (2004)
193. L.M. Huang, H.T. Wang, Z.B. Wang, A. Mitra, K.N. Bozhilov, Y.S. Yan: Adv. Mater. **14**, 61 (2002)
194. L.M. Huang, H.T. Wang, Z.B. Wang, A.P. Mitra, D. Zhao, Y.H. Yan: Chem. Mater. **14**, 876 (2002)
195. Y.S. Yan, L.M. Huang: Abstr. Pap. Am. Chem. Soc. **221**, U629 (2001)
196. H.M. Luo, J.F. Zhang, Y.S. Yan: Chem. Mater. **15**, 3769 (2003)
197. H.M. Luo, L. Sun, Y.F. Lu, Y.S. Yan: Langmuir **20**, 10218 (2004)
198. M. Li, H. Schnablegger, S. Mann: Nature **402**, 393 (1999)
199. C.L. Cheung, J.A. Camarero, B.W. Woods, T.W. Lin, J.E. Johnson, J.J. De Yoreo: J. Am. Chem. Soc. **125**, 6848 (2003)
200. S.J. Choi, S.M. Park: Adv. Mater. **12**, 1547 (2000)
201. S. Hoeppener, L.F. Chi, H. Fuchs: Nano Lett. **2**, 459 (2002)
202. H. Zhang, R.C. Jin, C.A. Mirkin: Nano Lett. **4**, 1493 (2004)
203. M. Toerker, R. Staub, T. Fritz, T. Schmitz-Hubsch, F. Sellam, K. Leo: Surf. Sci. **445**, 100 (2000)
204. R. Staub, M. Toerker, T. Fritz, T. Schmitz-Hubsch, F. Sellam, K. Leo: Langmuir **14**, 6693 (1998)
205. G.E. Poirier: Langmuir **15**, 3018 (1999)
206. G.E. Poirier: Langmuir **15**, 1167 (1999)
207. S. Baral, P. Schoen: Chem. Mater. **5**, 145 (1993)
208. S.K. Sinha: *Ordering in Two Dimensions*, (Elsevier North Holland, Amsterdam, 1980)
209. J.H. Fendler: Chem. Mater. **13**, 3196 (2001)
210. Y.S. Zhou, B. Wang, S.Q. Xiao, S.Q. Xiao, Y.L. Li, J. G. Hou: Appl. Surf. Sci. **252**, 2119 (2006)
211. Y.S. Zhou, B. Wang, M.Z. Zhu, J.G. Hou: Chem. Phys. Lett. **403**, 140 (2005)

212. A.K. Grag, A.K. Tripathi, T.C. Goel, M.M.A. Sekar, C.N. Sukenik: Mater. Sci. Eng. **87**, 87 (2001)
213. H. Lin, H. Kozuka, T. Yoko: Thin Solid Films **315**, 111 (1998)
214. M. Nagtegaal, P. Stroeve, J. Ensling, P. Gutlich, M. Schurrer, H. Voit, J. Flath, J. Kashammer, W. Knoll, W. Tremel: Chem. Eur. J. **5**, 1331 (1999)
215. O. Azzaroni, P.L. Schilardi, R.C. Salvarezza: Appl. Phys. Lett. **80**, 1061 (2002)
216. Y. Masuda, W.S. Seo, K. Koumoto: Thin Solid Films **382**, 183 (2001)
217. Y.F. Gao, Y. Masuda, T. Yonezawa, K. Koumoto: Chem. Mater. **14**, 5006 (2002)
218. H. Zhang, Z. Li, C.A. Mirkin: Adv. Mater. **14**, 1472 (2002)
219. J.C. Smith, K.B. Lee, Q. Wang, M.G. Finn, J.E. Johnson, M. Mrksich, C.A. Mirkin: Nano Lett. **3**, 883 (2003)
220. X.G. Liu, L. Fu, S.H. Hong, V.P. Dravid, C.A. Mirkin: Adv. Mater. **14**, 231 (2002)
221. L.M. Demers, S.J. Park, T.A. Taton, Z. Li, C.A. Mirkin: Angew. Chem. Int. Ed. Engl. **40**, 3071 (2001)
222. L.M. Demers, C.A. Mirkin: Angew. Chem. Int. Ed. Engl. **40**, 3069 (2001)
223. J. Tien, A. Terfort, G.M. Whitesides: Langmuir **13**, 5349 (1997)
224. I.W. Hamley: Angew. Chem. Int. Ed. Engl. **42**, 1692 (2003)
225. L.F. Yuan, J.L. Yang, H.Q. Wang, C.G. Zeng, Q.X. Li, B. Wang, J.G. Hou, Q.S. Zhu, D.M. Chen: J. Am. Chem. Soc. **125**, 169 (2003)
226. J.G. Hou, J.L. Yang, H.Q. Wang, Q.X. Li, C.G. Zeng, L.F. Yuan, B. Wang, D.M. Chen, Q.S. Zhu: Nature **409**, 304 (2001)
227. R.W. Zehner, W.A. Lopes, T.L. Morkved, H. Jaeger, L.R. Sita: Langmuir **14**, 241 (1998)
228. T. Lee, N. Yao, I.A. Aksay: Langmuir **13**, 3866 (1997)
229. J. Frommer, R. Luthi, E. Meyer, D. Anselmetti, M. Dreier, R. Overney, H.J. Guntherodt, M. Fujihira: Nature **364**, 198 (1993)
230. J.Y. Fang, C.M. Knobler: Abstr. Pap. Am. Chem. Soc. **214**, 1 (1997)
231. J.Y. Fang, C.M. Knobler: Langmuir **12**, 1368 (1996)
232. P. Moraille, A. Badia: Angew. Chem. Int. Ed. Engl. **41**, 4303 (2002)
233. M. Brust, D. Bethell, C.J. Kiely, D.J. Schiffrin: Langmuir **14**, 5425 (1998)
234. K.V. Sarathy, P.J. Thomas, G.U. Kulkarni, C.N.R. Rao: J. Phys. Chem. B. **103**, 399 (1999)
235. S.B. Lei, C. Wang, S.X. Yin, C.L. Bai: J. Phys. Chem. B. **105**, 12272 (2001)
236. H. Byrd, S. Whipps, J.K. Pike, D.R. Talham: Abstr. Pap. Am. Chem. Soc. **206**, 216 (1993)
237. H. Byrd, J.K. Pike, D.R. Talham: Chem. Mater. **5**, 709 (1993)

5 Carbon Nanotubes

Lisa Vaccari, Dimitrios Tasis, Andrea Goldoni, and Maurizio Prato

5.1 Introduction

Carbon nanotubes (CNTs) [1] are unique nanosystems with extraordinary mechanical and electronic properties, which derive from their unusual molecular structure. An ideal carbon nanotube can be thought of as a single graphite layer (graphene sheet), rolled up to make a seamless hollow cylinder (Fig. 5.1). These cylinders can be tens of microns long, with diameters as small as 0.7 nanometers and are closed at both ends by fullerene-like caps. CNTs having wall thickness of one carbon sheet are named single-wall carbon nanotubes (SWCNTs). In consequence of the van der Waals interactions between nanotubes, they often aggregate in large ropes: ordered arrays of SWCNTs arranged on a triangular lattice. SWCNTs can be considered as the building bocks of multi-wall carbon nanotubes (MWCNTs), which consist of a coaxial array of SWCNTs with increasing diameter. MWCNTs are also usually long many microns, with the external diameter that ranges from two to several tens of nanometers, providing very high aspect ratio structures.

5.2 Historical Background

The discovery of carbon nanotubes originates from the interest in the carbon materials that followed the synthesis of cage structures called fullerenes at Rice University by Kroto, Smalley and coworkers in the 1980's [2]. In 1991 Iijima [1], at NEC laboratory, in Japan, first observed carbon nanotubes using High Resolution Transmission Electron Microscopy (HRTEM) (Fig. 5.2). In the same period, Russian scientists at the Institute of Chemical Physics in Moscow independently discovered CNTs and nanotube bundles, but possessing a much smaller aspect ratio [3]. Both research groups observed MWCNTs. Soon after, in 1992, the Iijima group [4] and Bethune's at IBM laboratory [5] observed for the first time SWCNTs. In the same period, a lot of theoretical studies were performed, predicting that carbon nanotubes could be either metallic or semiconducting depending on their diameter and chirality (orientation of their hexagons with respect to the nanotube axis) [6–8]. These exciting theoretical predictions justify the extraordinary interest that CNTs stimulated inside the research community and the efforts made by the scientists to overcome the enormous problems that they had at the beginning

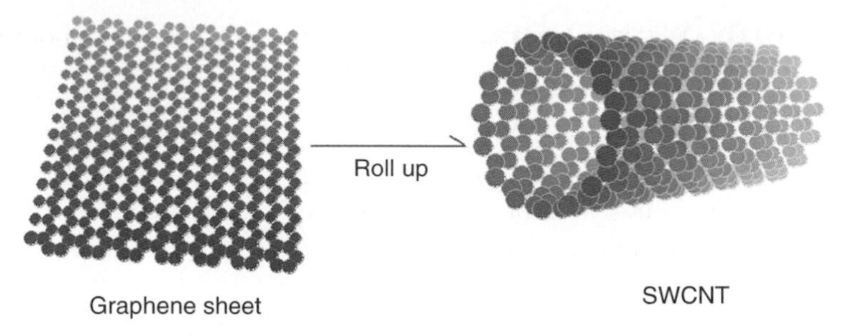

Fig. 5.1. Rolling up of a graphene sheet to make a SWCNT

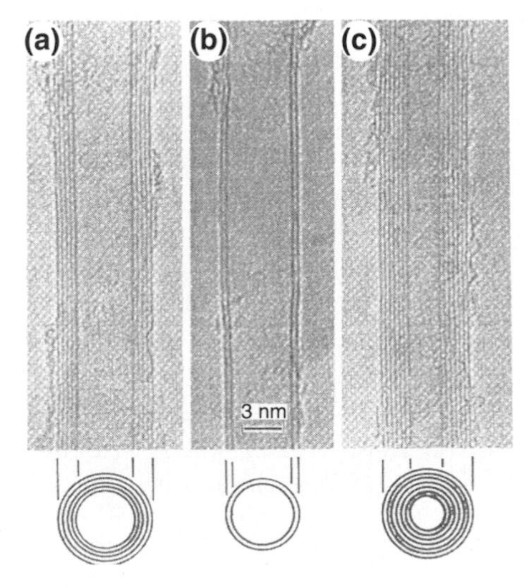

Fig. 5.2. Electron micrographs of MWCNTs from the original work of Iijima [1]. Parallel dark lines correspond to (002) lattice image of graphite: (**a**) Cross section of five graphite sheets, outer diameter 6.7 nm; (**b**) Cross section of a two-sheets tube, outer diameter 5.5 nm; (**c**) Seven-sheets tube, outer diameter 6.5 nm. Reprinted by permission from Macmillan Publishers Ltd: Nature [1], copyright 1991

to grow, purify and manipulate CNTs. It was only in 1998 that the theoretical prediction regarding electronic properties of carbon nanotubes were corroborated by experimental measurements [9, 10].

5.3 Atomic Structure of CNTs

Experimental studies on carbon nanotubes structure have mainly been carried out using HRTEM and Scanning Tunnelling Microscopy (STM). The information collected with these two techniques along with Atomic Force

Microscopy (AFM) gave a good understanding of the structure of both single-wall and multi-wall carbon nanotubes. Because of the very close relationship between CNTs and graphite, the description of the structure of CNTs will follow a short introduction on graphite structure.

The electronic structure of a free carbon atom is: $(1s)^2$ $(2s)^2$ $(2p)^2$. In order to form covalent bonds, one of the 2s electron is promoted to a 2p orbital giving rise to 2s, $2p_x$, $2p_y$, $2p_z$ configuration. The 2s valence orbital hybridizes with 1, 2 or 3 2p orbitals to form sp, sp^2 or sp^3 hybridized molecular orbitals, respectively. In the sp hybridization, the two sp orbitals are oriented at 180° along the bond axis with the two remaining 2p orbitals perpendicular to this. In the sp^2 hybridization, the three orbitals are arranged at 120°, with the third p_z orbital perpendicular to the plane. Finally, the four sp^3 orbitals are organized in a tetrahedral structure, typical of the carbon atom in a diamond crystal.

The ideal crystal structure of graphite (see Fig. 5.3) consists of layers of carbon atoms that are arranged in an open honeycomb network with two atoms per unit cell, labelled A and B. This structure is a consequence of the sp^2 hybridization of carbon atoms in graphite crystals. The sp^2 orbitals form strong σ bonds in the graphite plane while the p_z orbitals overlap to give π-orbitals that provide the weak van der Waals forces between the planes. Atoms A and A' on consecutive layers are aligned but atoms B and B' are shifted so that atom B is over the unoccupied center of the hexagon of the adjacent layer, and similarly for B'. Then, in graphite crystals there are two distinct planes, labelled **A** and **B**, that are arranged in the **ABAB** stacking, also known as Bernal graphite structure. The Bernal graphite structure belong to the $P6_3/mmc$ space group with in plane lattice constant a_0=2.462 Å and lattice constant along c direction c_0=6.708 Å. The in plane nearest-neighbour carbon–carbon distance, a_{c-c}, is 1.421 Å and the interlayer distance, $c_0/2$=3.354 Å.

Very often graphene sheets do not stack perfectly, giving rise to a deviation from the Bernal structure producing an increase of the interlayer distance to 3.440 Å. The resulting structure is made by uncorrelated graphite layers and is named turbostratic graphite.

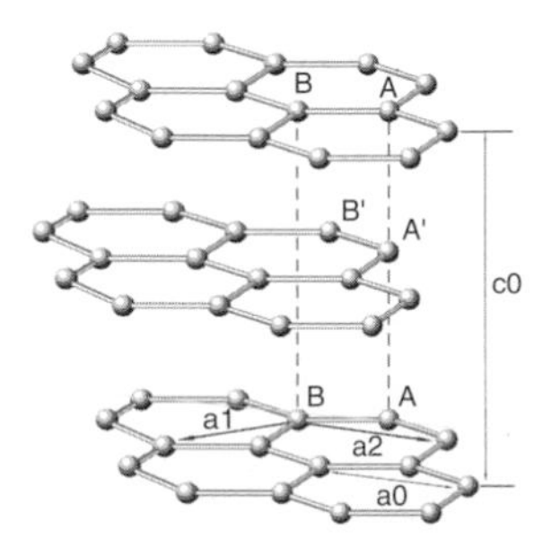

Fig. 5.3. Crystal structure of graphite with **ABAB** Bernal stacking

5.3.1 Structure and Symmetry of SWCNTs

The structure of a SWCNT can be described introducing two parameters: the chiral vector, $\boldsymbol{C_h}$, and the chiral angle, θ. The chiral vector $\boldsymbol{C_h}$ connects two equivalent sites on the original graphene sheet (see Fig. 5.4); the cylinder is produced by rolling up the graphite layer so that the two ends of the $\boldsymbol{C_h}$ vector superimpose. If $\boldsymbol{a_1}$ and $\boldsymbol{a_2}$ are the two unit cell base vectors of the honeycomb graphene sheet, the chiral vector can be geometrically expressed in terms of two integers, n and m, such that

$$\boldsymbol{C_h} = n\boldsymbol{a_1} + m\boldsymbol{a_2} \qquad (0 \leq |m| \leq |n|) \tag{5.1}$$

Since $|\boldsymbol{a_1}| = |\boldsymbol{a_2}| = \sqrt{3}a_{c-c}$, the carbon nanotube diameter is $d_t = |\boldsymbol{C_h}|/\pi$:

$$d_t = \frac{\sqrt{3}a_{c-c}}{\pi}\sqrt{m^2 + n^2 + mn} \tag{5.2}$$

The chiral angle, θ, defined as the angle between the chiral vector $\boldsymbol{C_h}$ and the $\boldsymbol{a_1}$ direction, is given by:

$$\theta = \arctan\frac{\sqrt{3}m}{2n + m} \qquad 0^\circ \leq \theta \leq 30^\circ \tag{5.3}$$

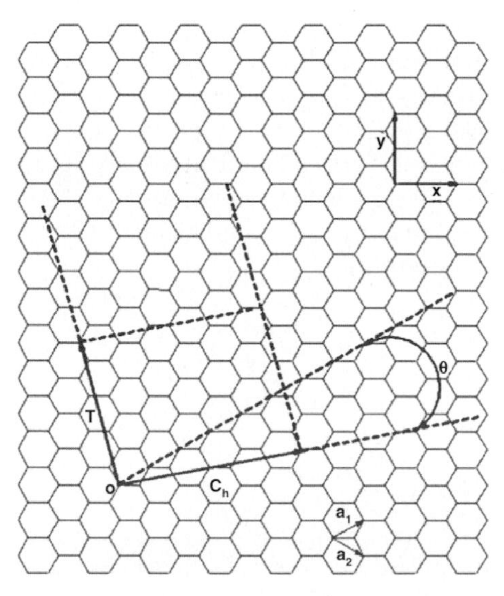

Fig. 5.4. 2-D representation of chiral vector, $\boldsymbol{C_h}$, translation vector, $\boldsymbol{T}$, and chiral angle θ, for a (4,2) carbon nanotube. $\boldsymbol{a_1}$ and $\boldsymbol{a_2}$ are the basic lattice vector for the honeycomb lattice of graphene sheet. Adapted from [11]

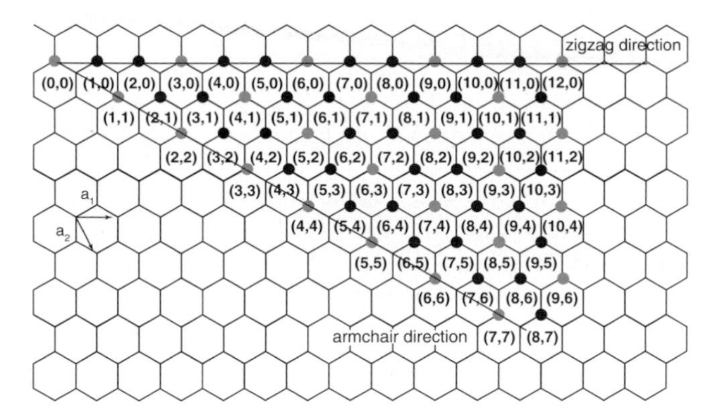

Fig. 5.5. Possible vectors for general carbon nanotubes, specified using (n,m) notation. Zigzag and armchair direction are visualized. The grey dots denote metallic nanotubes, the black ones semiconducting nanotubes. Adapted from [7]

Therefore, the structure of a SWCNT is uniquely determined by the pair of integers (n,m) that define the chiral vector and the chiral angle (see Fig. 5.5). Rolling up the graphene sheet as described before, three distinct nanotube structures can be obtained (see Fig. 5.6): the zig-zag nanotube, for which $\theta = 0$; the armchair nanotube, for which $\theta = 30°$; and the chiral nanotube for which $0° < \theta < 30°$. All (n,n) are armchair nanotubes; $(n,0)$ and $(0,m)$ are zig-zag nanotubes. (n,m) with $n \neq m \neq 0$ are chiral nanotubes.

A carbon nanotube can be thought as a one-dimensional crystal having a cylindrical translational unit cell. According with the method described by Jishi, Dresselhaus and coworkers [11, 12], the chiral vector $\boldsymbol{C_h}$ and the translation vector $\boldsymbol{T}$ define the unit cell for a SWCNT (see Fig. 5.4). The $\boldsymbol{T}$ vector is a vector starting from the origin, perpendicular to $\boldsymbol{C_h}$ and extending until it encounters an equivalent lattice point. If t_1 and t_2 are the components of $\boldsymbol{T}$ vector,

$$\boldsymbol{T} = t_1\boldsymbol{a_1} + t_2\boldsymbol{a_2} \tag{5.4}$$

Imposing the orthogonality condition between $\boldsymbol{T}$ and $\boldsymbol{C_h}$, from (5.1) and (5.4) it is possible to obtain the following relationships:

$$t_1 = \frac{2m+n}{d_r} \tag{5.5}$$

$$t_2 = -\frac{2n+m}{d_r} \tag{5.6}$$

$$|\boldsymbol{T}| = \frac{|\boldsymbol{C_h}|\sqrt{3}}{d_r} \tag{5.7}$$

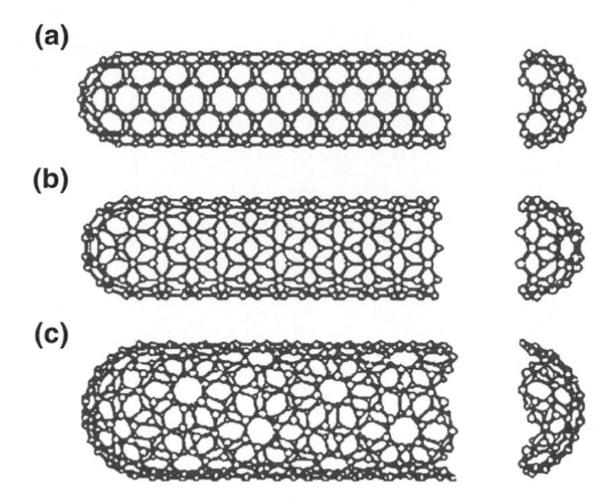

Fig. 5.6. Schematic models for SWCNTs. (**a**) Armchair carbon nanotube, $(n,m)=(5,5)$, $\theta = 30°$. (**b**) Zigzag carbon nanotube, $(n,m)=(9,0)$, $\theta = 0°$. (**c**) Chiral nanotube, $(n,m)=(10,5)$, $\theta = 19°$. Reprinted from [12], Copyright 1995, with permission from Elsevier

where d_r is the greatest common divisor between $2n+m$ and $2m+n$. d_r is related to the greatest common divisor d between n and m, in such a way:

$$d_r = \begin{cases} 3d & \text{if } (n\text{-}m) \text{ is multiple of } 3d \\ d & \text{if } (n\text{-}m) \text{ is not multiple of } 3d \end{cases} \tag{5.8}$$

The number of hexagons in the unit cell of a (n,m) carbon nanotube, N, is given by

$$N = \frac{2(n^2 + m^2 + nm)}{d_r} \tag{5.9}$$

Each hexagon in the honeycomb lattice contains two carbon atoms, then $2N$ is the number of carbon atoms in the carbon nanotube unit cell. In Fig. 5.7 we draw the unit cell for (5,5) armchair and (9,0) zigzag nanotubes.

Normally, carbon nanotubes are closed at both ends. Fujita, Dresselhaus and coworkers predicted possible caps of carbon nanotubes [6, 14, 15]. They suggested that the ends of capped carbon nanotubes have to obey the Euler's rule, where a closed structure made only by pentagons and hexagons has to contain 12 pentagons [16]. Then caps of carbon nanotubes are half a fullerene molecule. The smallest possible fullerene structure is the dodecahedron with twenty carbon atoms, C_{20}, but for the isolated pentagon rule (IPR) the smallest stable fullerene is the C_{60}. Assuming that CNT caps follow the IPR, we expect the smallest capped nanotube to have the same diameter of C_{60}

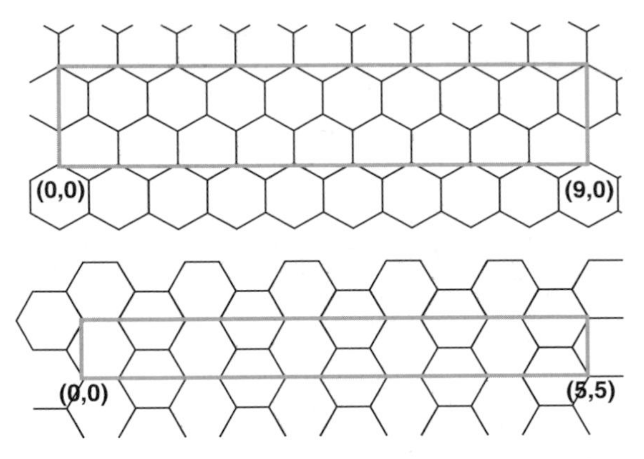

Fig. 5.7. Unit cells for (**a**) (9,0) zigzag nanotube, (t_1, t_2)=(1,-2), and (**b**) armchair (5,5) nanotube, (t_1, t_2)=(1,-1). Adapted from [13]

fullerene structure, 7 Å (see Fig. 5.6). Then, armchair (5,5) nanotube and zigzag (9,0) nanotube, whose diameter is around 7 Å, are supposed to be the smallest nanotubes closed at both ends with half a C_{60} fullerene molecule, bisected normal to five-fold axis and three-fold for (5,5) and (9,0) nanotubes respectively. (5,5) and (9,0) carbon nanotubes are often referred to as archetypal nanotubes. Armchair and zigzag nanotubes of larger diameters have bigger caps and the number of possible cap structures increases with the diameter, according with the increasing number of structural isomers for fullerenes larger than C_{60}. Very similar considerations can be drawn also for chiral nanotube (see Fig. 5.6).

According to theoretical predictions, the smallest free-standing stable SWCNT was observed by Ajayan and Iijima [17] in 1992. SWCNTs usually observed have diameters greater than 10 Å. This could be related to the higher number of ways that caps can be formed for larger CNTs. Furthermore caps with conical shape are very often observed. Caps shaped as cones can be formed from a hexagonal network introducing a number of pentagons inferior to six, that are needed to give a half sphere.

Being the diameter of carbon nanotubes much smaller than their length, tube caps can be neglected when discussing the symmetry properties of the nanotubes. Achiral CNTs belong to a symmorphic symmetry group, for which rotational and translation symmetry operation can be applied independently, while chiral CNTs are non-symmorphic, because the basic symmetry operation is a roto-translation. We will consider first the general case of the chiral nanotubes, referring to the work of Dresselhaus et al. [11,12,18]. For a (n,m) chiral nanotube the basic symmetry operation, $\mathrm{R}=(\phi,\boldsymbol{\tau})$, consist of a rotation around the tube axis by an angle ϕ given by

$$\phi = 2\pi \frac{\Omega}{Nd} \tag{5.10}$$

followed by a translation τ along the tube axis, given by

$$\tau = \mathbf{T}\frac{d}{N} \tag{5.11}$$

where $\boldsymbol{T}$ is the translation vector (see (5.5), (5.6) and (5.7)), N the number of hexagons in the CNT unit cell (see (5.9)) and Ω an integer defined by the relation

$$\Omega = [p(2n+m) + q(2m+n)]\frac{d}{d_r} \tag{5.12}$$

where p and q are integers defined as $(mp - nq = d)$ under the conditions $q < m/d$ and $p < n/d$.

For the case d=1, the symmetry group of a (n,m) chiral nanotube is a cyclic group of order N, given by:

$$C_{N/\Omega} = \left\{R_{N/\Omega}, R^2_{N/\Omega}, ..., R^{N-1}_{N/\Omega}, R^N_{N/\Omega} = E\right\}$$

where E is the identity element and the symmetry element $R_{N/\Omega} = (2\pi(\Omega/N), \boldsymbol{T}/N)$.

For the general case $d \neq 1$, the symmetry group of the nanotube is given by:

$$C = C_d \otimes C'_{Nd/\Omega}$$

where

$$C_d = [C_d, C^2_d, ..., C^d_d = E]$$

and

$$C'_{Nd/\Omega} = [R_{Nd/\Omega}, R^2_{Nd/\Omega}, ..., R^{N/d}_{ND/\Omega} = E]$$

The symmetry element C_d is a rotation around the tube axes by $2\pi/d$ and the symmetry element $R_{Nd/\Omega} = (2\pi(\Omega/Nd), \boldsymbol{T}(d/N))$.

Applying the above symmetry formulation to armchair (m=n) and zigzag nanotubes (m=0), they are found to belong to the symmetry group given by the product of cyclic group C_n and C'_{2n} that contain only two symmetry operations: the identity and a rotation by $2\pi/2n$ followed by a translation of $\boldsymbol{T}/2$. Armchair and zigzag nanotubes, however, have other symmetry operations, such as inversion center and reflection in planes parallel to the tube axis. According to the Dresselhaus notation [12], non-chiral nanotubes can be described by symmetry groups D_{nh} and D_{nd} for even and odd n respectively, since the inversion center is an element of D_{nd} group only for odd n, and of D_{nh} group only for even n.

5.3.2 Structure and Symmetry of MWCNTs

MWCNTs consist of more than one carbon nanotube arranged concentrically around the axis of the inner tube and having increasing diameters and different helicities. Structural properties of each tube can be described in terms of n and m integers. From geometrical consideration, each tube is expected to be lattice mismatched from the others [19]. In fact, consecutive layers of a tube are commensurate, i.e. present a stacking similar to AB stacking in Bernal graphite, only if the ratio between their unit cell lengths is a rational number. The probability to have commensurate MWCNTs obviously decreases with the increasing of the number of shells and of nanotube diameter. Commonly, adjacent layers are not-correlated, then we expect interlayer spacing in MWCNTs to be closer to turbostratic graphite (3.44 Å) rather than Bernal structure (3.35 Å). The outer diameter of MWCNTs depends on the technique used to grow them; multi-wall carbon nanotubes either with diameter of a few nanometers and bigger than 100 nanometers have been observed. Inner tubes with diameters of 5 Å [20] and 4 Å [21, 22] have been recently observed. They are related to C_{36} and C_{20} fullerene respectively. There are two nanotube structures with a diameter of about 5 Å: the (6,0) zigzag nanotube with a diameter of 4.7 Å, and the zigzag (7,0) nanotube with a diameter of 5.5 Å. Instead, three are the structures related to a 4 Å diameter nanotube: the zigzag (5,0), the armchair (3,3) and the chiral (4,2) nanotubes with diameters of 3.93, 4.07 and 4.14 Å respectively.

Isolated, not self-standing SWCNTs with 4 Å diameter were prepared by pyrolysis of tripropylamine in the channels of porous zeolite $AlPO_4^- - 5$ (AFI) single crystal (AFI-SWCNTs) [23, 24].

MWCNTs can be capped in many ways. Symmetric caps related to very big fullerenes are not observed very often [25]. Most common tubes have asymmetric structures [26, 27]. In Fig. 5.8 are shown two HRTEM images of MWCNTs capped with two very common asymmetric structures: the asymmetric cone and the bill-like, in Fig. 5.8 (**a**) and 5.8(**b**) respectively. Sometimes, CNTs grow without caps, ending with very complex structures [28].

5.4 Electronic Structure of CNTs

Experiments and theory have shown that carbon nanotubes can be metallic or semiconducting depending on their diameter and chirality. The remarkable electrical properties of SWCNTs grow out of the unusual electronic structure of the two-dimensional graphene sheet [29]. For this reason, the electronic properties of graphite will be briefly discussed first. Then, SWCNTs electronic properties will be taken into account, with some comments on the electronic properties of MWCNTs .

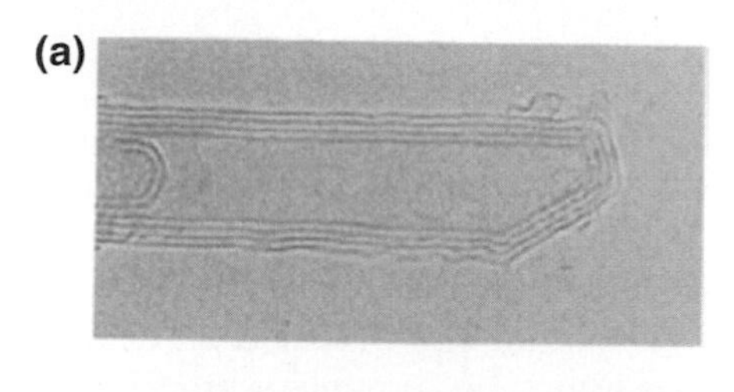

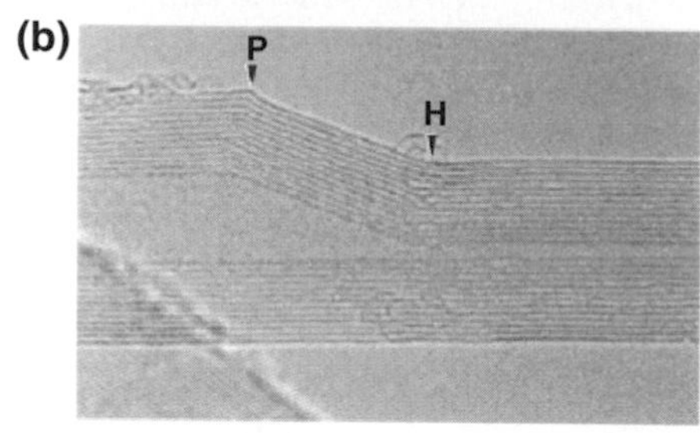

Fig. 5.8. MWCNT possible caps: (**a**) Tip structure of a MWCNT with asymmetrical conical end; (**b**) MWCNT ending with a bill-like structure. P and H are two folding points generated by insertion of a single pentagon at point P and a heptagon at point H. Reprinted in part from [26], copyright 1997, with permission from IOP Publishing Ltd

5.4.1 Electronic Structure of Graphite

The electronic structure of a 3D graphite crystal can be approximated to the electronic structure of a 2D graphene sheet. In fact, the interlayer interactions between two adjacent layers in graphite are small relative to the intra-layer ones, being the interlayer separation in graphite crystal, 3.4 Å, much larger than the distance between two carbon atom in the graphite layer, a_{c-c}=1.42 Å.

Figures 5.9(**a**) and 5.9(**b**) show the unit cell and the Brillouin zone of two-dimensional graphite; as usual, $\boldsymbol{a_1}$ and $\boldsymbol{a_2}$ are the unit vectors in the real space, $\boldsymbol{b_1}$ and $\boldsymbol{b_2}$ are the reciprocal lattice vectors. In the x,y coordinate system shown in Fig. 5.9, they are expressed in such way:

$$
\begin{aligned}
\boldsymbol{a_1} &= \left(\frac{\sqrt{3}}{2}a, \frac{a}{2}\right) & \boldsymbol{a_2} &= \left(\frac{\sqrt{3}}{2}a, -\frac{a}{2}\right) \\
\boldsymbol{b_1} &= \left(\frac{2\pi}{\sqrt{3}a}, \frac{2\pi}{a}\right) & \boldsymbol{b_2} &= \left(\frac{2\pi}{\sqrt{3}a}, -\frac{2\pi}{a}\right)
\end{aligned}
\tag{5.13}
$$

where $|\mathbf{a}_1| = |\mathbf{a}_2| = a = \sqrt{3}a_{c-c}$.

Γ, K and M are three high symmetry points in the Brillouin zone, its center, corner and center of the edge respectively.

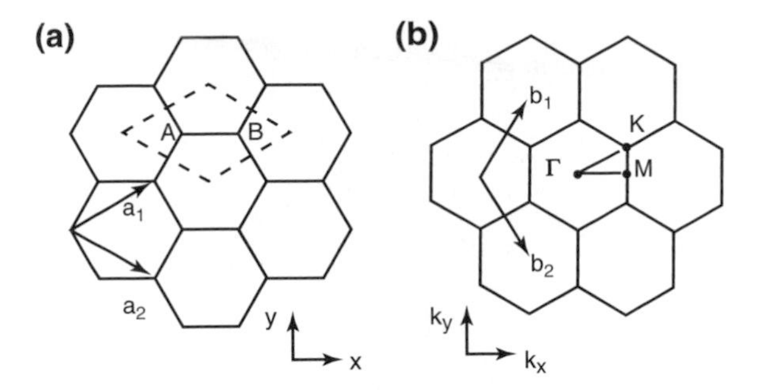

Fig. 5.9. (**a**) Portion of the graphene sheet lattice where $\boldsymbol{a_1}$ and $\boldsymbol{a_2}$ are the unit vectors. The dashed line rhombus is the unit cell containing two carbon atoms: A and B. (**b**) Portion of the reciprocal lattice of graphene sheet where $\boldsymbol{b_1}$ and $\boldsymbol{b_2}$ are the reciprocal lattice vectors. The central hexagon is the Brillouin zone and Γ, K and M are high symmetry points. Adapted from [30]

As previously discussed in Sect. 3.1, each carbon atom in the graphene layer has a sp^2 configuration; these orbitals are involved in the formation of three strong σ bonds. The forth valence orbital, $2p_z$, is perpendicular to the graphene plane and makes π covalent bonds. π-orbitals can be treated separately from σ orbitals because they do not interact owing to a different symmetry. Here we will briefly consider only the π-bands for 2D graphite because they are the most important in determining electrical properties of a graphene sheet. Since there are two inequivalent atoms per unit cell, graphene will be characterized by two π-bands, a bonding π and an anti-bonding π^*-band. The energy dispersion relation for the π-bands of graphene as a function of the wave vectors k_x and k_y, E_{2D}, obtained employing a Slater–Koster tight binding model [31, 32] is given by:

$$E_{2D} = \pm\gamma_0 \left[1 + 4\cos\left(\frac{\sqrt{3}}{2}k_x a\right)\cos\left(\frac{k_y a}{2}\right) + 4\cos^2\left(\frac{k_y a}{2}\right)\right]^{1/2} \tag{5.14}$$

where $a = 2.46$ Å and γ_0 is the nearest neighbour overlap integral [31].

In Fig. 5.10, the three dimensional energy dispersion relations for graphene and its 2D projection are shown [33, 34]. The upper π^*-band and the lowest π-band are degenerate (i.e. touch themselves) at the six K-point of the Brillouin zone, making the system an electrical conductor. At 0 K the bonding π-band would be completely filled, and the antibonding π^*-band completely empty, with a vanishing density of states at the Fermi level. Therefore, two-dimensional graphite can be seen as a zero-band gap semiconductor rather than a metal.

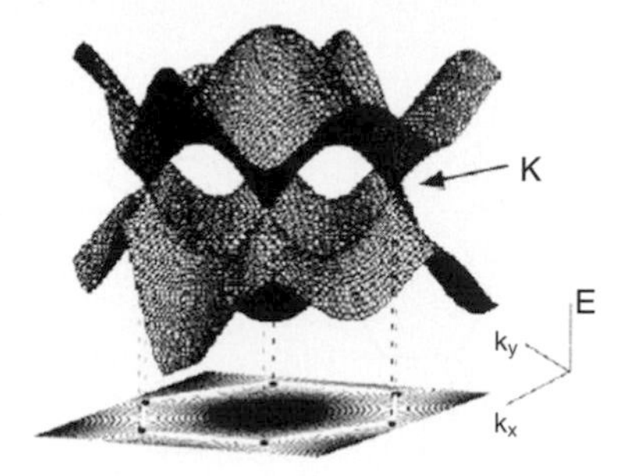

Fig. 5.10. 3-D dispersion energy relation of graphene and its projection. Reprinted in part with permission from [34]. Copyright 2002 American Chemical Society

5.4.2 Electronic Structure of SWCNTs

A SWCNT is a one-dimensional periodic structure along the tube axis. As a consequence of their high aspect ratio, SWCNTs are characterized by quantized wave vectors in the radial direction, while wave vectors associated with the axial direction are expected to be continuous, in the ideal case of a SWCNT having infinite length.

The lattice vectors, $\boldsymbol{C_h}$ and $\boldsymbol{T}$, and the real cell of a carbon nanotube have been discussed and defined in the Sect. 3.2. From (5.1), (5.5), (5.6), (5.7) and (5.13) it is easy to obtain the components of these two vectors in Cartesian coordinates:

$$\boldsymbol{C_h} = \left(\frac{a}{2}\right)\left[\sqrt{3}(n+m),(n-m)\right] \tag{5.15}$$

$$\boldsymbol{T} = \left(\frac{3a}{2d_r}\right)\left[-\frac{(n-m)}{\sqrt{3}},(n+m)\right] \tag{5.16}$$

where d_r is defined in the relation (5.8) and $a{=}\sqrt{3}a_{c-c}$.

For the reciprocal lattice, two reciprocal lattice vectors, $\boldsymbol{K_1}$, in the circumferential direction and $\boldsymbol{K_2}$, along the nanotube axis, are defined. $\boldsymbol{K_1}$ and $\boldsymbol{K_2}$ have to satisfy the following conditions:

$$\begin{aligned} \boldsymbol{C_h} \times \boldsymbol{K_1} &= 2\pi \qquad & \boldsymbol{T} \times \boldsymbol{K_1} &= 0 \\ \boldsymbol{C_h} \times \boldsymbol{K_2} &= 0 \qquad & \boldsymbol{T} \times \boldsymbol{K_2} &= 2\pi \end{aligned} \tag{5.17}$$

From (5.15) and (5.16) and the conditions (5.17), expression of $\boldsymbol{K_1}$ and $\boldsymbol{K_2}$ can be derived:

$$\begin{aligned} \boldsymbol{K_1} &= \left(\tfrac{2\pi}{Nad_r}\right)\left(\sqrt{3}(n+m),(n-m)\right) \\ \boldsymbol{K_2} &= \left(\tfrac{2\pi}{Na}\right)\left(\tfrac{-(n-m)}{\sqrt{3}},(n+m)\right) \end{aligned} \tag{5.18}$$

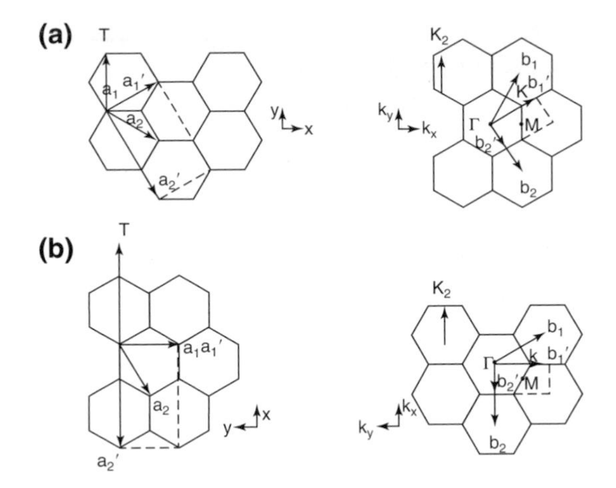

Fig. 5.11. Unit cell and Brillouin zone of **(a)** armchair and **(b)** zig zag SWCNTs (*dashed lines*). $\boldsymbol{a}_1'$, $\boldsymbol{a}_2'$, $\boldsymbol{b}_1'$ and $\boldsymbol{b}_2'$ are the unit vectors and recipocral lattice vectors. Unit vectors ($\boldsymbol{a}_1$, $\boldsymbol{a}_2$) and reciprocal lattice vectors ($\boldsymbol{b}_1$, $\boldsymbol{b}_2$) for graphene sheet are shown for comparison. The $\boldsymbol{T}$ vector and the corresponding $\boldsymbol{K}_2$ vector are also shown. Adapted from [36]

SWCNT electronic band structure can be derived from the graphene sheet one, imposing periodic boundary conditions in the circumferential direction and neglecting the finite curvature effects that produce a mixing of σ and π-orbitals. If we cut and unroll a SWCNT parallel to the cylinder axis, we get a graphite layer in which translation symmetry exists in the direction identified by $\boldsymbol{T}$ or $\boldsymbol{K}_2$, and periodic boundary condition have to be applied to the radial direction, identify by $\boldsymbol{C}_h$ or $\boldsymbol{K}_1$. For simplicity, we consider first armchair and zigzag nanotubes, and we will follow on the work of Mildred Dresselhaus and coworkers [33,35]. In Fig. 5.11(**a**) and 5.11(**b**) we report the unit cell and the Brillouin zone for armchair and zigzag nanotubes respectively. $\boldsymbol{a}_1'$, $\boldsymbol{a}_2'$, $\boldsymbol{b}_1'$ and $\boldsymbol{b}_2'$ are the lattice vectors for direct and reciprocal lattice respectively. Direct and reciprocal lattice vectors for a graphene sheet are also shown.

Imposing periodic boundary conditions to an $(n,\ n)$ armchair nanotube

$$\boldsymbol{C}_h \times k = 2\pi j \qquad \text{where } j \text{ is an integer}$$

we obtain the number of allowed wave vectors in the circumferential direction:

$$k_{x,j} = \frac{2\pi j}{na\sqrt{3}} \qquad (j = 1, \dots\dots, 2n)$$

$\{k_{x,j}\}_{j=1,\dots,2n}$ represents the group of allowed one-dimensional wave vectors for an (n,n) nanotube in the direction of $\boldsymbol{K}_2$ vector. Considering the

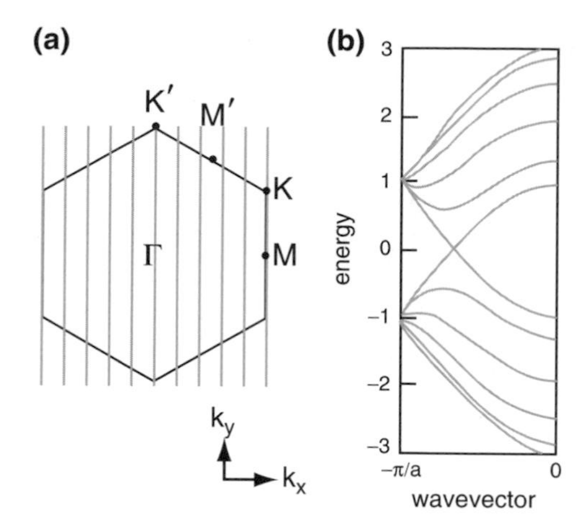

Fig. 5.12. Dispersion relation for a (5,5) armachair nanotube: (**a**) Projection of the allowed k values onto the first Brillouin zone of a (5,5) armchair nanotube; (**b**) 2D dispersion relation for the (5,5) armchair nanotube. Adapted from [36]

first Brillouin zone for graphene a layer, the $\boldsymbol{K_2}$ direction is parallel to the Γ–K segment. Therefore, the 1D energy dispersion relationship for an armchair nanotube can be obtained as sections of 2D energy dispersion curve of graphite in the first Brillouin zone along the $2n$ segments identified by $\{k_{x,j}\}_{j=1,\ldots,2n}$ and parallel to the direction identified by Γ/K. The explicit energy dispersion relation for an $(n,\ n)$ armchair nanotube can be obtained substituting the allowed $k_{x,j}$ vectors in (5.14). In Fig. 5.12(**a**) we report the allowed k values in the first Brillouin zone of 2D graphite for the archetypal (5,5) armchair nanotube and its normalized energy dispersion relation in Fig. 5.12(**b**). From this figure we can see that there are two energy bands that meet at the Fermi energy at E=0, since the corresponding energy bands in 2D graphene meet at the K point, where π and π* are degenerated. Because the energy bands for a carbon nanotube are one-dimensional, the density of the state at the K point is finite and different from zero, then the (5,5) archetypal nanotube is metallic. Calculations show that all the armachair nanotubes have a similar band structure, and then all armchair nanotubes are metallic. A different way to illustrate this concept is to affirm that all carbon nanotubes that have allowed k values that pass through K points of the first Brillouin zone for 2D graphite are metallic.

Considering now a generic $(n,0)$ zigzag nanotube. The allowed wave vectors are given by:

$$\boldsymbol{k_{y,j}} = \frac{2\pi j}{na} \qquad (j = 1, \ldots\ldots, 2n)$$

In Fig. 5.13 we report the normalized energy dispersion curve for two zigzag nanotubes: the archetypal (9,0) and the (10,0).

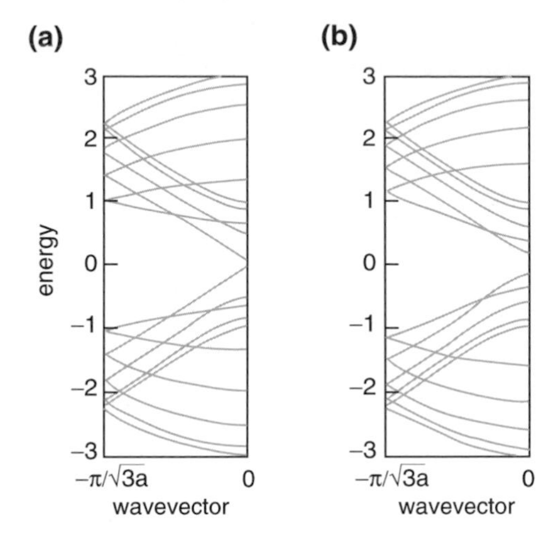

Fig. 5.13. 2D dispersion relation for two zig-zag nanotubes: (**a**) (9,0) zigzag nanotube; (**b**) (10,0) zig zag nanotube. Adapted from [36]

The conduction and valence bands of an (9,0) nanotube meet at k=0, then at the Γ point, therefore it is a metallic tube. At the same point there is an energy gap between conduction and valence bands for an (10,0) nanotube, which is then expected to be semiconducting. Calculations show that all the zigzag nanotubes such that

$$n{=}3q \qquad q = \text{integer}$$

are expected to be metallic, while the others are semiconducting.

Extending the discussion to chiral nanotubes, it is possible to conclude that all (n,m) nanotubes such that

$$n\text{-}m{=}3q \qquad q = \text{integer}$$

are expected to be metallic, while the others are semiconducting with a bandgap between valence and conduction band that depends on the carbon nanotube diameter.

The π-only tight binding approach here considered describes quite well the electronic structure of carbon nanotubes with big diameters. As predicted by theoretical calculations [35, 37–40] and demonstrated by STM investigation [34, 41, 42] the curvature effect really affects the electronic properties of nanotubes opening a small gap at the Fermi energy in metallic zigzag and chiral nanotubes. An indirect validation of the relationship between curvature effect and deviation from predicted electronic structure arises from the fact that the gap magnitude at the Fermi energy increases inversely with the radius of nanotube. In conclusion, all the nanotubes for which n-m=$3q$

with q different from zero, are not metallic but semiconducting with very small bandgaps. Then, carbon nanotubes can be metallic, semiconducting with very small bandgaps or semiconducting with large bandgaps.

5.4.3 Electronic Structure of MWCNTs

Multi-wall carbon nanotube have electronic properties more complicated despite SWCNTs. Though consecutive layers of MWCNTs are generally expected to be non commensurate, and then the interlayer coupling less effective than in graphite, interaction between adjacent layers cannot be neglected [43]. Many theoretical studies [44] have been carried out to try to explain the big variety of electrical transport phenomena observed for MWCNTs, but the situation is not yet well understood.

5.5 Handling of CNTs

The realization of all kind of CNT-based device, both for electron transport studies and technological applications, is based on the ability to put the right nanotube, single-wall, multi-wall, metallic or semiconducting, at the right place, for example between metallic electrodes. However, the difficult handling of carbon nanotubes is limiting the potential applications of this material. Due to strong van der Waals interactions, carbon nanotubes aggregate to form bundles or ropes, of up to tens of nanometres in diameter for SWCNTs, which are difficult to unbundle. Furthermore, these ropes are tangled like spaghetti.

At present, improvements in the fabrication of carbon nanotubes make it possible to selectively grow SWCNTs or MWCNTs with a relatively narrow distribution of their diameters [45]. Nevertheless, there is no, up to now, any growth technique to produce CNTs with the desired chirality. However, the aligned growth of CNTs on pre-patterned catalyst substrates or between metallic pads can be usefully employed to fabricate some CNT-based devices, as field emitters and displays (see Sect. 6 in this chapter) [46–50]. Parallel to the directional growth, many efforts have tried to implement the deposition techniques that permit to realize CNT-based devices with more general design and with characteristics of simplicity, reliability and reproducibility, essential for every engineered device for technological applications on large scale. Here, we present a brief description of the deposition approaches employed for the realization of CNT-based devices, especially for nanoelectronics and sensing applications, described in the following section of this chapter.

A brief overview of the purification methods will be also presented as introduction to this section. As a matter of fact, after growth, CNTs have to be purified from the impurities present in the sample as catalytic particles, amorphous carbon, fullerenes and other forms of carbon, and then dispersed in a solvent to be deposited.

5.5.1 Purification of CNTs

A common problem concerning applications of carbon nanotubes is the purification of raw material. The as-produced carbon nanotubes contain a lot of impurities, like amorphous carbon or carbon-coated metal nanoparticles. These impurities interfere with most of the physical properties of carbon nanotubes. In order to study these fascinating superstructures, the material has to be as pure as possible. The main techniques applied for purification of nanotubes are oxidation of contaminants, functionalization/defunctionalization, wrapping by macromolecules, filtration, chromatography, microwave irradiation and mechanical force treatment. However, considerable problems remain for most of the present purification methods. Some affect the structure of the tubes, while others tend to be slow and inefficient processes.

Oxidation

Oxidative treatment of nanotubes is a good way to remove carbonaceous impurities or metal nanoparticles [51–56]. There are two general approaches for oxidation of nanotubes, the chemical treatment with various mixtures of inorganic acids [51–55] or air oxidation at high temperatures [52–54, 56]. The main disadvantage of oxidation is that not only the impurities are eliminated, but also a relevant amount of nanotubes is destroyed. In addition, carboxylic functionalities are generated both at ends and at sidewalls. The efficiency and the yield of the pure material that is obtained depend on many factors such as metal content, oxidation time, oxidizing agent and temperature. The complete removal of metal impurities can be assisted by sonicating a suspension of raw carbon nanotubes in a mixture of inorganic acids [55]. This type of treatment produces the cutting of SWCNTs, reducing them in short "fullerene pipes". Dangling bonds present at the opened ends of pipes easily convert into oxygenated functionalities, as carboxylic acid, anhydrides, quinones and esters. These functionalities are also present on the sidewall of carbon nanotubes, particularly concentrated at defect sites. In a similar approach, raw carbon nanotubes (HiPco) in methanol have been subjected to ozonolysis at -78°C, followed by treatment with various reagents, to generate a higher amount of carboxylic acid functions [57].

Functionalization–Annealing

A completely different approach for the purification of single-wall carbon nanotubes (HiPco) has been reported [58]. The method consists of the following steps: a) organic covalent functionalization of raw material via 1-3 dipolar cycloaddition, b) purification of the soluble functionalized nanotubes by recrystallization and c) thermal detachment of the functional groups followed by annealing at high temperature. The combination of these steps leads to high quality material. This purification procedure has the advantage of not using oxidizing reagents. Moreover, the processing of the soluble nanotubes is much easier and the material can be deposited onto various substrates for application in nanoelectronics. Subsequenty, the functionalities can be removed by simple heating, leaving the nanotubes intact and patterned. Spectroscopic studies of the thermally recovered material have shown that the graphite network of nanotubes is restored totally as well as their electronic properties are regenerated.

A different approach of functionalization of nanotubes for purification is based on electrochemical reactions [59]. In situ produced halogen atoms were bonded to the nanotube surface. The resulting material was solvated easily in water or alcohols whereas the impurities remained insoluble.

Wrapping by Macromolecules

Bundles of single-wall carbon nanotubes helically wrapped with conjugated polymers were produced [60–62]. The backbone structure of these polymers are of the poly(phenylenevinylene) type and the wrapping arises probably from van der Waals interactions with carbon nanotubes. Hybrid systems have been characterized by UV–Vis and Raman spectroscopy, as well as electron microscopy. It is demonstrated that solutions of the polymer are capable of suspending nanotubes while the majority of the accompanying amorphous carbon impurities precipitate out of solution.

Similar results have been demonstrated with other organic polymers such as Arabic gum [63] or polyvinyl pyrrolidone [64]. The wrapping polymer can be easily removed from the nanotubes either by changing the solvent or by smooth oxidation.

Microfiltration

Microfiltration is based on size separation. With application of pressure, the metal catalyst, fullerenes and carbon nanoparticles pass through the filter whereas the nanotubes remain on the membrane [65–67]. Also, the impurities can be removed by ultrasonically-assisted microfiltration [66]. Sample sonication during the filtration prevents filter contamination and provides a fine nanotube–nanoparticle suspension throughout the purification process.

Chromatographic Methods

The purification of carbon nanotubes by gel permeation chromatography has been reported [68]. Generally, this method is a powerful tool for the separation of macromolecules. Raw material is dispersed homogeneously with the aid of a surfactant in water. This colloidal aqueous dispersion is used directly for chromatography. The packing of the column consists of controlled-pore glass which is chemically inert. Several fractions are gathered and examined by microscopy techniques. A statistical evaluation of the size distribution shows that longer nanotubes are eluted first whereas small particles follow. The main advantage of this technique is that the material is not destroyed but, on the other way, this is a slow process.

There were also some reports on the separation of soluble single wall nanotubes from amorphous carbon material by chromatography [69, 70]. Solutions of these short nanotubes were prepared in THF and volumes of the milliliters scale were injected into a chromatograph. Two main fractions were eluted of which the first one consisted mainly of nanotubes, while the second contained the nanoparticulate matter. The composition of the fractions was monitored by UV–Vis spectroscopy and atomic force microscopy. Again this method is a very slow process and is rather inefficient for full-length carbon nanotubes.

Capillary Electrophoresis – Field Flow Fractionation

Electrophoretic separations are based on charge and size-dependent mobility of solution phase species under the influence of an applied electric field. The size separation of carbon nanotubes has been demonstrated using capillary electrophoresis (CE) [71]. Raman and UV–Vis spectroscopies show the ability of CE to provide high resolution separation of nanotube fractions, whereas microscopy studies demonstrate that separations are based on tube length. The separation is suggested to be controlled by the alignment of the nanotubes along the separation field. Another elution technique, which has been used for purification of carbon nanotubes, is field flow fractionation [72]. The mechanism of this approach combines elements of chromatography and field-driven techniques, such as electrophoresis and ultracentrifugation. The field is applied at some angle to flow and serves to drive components into different laminar stream in a capillary channel.

Microwave Irradiation

This method has been used for the elimination of metal nanoparticles of various types of nanotubes such as HiPco and arc-discharge [73–75]. Single-wall nanotubes (HiPco) have been purified by heating raw material in a domestic microwave oven. The oxidized metal was then washed by a concentrated solution of hydrochloric acid. Microwave irradiation reaches high metal removal percentages and time is drastically reduced compared to conventional acid treatments.

Mechanical Purification

In this approach the raw nanotube material is mixed with inorganic nanoparticles in ultrasonic bath [76]. The basic principle of this method is based on the energy of elastic impact between the encapsulated metal catalysts and the inorganic particles. This causes the ferromagnetic particles to be removed mechanically from their graphite shells. At the same time these magnetic particles are attracted by permanent magnetic poles and this results in a high purity nanotube material.

5.5.2 Random Approach

Up to 2000, all the demonstrated CNT-based electronic devices have been fabricated using a random approach: the random deposition of CNTs on a pre-patterned substrate (direct random approach) or the pattern of metallic contacts onto randomly deposited CNTs after their observation (indirect random approach).

The first examples of this kind of devices were realized employing MWCNTs and an indirect random approach, being the imaging of MWCNTs easier also with conventional imaging systems, as Scanning Electron Microscope (SEM). Langer and coworkers in 1996 first reported electrical resistance measurements on individual MWCNT [77]. In order to contact a CNT, they dispersed the pre-growth material, MWCNTs synthesized using arc-discharge method with average diameter of 18 nm, on a Si/SiO_2 wafer, onto which an array of large square gold pads had previously been lithographed. After evaporation of a thin gold layer and deposition of a negative electron resist, the sample was introduce in a Scanning Tunnelling Microscope to localize a CNT, then the STM tip was used to expose the resist from the CNT towards the pads. Unexposed resist was removed in ethanol and the underlaying thin gold layer etched by Ar ion milling, giving a CNT electrically connected. Imaging of MWCNTs and electrical contacts with metallic pads could also be obtained in a different manner: a Focus Ion Beam System was employed first to image and then to contact MWCNTs by ion induced tungsten deposition. In Fig. 5.14 we show a SEM image of the electrical device obtained in such a way [78]. In the same year, another laboratory developed a general approach that permited to determine the conductivity of a CNT simultaneously to its structure [79]. In the caption of Fig. 5.15 a description of the technique is given, which exploits both imaging and lithographic capabilities of Atomic Force Microscopy (AFM). Kasumov and coworkers developed a different approach for the simultaneous determination of the electronic properties and the atomic structure of individual MWCNTs based on High Resolution Transmission Electron Microscopy (HRTEM) [80]. A metal film was deposited on a Si_3N_4 membrane onto which a 100 μm width slit had previously been cut using a focused ion beam. IEmploying a focused ion laser, a nanotube was removed from a target onto the sample between the edges

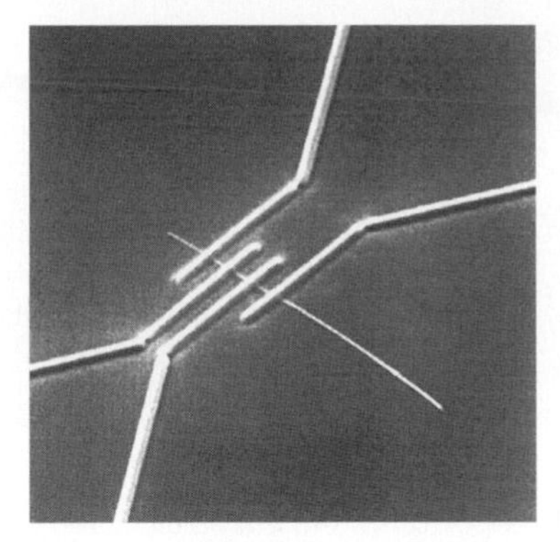

Fig. 5.14. Contacting tungsten wires deposited on a MWCNT using a FIB system, imaged by SEM. Reprinted by permission from Macmillan Publishers Ltd: Nature [78], copyright 1996

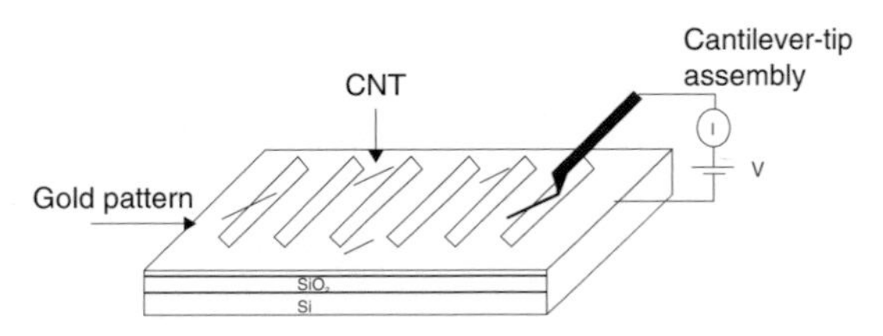

Fig. 5.15. Schematic diagram of the device employed for electrical measurements by Dai, Wong and Lieber. MWCNTs are ultrasonically dispersed in ethanol and then deposited on Si/SiO_2 substrate. After deposition of a gold layer by sputtering, rectangular openings ($4 \times 1000\,\mu m$, $15\,\mu m$ step) are obtained combining electron beam lithography and wet chemical etching of exposed gold after e-beam irradiation. A CNT having one end covered by gold and the other extending into the opened slot is imaged with AFM. Using a conductive tip, the free end of the nanotube is electrically contacted and the axial conductance of the nanotube is measured. Adapted from [79]

of metallic slits thus obtaining a suitable device for HRTEM investigation because the nanotube was suspended on a transparent medium.

In 1999 Ahlskog et al. developed a direct random approach based on electron beam lithography [81]. In a first lithographic step they defined alignment markers and interdigitating gold electrodes. In a second one, square windows with different areas were opened in correspondence to electrodes. Then, a drop of a suspension of MWCNTs in isopropanol was deposited on the substrate. After CNTs deposition, a lift-off process in boiling acetone was employed to remove not developed resist and CNTs on top of this, leaving unaffected CNTs deposited on gold interdigitating electrodes. With a good tailoring of window dimensions and concentration of the suspended CNTs, this technique allowed to deposit one or a few nanotubes in each window, reducing also the contamination deriving from the solvents, as resist developers.

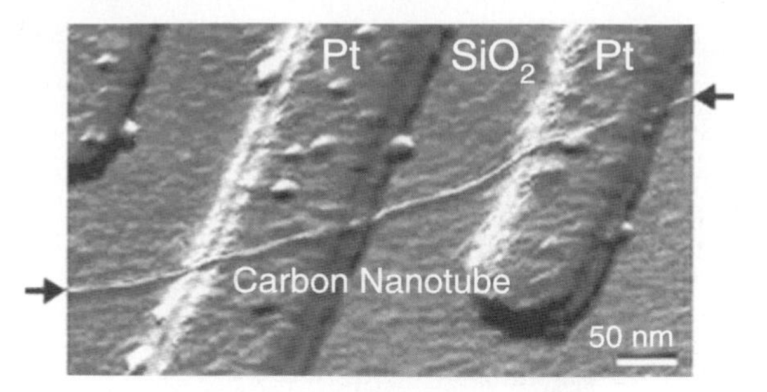

Fig. 5.16. AFM tapping mode image of a SWCNT, length 3 μm and height 1 nm, onto 15 nm high Pt-electrodes 140 nm spaced. Reprinted by permission from Macmillan Publishers Ltd: Nature [84], copyright 1997

It was only after 1996 that SWCNTs have been obtained with yields and structural uniformity sufficient to permit the realization of SWCNT-based electrical devices [82]. In March 1997 a research group reported transport measurement on individual ropes made of about 30% of (10,10) SWCNTs [83]. Ultrasonically dispersed nanotubes in acetone were deposited on a Si/SiO_2 substrate and localized relatively to position markers with an AFM operating in tapping mode. Then, on a single rope, four metallic electrodes were first defined by electron beam lithography, deposited by evaporation (Cr/Au), and finally the surface was cleaned by lift-off process. One month later, the first electrical devices based on a SWCNTs was nanofabricated [84] with the aim to demonstrate the quantum wire nature of CNTs. SWCNTs produced by laser evaporation with an average diameter of 1.38 nm were suspended in a solvent and then a drop was deposited by spin coating on top of a Si/SiO_2 substrate, onto which Pt-electrodes 15 nm thick had been previously nanofabricated. An AFM image of the device is show in Fig. 5.16. With the same direct approach one year later the first device active element was produced: a room-temperature transistor [85]. Other important results following this nanofabrication method were obtained by Antonov and Johnson, that observed current rectification in a molecular diode consisting of a semiconducting SWCNT and impurities [86] and by Yao and coworkers that reported electrical transport measurements on SWCNTs with intramolecular junctions [87] (see Fig. 5.17).

Crossed nanotubes junctions, metal–metal, metal–semiconductor and semiconductor–semiconductor, were studied by Fuhrer and coworkers in 2000 [88]. SWCNTs produced by laser ablation, dispersed in CH_2Cl_2 using ultrasonication, were spread onto a Si/SiO_2 substrate on which Cr/Au aligner markers had been previously defined. After the exact location of CNTs having the right geometry, the Cr/Au contact pads were fabricated employing e-beam lithography followed by lift-off process.

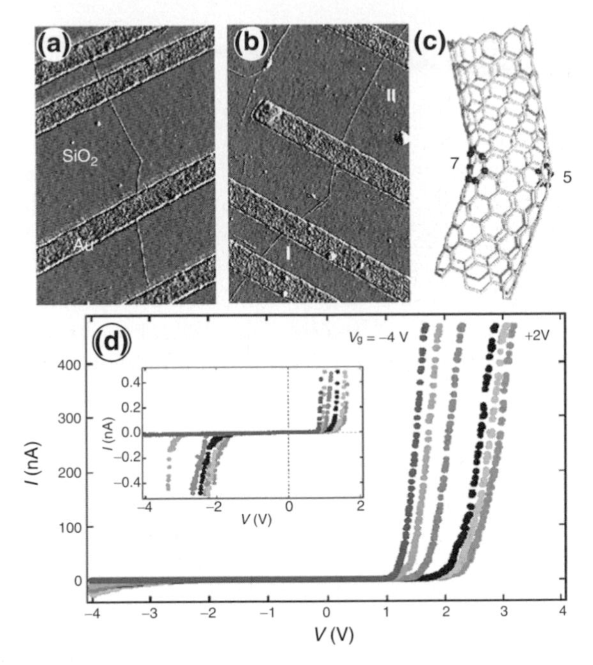

Fig. 5.17. On a Si/SiO_2 substrate, Ti-Au electrodes 250 nm wide and 20 nm thick are nanofabricated combining electron beam lithography, evaporation and lift-off techniques. (**a**), (**b**) Two CNTs characterized by a single kink junction of 36° and 41° respectively are imaged by tapping mode AFM. (**c**) Two segments of nanotubes having different atomic and electronic structures can be fused together introducing pentagon and heptagon in the hexagonal lattice, 5-7 (pentagon–heptagon) defects. In (**a**) we have a semiconductor–metal hetero-junction, in (**b**) a metal–metal hetero-junction. (**d**) I-V characteristic of a CNT Y-junction. Reprinted by permission from Macmillan Publishers Ltd: Nature [87], copyright 1999

In the case of the direct random approach, a very low surface coverage is needed to ensure that only one single-wall carbon-nanotube contacts more than one electrode, two for example, a source and a drain, in the case of Field Effect Transistor, FET. Under these conditions, many circuits remain disconnected while the connected ones can incorporate a metallic or semiconducting nanotube. This, of course, implies a waste of time and material. A possible way to improve this method is its combination with scanning probe manipulation of nanotubes. A Scanning Probe Microscope (SPM) can be used to move CNTs, both multi wall and single wall, not only to study their adhesion properties to a substrate [89, 90] or their mechanical characteristics [91, 92] but also to place a nanotube in the right place on a prepatterned substrate.

An example of single-electron transistor of MWCNT fabricated using scanning probe manipulation was presented by Roschier and coworkers in 1999 [93]. They moved a MWCNT, 410 nm long and 20 nm in diameter, onto

Cr/Au electrodes 16 nm thick prepatterned onto an oxidized silicon surface, operating an Atom Force Microscope (AFM) in non-contact mode. Individual nanotubes were moved on similar prepatterned substrates by Avouris and coworkers but operating an AFM in contact mode [91].

SPM manipulation of carbon nanotubes allows a full control of the direct random approach but the motion of an individual tube on a surface is a very slow process. For this reason, this technique, useful for scientific investigations, is not applicable to technological purposes.

A more powerful extension of the random approach is represented by electrical breakdown in MWCNTs and SWCNTs ropes proposed by Collins et al. in 2001 [94,95]. Current-induced electrical breakdown was demonstrated to be a simple and reliable method to permanently modify MWCNT and SWCNT ropes to tailor their properties. Considering a rope of SWCNTs: each tube of a rope can connect independently two metal electrodes and then the pore can be modelled as an ensemble of parallel conductors with an overall conductance, G(Vg)= Gm + Gs(Vg), where Gm is the metallic nanotubes contribution independent of the gate voltage, Vg, and Gs is the gate-dependent semiconducting nanotubes contribution. Applying a gate voltage high enough to deplete semiconducting-SWCNTs (s-SWCNTs), a bias voltage of several volts induces the selective electrical breakdown of metallic-SWCNTs (m-SWCNTs). The Gm component of conductance decreases to zero while the Gs(Vg) component does not change. This is also a direct evidence that depleted s-SWCNTs are protected from damage and suggests that little electronic interactions between nanotubes exist in a rope. A practical consequence of these experiments is that densely packed FETs that require s-SWCNTs at enough density to reduce the number of chip non-connected, can be fabricated. An array of electrodes (source, drain and gate) can be lithographed on a film of SWCNTs having a density tailored in such a way that at least one rope shorts every set of electrodes. Afterwards, each set of electrodes can be converted to a FET by selective electrical breakdown. Preliminary results indicate that more than 90% of the circuits is connected and present the desired characteristics (Fig. 5.18).

5.5.3 Microfluidic Approach and AC-Electrophoresis

More general techniques to assembly CNTs in a controlled way were proposed recently. Two of the most simple and reliable ways are the microfluidic method and the approach based on Alternating Current electrophoresis (AC-electrophoresis).

Huang, Lieber and coworkers demonstrated the possibility to realize predictable and controlled well-ordered structures made by GaP, InP and silicon nanowires (NW) [96]. NWs can be aligned in a fluidic flow with a degree of alignment along the flow direction that increases with increasing of the flow rate and with an average NW-NW separation that decreases with increasing of the flow duration. The length control of NWs involved in the assembly

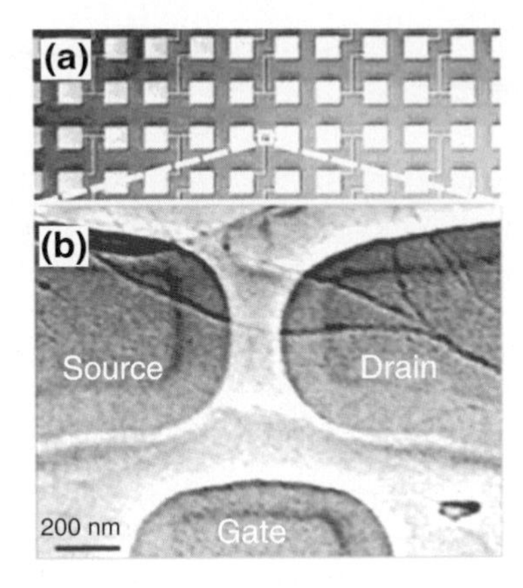

Fig. 5.18. (**a**) Employing conventional lithographic techniques, an array of three-electrode system was lithogrphed on a dispersion of SWCNTs-ropes (**b**) Selective electrical breakdown to remove m-SWCNTs was applied to obtain three-terminated FET with good characteristics. Reprinted with permission from [95]. Copyright 2001 AAAS

process permits to generate aligned structures without NW-NW contacts. Surfaces chemically modified introducing functional groups that anchor NWs, as -NH_2, allow an absolute control of the process (See Fig. 5.19(**a**)). Multiple crossed NW arrays (See Fig. 5.19(**b**)) and more complex structures (See Fig. 5.19(**c**)) are fabricated in a layer by layer process, changing the flow direction at each step and balancing the flow rate in such a way not to perturb previously generated geometries. The authors believe in the possibility to extend this approach to CNTs, on the base of the strict relation between CNTs and NWs. Actually, CNTs are flexible, have a very low solubility and techniques to control CNTs length are not well developed. Consequently, important results on handling of CNTs with fluidic approaches have not been presented yet.

On the contrary, AC-electrophoresis, or dielectrophoresis, has been demonstrated to be a valuable method to purify and assemble at desired locations multi-wall and single-wall CNTs, both in bundles and individual. AC electrophoresis is a phenomenon where a lateral motion is imparted on uncharged particles as a result of polarization induced by a non-uniform electric field. As a matter of fact, an AC electric field induces on a particle a dipole moment that, in the presence of a field gradient, experiences a force towards either the strong field region (positive dielectrophoresis) if the particle is more polarizable of the surrounding medium, or away from it (negative dielectrophoresis) in the opposite situation. Without entering in detail of dielectrophoretic phenomena [97], we have to stress that it is controlled by many parameters: AC-frequency, intensity of the electric field, shape of the particle, shape and dimension of electrodes, dielectric constants of the medium and the particle.

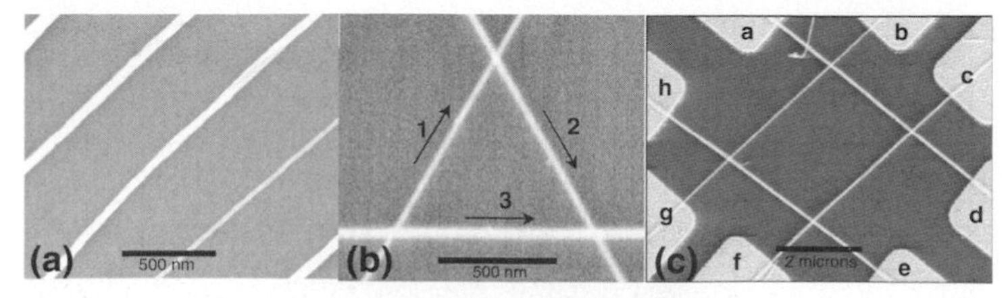

Fig. 5.19. (**a**) Parallel array of GaP Nws, 500 nm spaced, assemled on a Si/SiO_2 surface chemically patterned with a NH_2-terminated monolayer. A self-assembled monolayer (SAM) methyl-terminated was first obtained by immersion of Si/SiO_2 in pure hexamethyldisilazane (HMDS) for 15 minutes at 50° and 10 minutes at 100°. Then, a pattern of lines spaced 500 nm was obtained using e-beam liyhography. Soaking the sample in a solution of 3-aminopropyltriethoxysilane NH_2-terminated lines were obtained (**b**) Equilateral triangle geometry of GaP NWs obtained in a three step process. directions of flows are shown in the image. (**c**) SEM image of a crossed array of InP NWs. Ni/In/Au contact pads are defined using e-beam lithography and thermal evaporation. Reprinted with permission from [96]. Copyright 2001 AAAS

By tailoring all these parameters, many experiments with different goals can be addressed. In 1998 Yamamoto and coworkers obtained the first results in orientation and purification of carbon nanotubes using AC-electrophoresis [98]. MWCNTs produced by arc discharge, 1–5 μm long and 5–20 nm in diameters, were separated from carbon particles and aligned on Al electrodes. A suspension of MWCNTs in isopropyl alcohol (IPA), obtained after ultrasonication and centrifugation to disperse CNTs bundles and remove larger particles, was dropped onto coplanar Al electrodes, separated by 400 μm. When applying an AC electric field of $2.2 \times 10^3\ V_{rms}cm^{-1}$ at 10 MHz until evaporation of IPA, carbon particles remained in the middle of the two electrodes whereas CNTs aligned along the lines of the inhomogeneous electric field. The separation was based on different dipole moments between CNTs and carbon particles. Dielectrophoresis conditions could be tuned to form bulk of MWCNTs between gold electrodes [99]. These systems can potentially serve as sensing elements for thermal and anemometrical sensors, operating at mW power range and with very fast frequency response (See Fig. 5.20). Using a combination of AC electric field with a DC one (biased AC field), Chung and Lee [100] were able to deposit a single MWCNT between two Al electrodes forming a nanoscale gap. Such system is potentially useful as nanoscale resonator for chemical sensing applications (See Fig. 5.21).

Dielectrophoresis is also useful when employed for the purification and alignment of SWCNTs between electrodes with a micro [101] or a nano gap [102]. Crossbar structures made of small ropes of carbon nanotubes was also obtained by Diehl and coworkers [103]. They achieved a controllable aggregation in ropes of SWCNTs suspended in ortho-dichlorobenzene (ODCB).

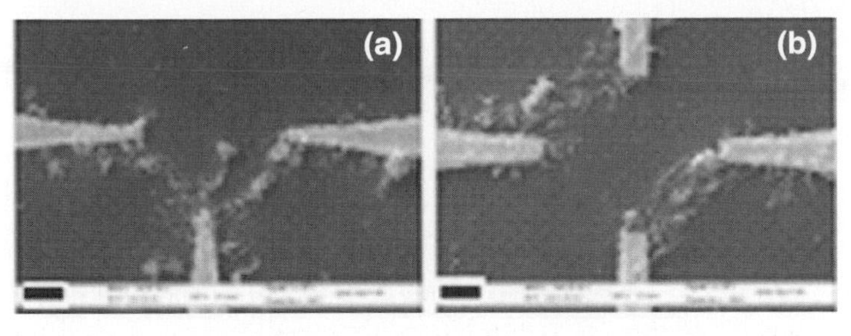

Fig. 5.20. SEM images showing a right angle (**a**) and a parallel (**b**) alignment of bulk of MWCNTs between gold electrodes. Reprinted with permission from [99]. Copyright 2003 IEEE

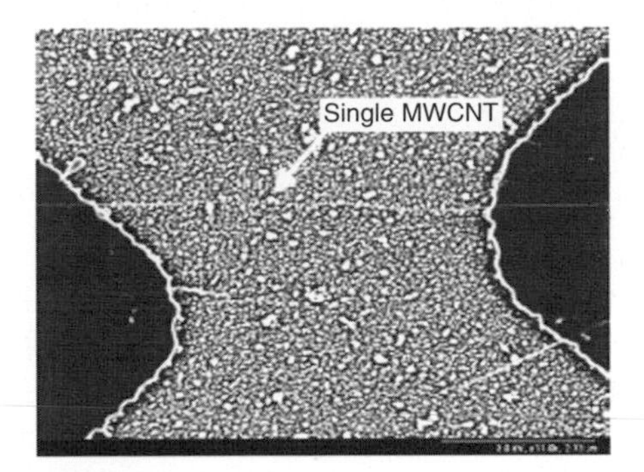

Fig. 5.21. Single MWCNT deposited between Al electrodes using the biased AC method working at 0.544 V_{rms}µm and 5 MHz. Reprinted from [100], copyright 2003, with permission from Elsevier

Crossbar structures were fabricated applying a 20 V peak-to-peak AC voltage at 4 MHz between two parallel electrodes separated by 10 micron gap and immersed into a SWCNTs suspension in ODCB/$CHCl_3$. These electrodes were then chemically etched and a new pair was fabricated orthogonal to the first one, and the process repeated. Multiple crossed SWCNT rope arrays, similar to the NW multilayer structures previously presented, were realized. The pitch of the structure is related to the length distribution of the ropes: narrow distribution allows to achieve more ordered and dense structures.

The most important achievement of dielectrophoretic approach for handling of CNTs was obtained by Krupke et al. in 2003: the separation of metallic from semiconducting carbon nanotubes from a suspension using AC electrophoresis [104]. Following the method proposed by M.J. O'Connel et al. [105], Krupke et al. [104] prepared an individual suspension of SWCNTs in D_2O adding sodium dodecylsulfate (SDS) as a surfactant. A drop of this suspension, having a mass concentration of roughly 10 mg/mL, was put on an

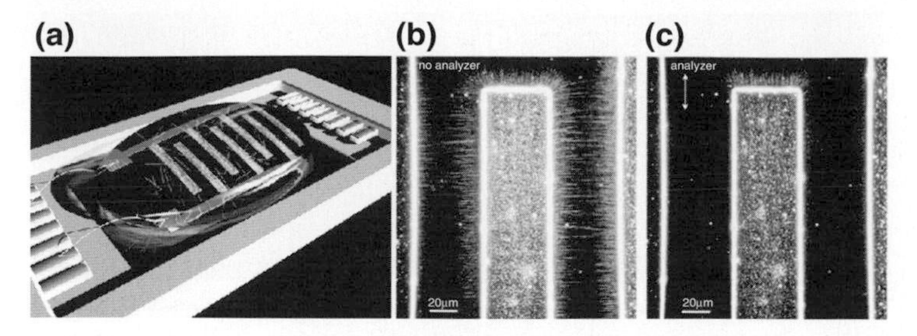

Fig. 5.22. Illustration of the experimental apparatus used to separate metallic from semiconducting nanotubes. Gold electrodes separated by 50 μm, 50 μm wide and 30 nm thick, are fabricated on a silicon substrate with 600 nm of thermal SiO_2 and a thin titanium layer to improve adhesion of gold. The chip is mounted on a chip carrier for the connection with an external power supply. (**b**) Dark field micrography of gold electrode and SWCNTs deposited during AC-electrophoresis. Due to strong Rayleigh scattering of SWCNTs in the green region, they appear green (bright filaments in Figure). (**c**) Using a light polarized perpendicularly to the nanotube axis and introducing an analyzer, uniformity of carbon nanotube deposited is visible. Reprinted with permission from [104]. Copyright 2003 AAAS

array of comb like electrodes and a peak-to peak voltage of 10 V at a frequency of 10 MHz was applied (see Fig. 5.22(**a**)). Tubes deposited onto the electrodes were then observed by microscopic resonance Raman spectroscopy. By comparing the Raman spectra of the sample with that of a reference prepared depositing a drop of SWCNTs suspension without application of any electric field, the researchers concluded that CNTs deposited were well aligned with their axes along electric field lines, data corroborated also by dark-field micrographs of the sample (see Figs. 5.22(**b**) and 5.22(**c**)). Furthermore, they concluded that the most part of CNTs aligned were metallic, about 80%. The authors justified this result by postulating that in the experimental condition s-SWCNTs exhibited a negative dielectrophoretic behavior, while metallic or quasi-metallic nanotubes showed a positive behavior. Actually, the yield of the process was very low, only 0.1% of the nanotubes in the sample were deposited, but, in principle, the method could be scaled up using microfluidic dielectrophoretic separation cells.

In the context of separation of metallic carbon nanotubes from semiconducting, three other methods have been reported [106]. Researchers at DuPont, USA, proposed a new method for the dispersion and separation of carbon nanotubes based on a biochemical approach [107]. They dispersed bundles of HiPco-SWCNTs in water by sonication in the presence of single stranded DNA (ssDNA). Experimental evidence of the dispersion came from absorption and fluorescence spectroscopy and from AFM analysis. DNA-assisted dispersion of CNTs is really efficient: very stable CNTs solutions with concentrations of about 0.4 mg/mL were obtained. Molecular modelling suggested that the flexibility on bond torsion of the sugar-phosphate backbone allows ssDNA to find the low energy conformation that maximizes the π-stacking interactions between the aromatic nucleotide base in ssDNA and the side-wall of carbon nanotubes and exposes the sugar phosphate groups to water. The phosphate groups of the DNA-CNT hybrid system confer a superficial negative charge. The interactions are so stable that the removal of free ssDNA by anion-exchange liquid chromatography leaves the DNA-CNT hybrids stable for months at room temperature. DNA-CNT hybrids emerge from the column as a single broad peak. Different fractions were analysed by absorption spectroscopy. The authors established that the first fractions were richer in metallic tube than the last fractions and explained these separation phenomena as a consequence of the less surface charge of metallic nanotubes as compared to the semimetallic nanotubes.

A pure chemical method for the bulk separation of semiconducting from metallic SWCNTs was proposed by Chattopadhyay et al. [108]. This method was based on the selective precipitation of metallic SWCNTs from a suspension of SWCNTs dispersed with octadecylamine (ODA) in tetrahydrofuran. The physisorption of ODA on the side walls of SWCNTs was proposed to confer extra stability to s-SWCNTs, due to stronger interactions of the semiconducting nanotube with the -NH_2 groups of the alkylamines [109]. Raman spectroscopy investigations of the supernatant solution and precipitate validated this hypothesis, being the supernatant solution enriched in s-SWCNTs and the precipitate in m-SWCNTs.

Rinzler and colleagues at the University of Florida proposed a similar methodology based on selective precipitation of m-SWCNTs. Carbon nanotubes dispersed in water with surfactants were treated with bromine. After centrifugation, to separate the heavier bromine complexes from the uncomplexed form, and a thermal treatment to remove not-reacted bromine, optical spectra revealed that the precipitate was richer in m-SWCNTs [110].

Although all the separation methods await further investigations for their validation, chemical and biochemical approaches seem to be good choices for the separation of large quantities of CNTs. On the contrary, the dielectrophoretic method should be a valid alternative to electrical breakdown being non-destructive and non-limited to transistor geometry.

5.5.4 Self-Assembly of CNTs

Ordered organization of CNTs onto a substrate can be achieved exploiting also chemico-physical properties of nanotubes. Two main directions have been followed: the first one is the self-organization of pristine carbon nanotubes (p-CNTs) on a prepatterned-chemically modified substrate, the second is the self-assembly of chemically modified carbon nanotubes on a prepatterned substrate. We will first review the approach on pristine nanotubes and then will give a brief overview of the techniques to chemically modify CNTs before to present some results obtained with the second approach.

Self-Assembly of p-CNTs on Prepatterned-Chemically Modified Substrates

The first controlled deposition of both multi-wall and single-wall nanotubes onto chemically modified Si/SiO_2 surface was reported by Burghard and coworkers in 1998 [111,112]. A sodium dodecylsulphate-stabilized CNT (SDS-CNTs) water suspension, purified via size exclusion chromatography [113], was prepared. SDS molecules organize around the tube giving a negativelly charged micellar surface. Electrostatic interactions between the negative shell of CNTs and the chemically modified surfaces, amino, methyl or carboxy-functionalized, was exploited to organize CNTs. For more details see to Fig. 5.23. However, the degree of organization described by the authors was not sufficient for fabrication and characterization of CNT-based device. Choi et al. [114] improved the Burghard's method by optimising the procedure for the amino-fuctionalization of silicon dioxide surface. A scheme of the deposition method is reported in Fig. 5.24. AFM investigations of the samples prepared with this procedure revealed that SWCNTs were deposited only on silane monolayer with a density that increased with increasing the solution

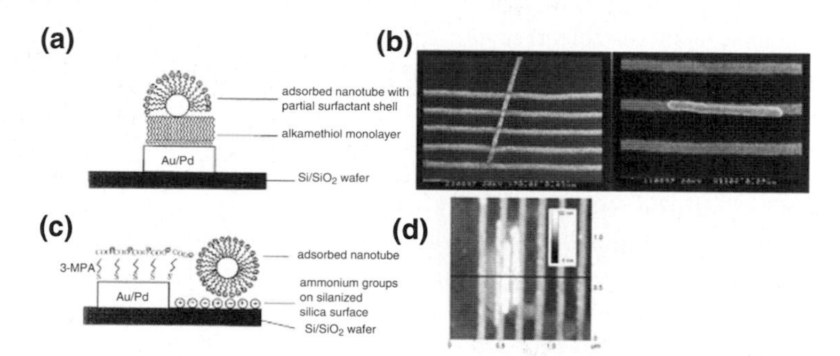

Fig. 5.23. (**a**) Organization of CNTs on Pd/Au electrodes: an octadecanethiol monolayer is first assembled onto electrodes exploiting the affinity of thiols for gold. Partially shelled CNTs prefer to adsorb on hydrophobic metal eletrodes instead of on silicon dioxide surface, negatively charged at pH conditions of experiment. (**b**) SEM images of a nanotube between CH_3 terminated electrodes and another one on a CH_3 terminated electrode. (**c**) Organization of CNTs on a SiO_2 substrate: SiO_2 surface is positively charged by amino silanization while gold electrodes negatively charged with 3-mercaptopropionic acid. Fully shelled CNTs assemble onto positively charged surface via electrostatic interaction. (**d**) AFM image of CNTs between electrodes on amino-silanized surface. Reprinted with permission from [111], copyright 1998 WILEY-VCH Verlag GmbH

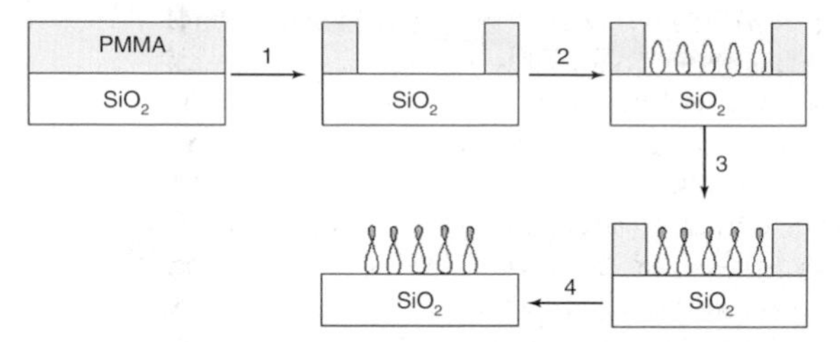

Fig. 5.24. Schematic representation of the deposition method proposed by Choi [114]. In the first step, 300 nm of silicon dioxide on silicon wafer are cleaned via sonication in acetone and methanol, immersion into Piranha solution, exposition to UV-Ozone treatment and rinsing in D.I. water. Then, a 100 nm thick PMMA resist layer is spun onto the substrate. Finally, lines with width ranging from 50 to 200 nm are patterned using e-beam lithography and developed. In the second step, after cleaning of the surface with oxygen plasma, a monolayer of 1,2-aminopropyltriethoxysilane (APTS) is formed by vapour deposition. In the third step, terminal amino groups are converted in NH_3^+ exposing the substrate to HCl vapours. In the forth step, SDS-SWCNTs purified via size exclusion chromatography are then deposited on the substrate and the PMMA layer is removed in acetone

concentration, deposition time and pattern line-width. CNTs were aligned along the silane lines with a quality of alignment that depended on the tube length and line-width: longer nanotubes were better aligned and, in general, CNTs were better aligned on smaller lines. This technique can be used with profit to fabricate crossed nanotube structures or to contact CNTs with metallic electrodes. In fact, if the prepatterned structure contains alignment markers, metallic electrodes can be patterned on the substrate after CNTs deposition without inspecting the sample. This overcomes the time-limiting step of the indirect random approach. Alternatively, metallic electrodes can be patterned before the assembling of CNTs onto chemically modified surface reducing in such a way possible sources of contamination derived from post-patterning procedure [115]. An accurate choice of process parameters as type and concentration of surfactants, electrode spacing and height, purity and length of nanotubes, allows to achieve excellent results.

In 1999, the group of Rick Smalley at Rice University proposed a procedure to prepare very stable suspensions of SWCNTs not based on surfactant-stabilization [116]: SWCNTs purified by reflux in 2.6 M HNO_3 were suspended in N,N-dimethylformamide (DMF) by sonication. They also proposed an alternative way to chemically modify a Si/SiO_2 surface for controlled deposition of CNTs. The most important steps of this procedure are shown in Fig. 5.25.

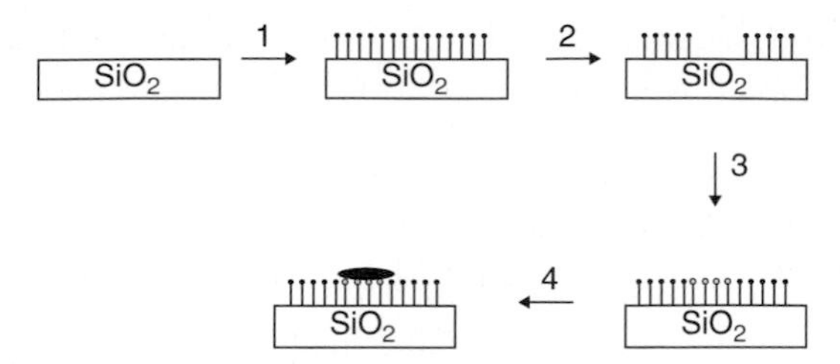

Fig. 5.25. Schematic representation of the deposition method proposed by Smalley and coworkers [116]. In the first step, an oxidized silicon surface is heated at 150°C for 5 hours with hexamethyldisilazane (HMDS). The surface results covered by a hydrophobic trimethyllsilyl (TMS) self-assembled monolayer (SAM). In the second step, by means of electron beam lithography, a pattern is obtained destroying the TMS-SAM at specific positions. The opened zones are filled with an NH_2 terminating monolayer (third step) by soaking the sample in a chloroform solution of 3-aminopropyltriethoxysilane (APS). At the end, SWCNTs are deposited from DMF suspension. They exhibit a strong affinity to NH_2-functionalized pattern

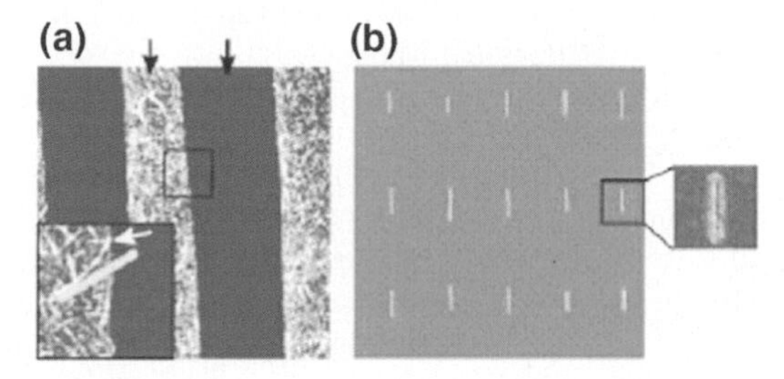

Fig. 5.26. (**a**) Atomic force micrograph of SWCNTs (white filaments in the inset) assembled on gold lines chemically functionalized with polar cysteamine. Lines free of nanotubes are chemically modified with apolar 1-octadecane thiol (ODT). (**b**) Atomic force micrograph of very thin gold lines functionalized with 2-mercaptoimidazole on which SWCNTs assembled (see inset). Reprinted by permission from Macmillan Publishers Ltd: Nature [117], copyright 2003

In 2003 Rao and coworkers proposed an approach for large-scale assembly of carbon nanotubes based on the same principle of that proposed by Smalley's group [117]: the selective self-assembly of carbon nanotubes on polar chemical groups, such as amino groups (NH_2/ NH_3^+) or carboxylic groups ($COOH$/ COO^-), relative to nonpolar groups, such as methyl (-CH_3). The innovation is in the use of dip-pen lithography or micro contact stamping instead of conventional lithographic techniques minimizing surface contamination. An example of the structures obtained is shown in Fig. 5.26.

In a different kind of chemical approach, the organization of carbon nanotubes onto many surfaces (silicon with native oxide, glass, quartz crystal microbalance resonator) was shown to proceed through metal-assisted assembly [118]. The substrate was first dipped into an acid solution of $FeCl_3$ and then into a basic solution of DMF to anchor iron to the surface, transforming it in its base hydroxide form (surface-$Fe(OH)_2$). Carboxy-functionalized nanotubes in DMF, shortened by oxidation, were assembled by acid-base neutralization reaction. While CNTs assembled on hydrophilic functionalized substrates, for example -NH_2 terminated, tended to orient parallel to the substrate, Raman studies on metal-assisted shortened nanotube assembly indicated a significant orientation perpendicular to the substrate, probably related to the high number of carboxylic groups at cut ends of oxidized CNTs and to strong van der Waals forces between adjacent nanotubes. Performance enhancements of devices such as sensors and optoelectronic devices should be achieved for these dense, normal-oriented structures.

Arrays of SWCNTs aligned normal to gold surfaces were obtained recently [119,120]. The assembling reaction proceeds in two steps. First a self-assembled monolayer of α,β-aminomercaptans on gold is produced following standard procedure, then a surface condensation reaction between the amino group of aminomercaptan and the carboxylic groups of oxidatively shortened carbon nanotubes is induced using dicyclohexylcarbodiimide as condensing agent. These systems can work as microelectrode array. The possibility to anchor a chemical species at the free end of the nanotubes assembled in such a way can be exploited to design bioelectrochemical systems to study protein electrochemistry or to develop ionic, molecular or biomolecular sensors.

Recently, the direct deposition of carbon nanotubes onto pyrene-modified oxide surfaces (Si/SiO_2, quart and Indium Tin Oxide-ITO) has been develop [121]. The driving force of the assembling process is the strong interaction between the pyrene groups and the sidewalls of CNTs via π-stacking [122]. Xin and Wolley [123] exploited the same type of interactions to selectively immobilize SWCNTs on double stranded λ-DNA pre-aligned on silicon surface and chemically modified via interaction with 1-pyrenemethylamine hydrochloride (PMA). The amino group of PMA is believed to interact with the negatively charged phosphate backbone of DNA, exposing then the pyrene-moiety for anchoring of CNTs. The possibility to use DNA as template for SWCNTs anchoring and localization opens a new fascinating way for the fabrication of nanodevices via bottom-up approach.

Self-Assembly of Functionalized Carbon Nanotubes

Chemical modification of carbon nanotubes is a very attractive research field as functionalization can induce the exfoliation of nanotubes bundles and then improve their solubility and processibility. Soluble carbon nanotubes can be easier purified, from amorphous carbon and/or catalyst particles; moreover, modification of chemical-physical properties of CNTs via functionalization can guide their assembly with any substrate. Several excellent reviews dealing with the many aspects of chemistry of nanotubes have already appeared in the literature [124–128].

Chemical reactivity of carbon nanotubes comes from their local-strained structure. Rolling of graphene sheets to give 1D tubes induce pyramidalization (deviation from ideal planar geometry of sp^2 carbon) and misalignment of the π-orbitals of the carbon atoms. For ideal CNTs, that is to say not defective, it is probable that the end caps reactivity is mostly guided by pyramidalization while the sidewall reactivity by π-orbitals misalignment. Furthermore, because pyramidalization of CNTs decreases with increasing of the diameters contrary to π-orbitals misalignment, nanotubes with different diameter are expected to exhibit a different chemical reactivity. Intrinsic or induced defects on carbon nanotubes also play an important role in their chemistry: five- or seven-atom rings that destroy the symmetry of the hexagonal lattice of graphene increasing CNT's curvature, not intact end caps presenting dangling bonds, or reactive functionalities introduced by hard treatment, like oxidation in acidic media, all create defect sites that are thought to react more easily.

There are three general categories of chemical modification of carbon nanotubes: functionalization of oxidized nanotubes, sidewall covalent functionalization and non-covalent functionalization. The following subsections contain many concepts already examined by Professor Prato and coworkers in [128].

Functionalization of Oxidized Nanotubes

One of the purification methods of raw CNT material is based on oxidation in acidic media [51]. Under these conditions, the end caps of CNTs are opened and they are cut in short tubes (fullerene pipes) with lengths of few hundreds of nanometers. At the defect sites, especially at the level of dangling bonds, acidic functionalities suitable for further derivatization are formed. Numerous amidation or esterification reactions to oxidized CNTs have been reported, resulting in soluble functionalized materials (See Fig. 5.27) [126, 129–132].

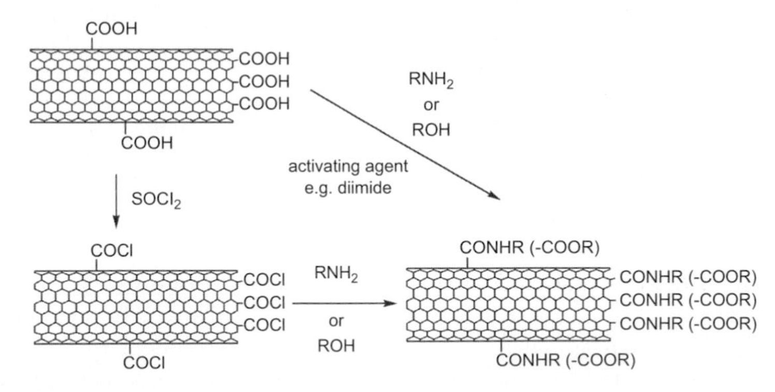

Fig. 5.27. Schemes of amidation and esterification of oxidized CNTs

Long-chain alkylamines were condensed with the carboxylic groups present on the surface of the CNT [133, 134]. Activation of the carboxy moieties with thionyl chloride and subsequent reaction with amines is preferred to direct condensation of oxidized nanotubes with amines [135,136], producing in general more soluble products. In fact, exfoliation of individual CNTs from large ropes has been achieved during the functionalization process. This is extremely important as it allows easy characterization of the soluble CNTs and facilitates their manipulation. The presence of the functional moieties at the defect sites of oxidized CNTs has been mainly monitored by IR spectroscopy, allowing an easy identification of the amide or ester bonds. Recently, SWCNTs functionalized in this manner have been purified by HPLC [70]. One of the more important goals of this method is that the electronic properties of such functionalized CNTs remain intact [51], being the functionalization almost completely restricted to nanotube ends.

Along the issue of controlled deposition of carbon nanotubes on gold surface for the construction of microelectrode arrays, important progress was achieved by the thiolization reaction of carboxyl-terminated carbon nanotubes [51, 137]. Short-length oxidized carbon nanotubes were treated with the appropriate thiol derivative and the resulting material was tethered chemically to a gold substrate (See Fig. 5.28).

Deposition of oxidatively shortened nanotubes on silver surface has been studied and the technique is based on spontaneous adsorption of the -COOH groups onto the surface [138]. Various techniques have been used to characterize the assembly such as Raman Spectroscopy, Atomic Force Microscopy and Transmission Electron Microscopy.

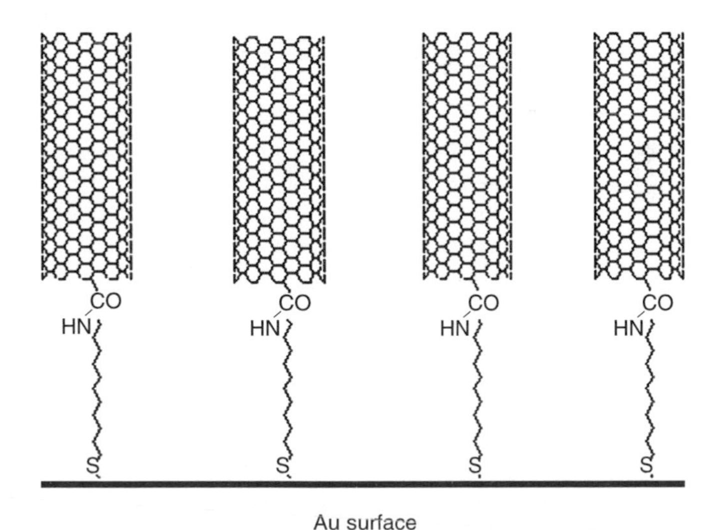

Fig. 5.28. Scheme of the assembling of thiol-derivatized CNTs on gold substrate

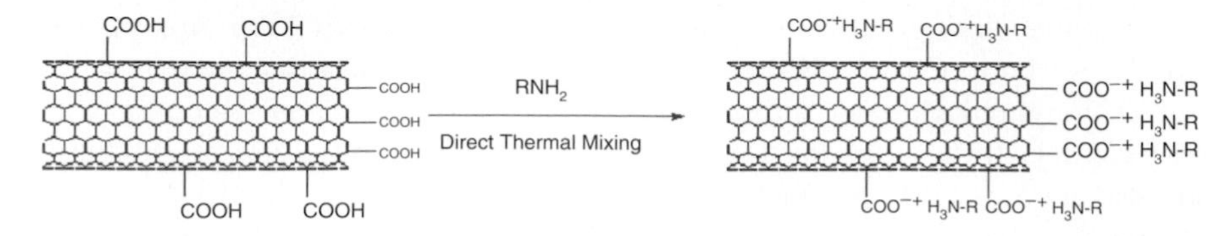

Fig. 5.29. Scheme of zwitterionic thermal assisted functionalization of long oxidized SWCNTs

Longer single-wall oxidized CNTs can be solubilized via acid-base direct reaction with long chain amine by formation of zwitterions (See Fig. 5.29). This simple reaction is very attractive as the cation can be easily exchanged with another one, organic or inorganic, opening a new way for electrostatic interactions of nanotubes with other charged species [135].

Sidewall Covalent Functionalization

Covalent chemistry of carbon nanotubes is mainly limited by two factors. The first problem is represented by the low solubility and dispersibility of CNTs. Secondly, the very poor reactivity of CNTs can be addressed only using very reactive species and/or hard reaction conditions. For these reasons it is not surprising that the first report of this kind of approach involved reaction of SWCNTs with well known reactive organic species such as dichlorocarbene [134, 139]. Moreover, SWCNTs have been also found to react with molecular fluorine at temperatures between 150–600°C [140]; different degrees of fluorination can be obtained depending on the reaction temperature. A hydrazine treatment of such derivatized CNTs results in their defunctionalization, so that intact CNTs can be recovered. Although fluorinated carbon nanotubes can be well dissolved in alcohols [141], the fluorine atoms can be further substituted by alkyl groups for achieving better dissolution [144]. However, fluorination applied to small-diameter HiPco-SWCNTs detrmines their shortening to an average length of less than 50 nm [142].

Extensive studies of organic reactions leading to sidewall functionalization of SWCNTs were performed [124, 143]. Reactions considered, already successfully applied to fullerenes, include: a) addition of nitrenes, b) addition of carbenes and c) addition of radicals. In the first case, alkyl azides were used as precursor compounds of nitrenes that, reacting with SWCNTs, allow to obtain functionalized SWCNTs soluble in dimethyl sulfoxide (DMSO). A dipyridyl imidazolium system, that was found to link to the surface of nanotubes, was utilized for the functionalization of SWCNTs with carbenes. Finally, perfluoroalkyl radicals were photochemically added on the sidewalls of SWCNTs (See Fig. 5.30).

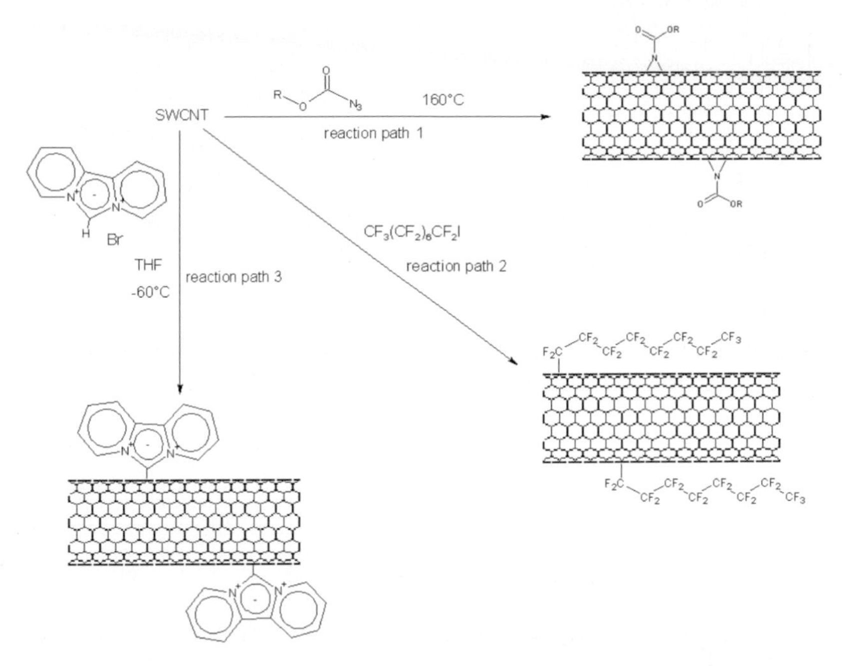

Fig. 5.30. Scheme of reaction paths of CNTs with highly reactive species such as nitrenes (reaction path 1), radicals (reaction path 2) and carbenes (reaction path 3)

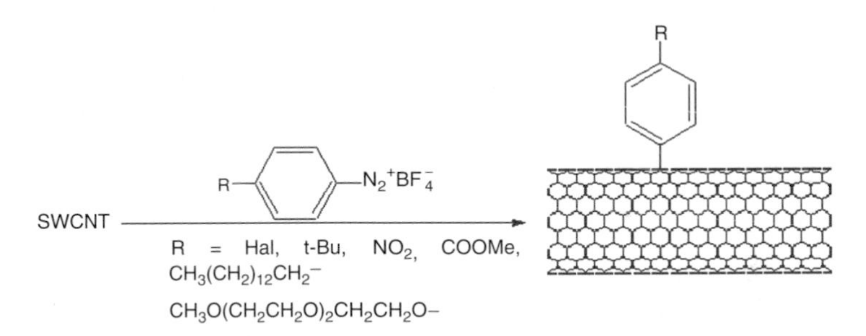

Fig. 5.31. Scheme of CNT aryl diazonium chemistry

In another approach for covalent side-wall functionalization of CNTs, aryl diazonium chemistry has been employied [125, 145–147]. Derivatization of HiPco-CNTs was achieved via electrochemical reduction of a variety of diazonium salts (See Fig. 5.31) [145], but diazonium compounds chemically generated in-situ were also reported [146, 147]. The electrochemically reductive coupling of aryl diazonium salts to CNT resultes in C–C bond formation; otherwise in the oxidative coupling of aromatic amines, the amines attaches directly to the surface of nanotubes (See Fig. 5.32) [148].

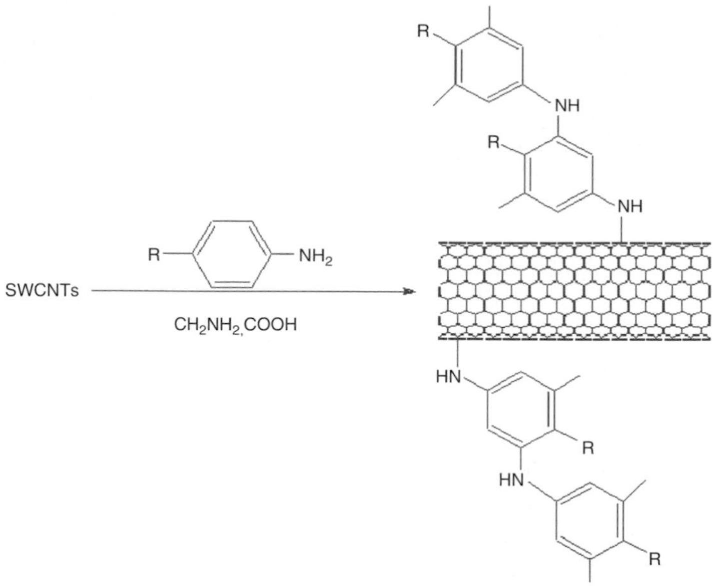

Fig. 5.32. Scheme of oxidative coupling reation of CNTs with aromatic amines

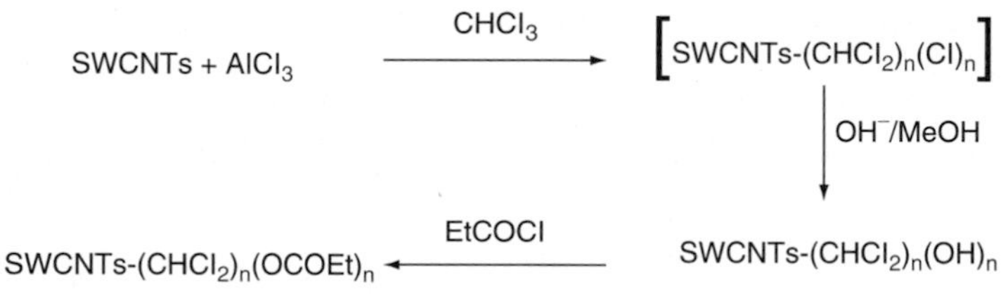

Fig. 5.33. Scheme of eletrophilic addition to SWCNTs of choloform molecules and following esterification via clorine-hydroxyl exchange

An unprecedented electrophilic addition to HiPco SWCNTs was recently reported [149]. In the presence of aluminum trichloride ($AlCl_3$), chloroform molecules were found to add to the sidewalls of CNT. It was also possible to replace the chlorine atoms with hydroxyl groups, esterified in a second step to yield the corresponding esters (See Fig. 5.33). This one-pot-two-step functionalization of HiPco-SWCNTs not only resulted in soluble functionalized material but also showed the way for multi-step synthesis of CNTs-based materials.

The powerful methodology of 1,3 dipolar cycloaddition of azomethine ylides, well established in the fullerene field [150, 151], has been applied for functionalizing and solubilizing CNTs [152, 153]. The main advantage of this reaction is the easy attachment to the sidewalls of CNT of pyrrolidine rings

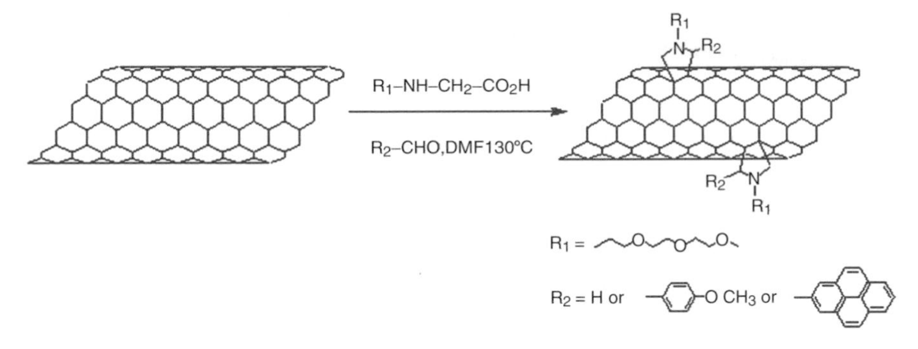

Fig. 5.34. Schematic diagram of 1,3-dipolar cycloaddition of azomethine ylides to SW or MWCNTs

substituted with chemical functions leading to the synthesis of novel materials with diverse properties. The generation of azomethine ylides is based on the decarboxylation of immonium salts derived from in situ condensation of α-amino acids with aldehydes. Functionalized aldehydes can give 2-substituted pyrrolidine moieties on the sidewalls of CNTs, while modified α-amino acids could also lead to numerous diverse functionalized materials (Fig. 5.34). CNTs functionalized in such a way are very soluble in $CHCl_3$, CH_2Cl_2, acetone, methanol and ethanol and practically insoluble in diethyl ether and hexane. Following this strategy, functionalized water soluble carbon nanotube were prepared when N-substituted α-aminoacid with a terminated amino tert-butoxycarbonyl (Boc) protected group [154] and para-formaldehyde were used as reagents for the functionalization of both SWCNTs and MWNTs [153] (See Fig. 5.35). There is great interest in developing water soluble CNT to be employed as nanoscale biosensors [122, 155–160].

Extensive covalent functionalization of CNTs induces substantial modifications of electronic properties of carbon nanotubes as proved by many investigation techniques like UV/vis/NIR absorption and Raman spectroscopy [145, 152]. Original electronic properties can be almost completely recovered defunctionalizing CNTs via chemical or thermal treatment [144, 152]. Full-length functionalized CNTs, soluble in a big variety of organic solvents, can be treated as conventional macromolecules, easily purified and/or anchored to desired substrates, and then the electronic properties of pristine CNTs can be fully recovered via defunctionalization.

Non-Covalent Functionalization

This type of functionalization is particularly attractive because of the possibility of attaching various groups without disrupting the conjugation among the benzene rings. CNTs can be solubilized in aqueous media in the presence of amphiphilic molecules (surfactants) [65, 68] whose hydrophobic moiety is

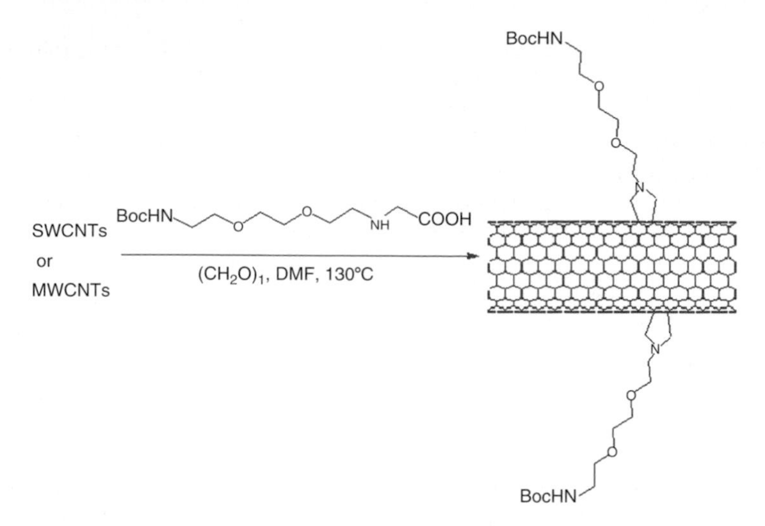

Fig. 5.35. Amino modified CNTs derivarized with Boc-protected aminoacid

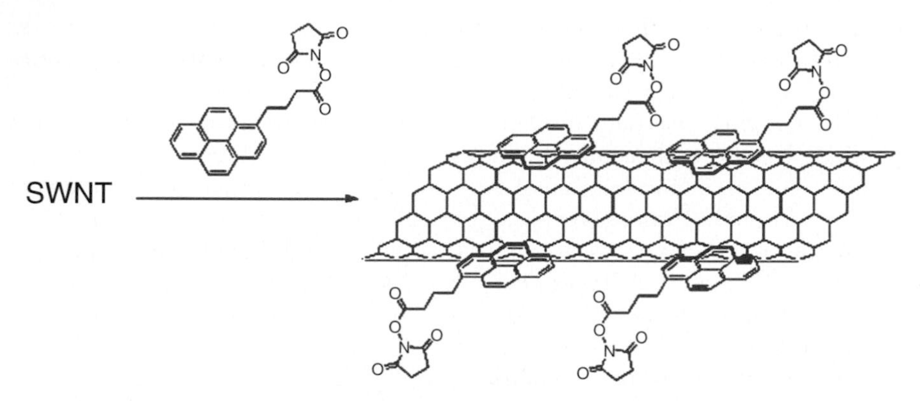

Fig. 5.36. Pyrene derivative adsorption to the CNT surface via π-π stacking

oriented towards the surface of carbon nanotubes and the polar part interacts with solvent molecules.

Recently, aromatic molecules, containing a planar pirene moiety, were non-covalent anchored to the side-walls of SWCNTs. The anchoring process is driven by strong π-π stacking interactions (See Fig. 5.36). The pyrene moiety was specifically designed to further couple via amide linkages [122] specific molecules, as proteins, DNA filaments or other biomolecules. The power of this methodology is in the possibility to couple the electronic properties of CNTs with the specific recognition properties of the anchored biomolecules leading to the concept of miniaturized bio-sensors. Pyrene-carrying ammonium ions were also used for the solubilization of CNTs in water [161].

In a different approach, bundles of SWCNTs helically wrapped with a conjugated polymer were produced [162] adding SWCNTs to a solution of poly-(m-phenylenevinylene)polymer, substituted with octyloxy alkyl chains. A stable suspension of CNTs was obtained upon sonication. Once again, the SWCNTs-polymer complex is the result of π-π stacking interactions. Due to their well defined dimensions, stilbenoid polymers have been used extensively in this type of functionalization [163, 164]. Soluble nanotube-polymer composites can be also prepared by in situ polymerization of phenylacetylene [165]. In the cited cases, the composites formed are soluble in organic solvents (chloroform, toluene etc.) but polymer wrapping around nanotubes has also been achieved in aqueous media. In particular, has been studied the coniugation in water of nanotubes with linear polymers like polyvinyl pyrrolidone and polystyrene sulfonate [64]. CNTs were found to unwrap by changing the medium to less polar solvents.

As solubilizing agents, short rigid conjugated polymers, such as polyarylene-ethynylenes, have also been used. These polymers have been attached by a non-wrapping approach onto the CNTs (the major interaction is stacking by van der Waals forces) while some pendant alkyl groups contribute to the solubility [166]. The non-wrapping approach allows the solubilization of SWCNTs in organic solvents (chloroform) by simply mixing with polymers.

Although polymers and surfactants can improve the solubility of CNTs in aqueous solutions, the main drawback of this class of materials is their nonbiocompatibility. Therefore, in order to exploit the unique properties of CNTs in a biological setting, the possibility of solubilizing SWCNTs in aqueous solutions of amylose was tested [167]. In the presence of iodine, which causes pre-organization of amylose from its linear conformation to a helical one, very stable and soluble composites of nanotubes-amylose were found to form. The investigation of the obtained composites, performed with atomic force microscopy (AFM), shows small bundles of CNT covered with amorphous polysaccharide. Moreover, the addition of amyloglucosidase (an enzyme that hydrolyzes starch) to the stable CNTs solution, led to precipitation of CNTs, suggesting a new method for purifying CNTs.

5.6 Applications of Carbon Nanotubes

In the last decade, an enormous amount of work has been devoted to reveal the, sometimes unique, structural, electrical, mechanical and chemical properties of carbon nanotubes and to explore what might be the most interesting applications of these novel materials [168]. First of all, CNTs are promising materials for building electronic devices, in particular field effect transistors [85, 169]. Moreover, the geometric properties of nanotubes, such as the high aspect ratio and small curvature radius of the tip, coupled with the extraordinary mechanical strength and chemical stability, make them ideal candidates for electron field emitters [170, 171], having several applications like flat panel displays [172] and gas sensors [173]. Sensing is, indeed, another interesting field of application of CNTs, which mainly exploits the strong sensitivity of the SWCNT electronic and transport properties to atomic structure and mechanical deformations [174, 175] or chemical doping [176, 177].

In the following subsections we will discuss in detail the possible applications of CNTs in electronic and field emission devices, sensing and microscopy applications. In addition to these fields, CNTs are actually investigated as useful high surface area and high conductivity electrodes in a variety of other applications. For example, the possibility to store hydrogen for fuel cell development has received much attention [178–184]. An H_2 storage capability of 6.5%-8% by weight appears to be the target for use in automobiles. Though interesting basic science of hydrogen storage is emerging, no technological breakthrough has been reported yet since several claims of high hydrogen storage levels have been demonstrated to be incorrect.

Filling nanotubes with a variety of metals has also been attempted [185–188]. The most noteworthy of these attempts involves lithium storage for battery applications [189, 190], and gallium filling for nano-thermometers [191]. The thermal conductivity of SWCNTs is reported to be about 3000 W/mK in the axial direction which is second only to epitaxial diamond [192]. This property can be exploited for cooling of semiconductor chips and heat pipes. Other interesting applications may include nanoscale reactors [193], supercapacitors [190–194], ion channels [195, 196], and drug delivery systems [197–199].

5.6.1 CNT Electronic Devices

The possibility of using carbon instead of silicon in nano-electronics has generated considerable interest. The ability to make metallic contacts to and to connect together nanotubes has allowed the investigation of the metallic and semiconducting behavior, as well as of the electronic transport through individual nanotubes and the realization of nanotube-based devices.

A major effort has been addressed to see if individual (or bundles of) nanotubes could be used as molecular wires for interconnections in electronic circuits. The room temperature two terminal resistance of a contacted metallic SWCNT can be as low as $6.5\,\mathrm{K}\Omega$ (the quantum limit for a balistic 1D conductor with 4 conducting channels) or as high as several $\mathrm{M}\Omega$, depending on the nature of the metal-tube contact and the length of the tube. Under certain conditions, metallic nanotubes have been found to carry high current densities (in the order of $10^9\,\mathrm{A/cm^2}$) which are believed to be due to minimal scattering of traversing electrons, that is, to ballistic conductance. In 1D systems like nanotubes, electrons can be scattered back in only one direction as opposed to several directions in 2D and 3D systems. Scattering of electrons can occur due to intrinsic defects in the nanotube or due to the interaction with phonons at high voltages and temperature. At high voltages, electrons have very high energy and they produce phonons. In order to conserve momentum, the momentum of the phonons has to be balanced by that of the electrons thereby resulting in backward electron scattering. At low voltages, electrons do not have sufficient energy to scatter back from defect-scatterers and at low temperature the phonon population is reduced, which limits the interaction probability. At low voltages and temperature, therefore, the electrons have a good probability to travel all along the nanotube length without suffering any scattering (ballistic).

Interestingly enough, other quantum transport phenomena, like Coulomb blockade, have been observed at liquid helium temperature in μm long tubes, when the nanotube-metal connections are weak (large contact resistance) and the energy needed for adding an electron to the system is larger than the thermal energy [83, 84]. In the Coulomb blockade regime the transport occurs via single electrons that occupy discrete energy levels arising from quantum confinement along the length of the tubes (well described by the particle-in-a-box picture with the box being the nanotube of finite length).

By improving the metal-tube contacts, Coulomb blockade is no more observed, and the "intrinsic" electrical properties of nanotubes can be elucidated, confirming that metallic nanotubes behave as almost perfect 1D ballistic conductors at low temperature and voltages [200–202]. Earlier measurements [200] showed that the resistance of a μm long metallic SWCNT can be about 12 kΩ at room temperature. As expected for a metal, the conductivity of the sample increases as the temperature is decreased, due to reduced phonon scattering, approaching the quantum limit at low temperatures and low voltages [201, 202].

In this regard, new results having important implications to high performance nanotube interconnections have been reported by Javey et al. [203, 204]. Contacting metallic single-wall carbon nanotubes by palladium, a noble metal with high work function and good wetting interactions with nanotubes, affords highly reproducible ohmic contacts and allows for the detailed elucidation of ballistic transport in metallic nanotubes. The estimated mean-free path for acoustic phonon scattering at room temperature in SWCNTs, with Pd contacts and lengths ranging from several microns down to 10 nm, is about 300 nm while that for optical phonon scattering is in the order of 15 nm. Therefore, transport through very short ($\leq$ 15 nm) nanotubes is free of significant acoustic and optical phonon scattering and thus ballistic and quasiballistic at the low- and high-bias voltage limits, respectively. Under these conditions, high currents of up to 70 μA can flow through a short nanotube [203, 204]. This corresponds to a current density several orders of magnitude higher than those found in present interconnections.

Good metal-tube contacts allow the measurement of low resistance (in the range of several tens of kΩ) also in semiconducting SWCNT samples [205], for which room temperature demonstration of conventional switching mechanism, such as in field effect transistors (FET), first appeared in 1998 [84, 169]. As already shown in Figs. 5.16–5.18, a FET device can be realized by placing a SWCNT to bridge a pair of metal electrodes serving as source and drain. The electrodes are defined using lithography on a layer of SiO_2 in a silicon wafer, which acts as the back gate. The variation of drain current with gate voltage at various source-drain biases demonstrates that the gate controls the current flow through the nanotube and the characteristic curve of FETs is obtained. In this device, the holes are the majority carriers of current, as evidenced by an increase in current at negative gate voltages. Single-wall carbon nanotube FETs built from as-grown tubes are found to be unipolar p-type, i.e., no electron current flows even at large positive gate biases. Positive gate voltages caused the Fermi level shifting away from the valence band into the band-gap, depleted the holes and turned the system into an insulating state. This behavior suggests the presence of a Schottky barrier at the metal–nanotube contact. It has beeen shown that in air the barrier for electrons is high because of the pinning of the Fermi level close to the valence band maximum at the nanotube–metal interface caused by adsorbed oxygen [206]. Kim et al. [207] have shown that carbon nanotube FETs, which commonly comprise nanotubes lying on SiO_2 surfaces, exposed to the ambient environment, exhibit hysteresis in their electrical characteristics because of charge trapping by water molecules around the nanotubes (mainly substrate-bound water proximal to the nanotubes). Hysteresis persists for the transistors in vacuum since the SiO_2-bound water does not completely desorb in vacuum at room temperature, a known phenomenon in SiO_2 surface chemistry. Heating under dry conditions significantly removes water and reduces hysteresis in the transistors. Nearly hysteresis-free transistors are obtainable by passivating the devices with polymers that hydrogen bond with silanol groups on SiO_2 (e.g., with poly-methyl methacrylate (PMMA)) [207].

Although the performances of CNT-FETs are far from being optimized, and significant improvements are expected soon, carbon nanotube-based field effect transistors have already shown excellent operating characteristics that are as good as, or better than, silicon devices [208–210]. Extraordinary transport properties at room temperature, such as field-effect mobility (79000 cm^2/Vs) and intrinsic mobility (>100000 cm^2/Vs), have been reported recently for semiconducting carbon nanotube transistors with channel lengths exceeding 300 microns. In these long transistors, carrier transport is diffusive and the channel resistance dominates the transport. These values exceed those for all known semiconductors. Moreover, CNTs do not have interface states that need passivation, as in the case of the Si-SiO_2 interface, which makes it easier to integrate the CNTs with high-K dielectrics. The integration of materials having a high dielectric constant (high-K) into CNT transistors promises to push the performance limit for molecular electronics. It has been demonstrated that zirconium oxide thin-films (about 8 nm) can be formed on top of individual single-wall carbon nanotubes by atomic-layer deposition and used as gate dielectrics for nanotube field-effect transistors [209]. The p-type transistors so obtained, exhibit subthreshold swings of about 70 mV per decade, approaching the room-temperature theoretical limit for field-effect transistors. Key transistor performance parameters, as transconductance and carrier mobility, reach values better than silicon devices. Once again, contacting semiconducting single-wall nanotubes by palladium greatly reduces or eliminates the barriers for transport through the valence band of nanotubes. With Pd contacts, the 'ON' states of semiconducting nanotubes can behave like ohmically contacted ballistic metallic tubes, exhibiting room-temperature conductance near the ballistic transport limit, high current-carrying capability (about 25 μA per tube), and Fabry–Perot interferences at low temperatures. Under high voltage operation, the current saturation appears to be set by backscattering of the charge carriers by optical phonons. High-performance ballistic nanotube field-effect transistors with zero or slightly negative Schottky barriers have been realized [208].

Since the demonstration of nanotube FETs, CNT logic circuits and complex electronic devices have been successfully fabricated exploiting various approaches. These include intra-tube heterojunctions containing sharp kinks at the junctions [87, 211], SWCNT crosses on substrates [88, 212] as well as doping of CNT structures. The capability to produce n-type transistors is important technologically, as it allows the fabrication of CNT-based complementary logic devices and circuits. Experiments have shown that p- to n-type conversion of the CNT-FETs can be made either by doping the surface of the tube using alkali metals or by simply annealing the device in vacuum or in an inert gas. By masking a portion of the nanotube and doping the remaining part with potassium vapor creates an n-type transistor. By combining the p- and the n-type transistors, complementary inverters are created. n-type FETs obtained by annealing in hydrogen the CNT-devices having zirconium oxide thin-films as gate dielectric, exhibit subthreshold swings of 90 mV per decade. High voltage gains of up to 60 are obtained for complementary nanotube-based inverters [209].

The natural p-doping of SWCNT based field effect transistors [85, 169], as well as the positive thermopower of SWCNT samples [213], gave rise to speculations regarding possible doping by atmospheric gases including oxygen and water. Although earlier experimental and theoretical investigations supported the charge transfer by adsorbed oxygen as a potential source of p-doping for single-wall carbon nanotubes [213–215], recent studies indicate that the main role of oxygen is not to dope the CNT but to modify directly the line-up of the CNT bands at the metal–nanotube junction [206, 216]. In contrast to doping, that changes the barrier thickness and introduces significant shifts of the threshold voltage of the device, absorption at the metal–nanotube junction or removal of the adsorbed oxygen by vacuum annealing results in the direct modification of the Schottky barrier height at the contacts. However, the electrical characteristics of n-type CNT-FETs produced

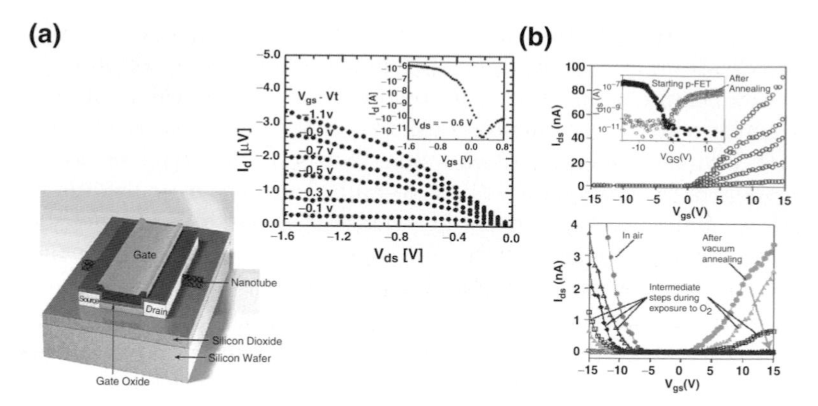

Fig. 5.37. (**a**) Left: schematic drawing of a top-gated CNT-FET architecture. Right: characteristic curves of a p-type CNT-FET. (**b**) Top: transformation of a p-type CNT-FET into a n-type CNT-FET by annealing in vacuum and t he eventually corresponding characteristic curves. Bottom: the inverse transformation, from n-type to p-type, by exposing the vacuum annealed n-type FET to oxygen. Reprinted in part with permission from [206]. Copyright 2002 American Chemical Society

by both methods, i.e. alkali metal doping or vacuum annealing, are rather similar although the mechanisms involved are quite different.

Chico et al. [217] proposed that by joining SWCNTs having different diameters and helicities via topological defects novel intramolecular devices may be obtained. The simplest way is to introduce pairs of heptagons and pentagons in an otherwise perfect hexagonal lattice structure. The resulting junction present a bend angle and can act like a rectifying diode. Yao et al. [87] investigated such device experimentally and observed a sharp kink on an individual SWCNT. Electron transport across the kink exhibited rectifying behavior (See Fig. 5.17). The system was suggested to contain a metallic tube joining a semiconducting tube with a Schottky junction formed at the kink [87]. This result demonstrated that extremely small electrical devices are obtainable on an individual molecular wire.

Heterojunctions with more than two terminals can be created as above by connecting different nanotubes through junctions welded by topological defects [211] or by crossing nanotubes over each other in order to simply form physically touching junctions [212]. The difference in the two approaches is the nature of the junction forming the device. In the first case, a stable junction is formed between the nanotubes which are welded through chemical bonds, giving rise to a variety of possible networks and structural shapes for switching, logic and transistor applications. Indeed, novel structures of carbon nanotube "T-" and "Y-junctions," as models of 3-terminal electronic devices, have been proposed [211, 218], while template-based chemical vapor deposition (CVD) [219] and pyrolysis of an organometallic precursor with nickelocene and thiophene [220] for the reproducible and high-yield fabrication of multi-wall carbon nanotube Y-junctions have been developed. Electrical measurements on Y-junction CNTs show intrinsic nonlinear and asymmetric I-V behavior with rectification at room temperature. These Y-junction nanotubes are nanoscale molecular rectifying switch with a robust behavior that is reproducible in a high-yield fabrication method [218, 220, 221]. Moreover, simulations propose that these molecular devices can function as 3-terminal bi-stable switches controlled by a "gate" voltage applied at a branch terminal [222], working as logic "OR" or "XOR" as well. The possible reasons for rectification in Y-junctions include constructive or destructive interference of the electronic wave functions through two different channels at the location of the junction; hence the rectification is strongly influenced by the structural asymmetry across the two branches in a junction [222].

In the second case, when two tubes cross, the junction is simply through a physical contact and will be affected by changes in the nature of the contact. The main applications of this second category will be in electromechanical bistable switches and sensors [212]. The bistable switches are bits in a CNT-based computing architecture. Fuhrer and coworkers have studied the electrical properties of crosses of SWCNTs [88]. Electron transport from metallic to metallic, metallic to semiconducting, and semiconducting to semiconducting tubes was investigated. Metallic and semiconducting tubes were found to form Schottky junctions at the crosses. A Schottky junction is typically formed at the interface between a bulk semiconductor and a pure metal. The difference in work function between the two materials causes energy band-bending in the semiconductor side. Characteristic of a Schottky junction is rectifying behavior in the current vs. voltage curves. Current flows across the junction only when the metal side is negatively biased. Tombler et al. [92] studied SWCNT crosses by using scanning probes as gates to identify the metallic or semiconducting nature of the crossing nanotubes.

Bachtold et al [224] reported logic circuits with CNT transistors which show high gain (>10), a large on-off ratio (>105) and room temperature operation. Their demonstration included operations such as an inverter, a logic NOR, a static random-access memory cell and an ac ring oscillator. Inter- and intra-molecular logic gates based on SWCNTs have been obtained also by Derycke et al. [216]

From the above discussion is clear that CNTs are currently one of the most promising materials for molecular electronics, however, many challenges remain before they can become a successful technology. Most of these challenges involve the synthesis, separation, and self-assembly of CNTs. Local catalytic growth and alignment of CNTs by electric fields or microfluidic techniques have already been demonstrated in the laboratory as described in Sect. 5.5.3. New types of synthetic approaches that operate at lower temperatures may be more selective. Selectivity could also be obtained using seeded growth.

Fabrication of nanotube-based 3-terminal devices requires placing nanotubes between the metal electrodes. While earlier works simply transplanted

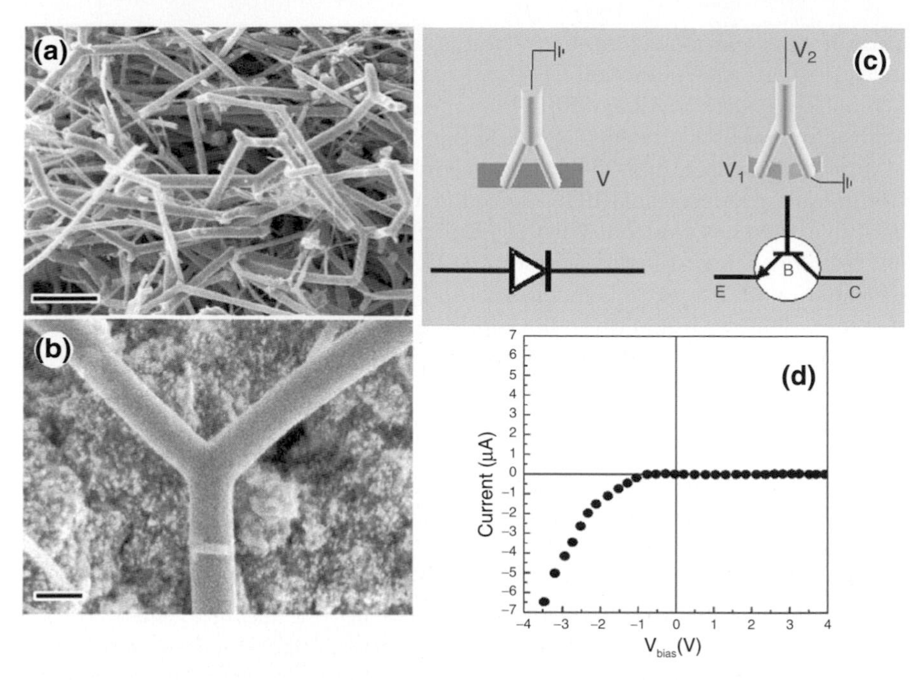

Fig. 5.38. (**a**) Low-magnification SEM image showing high yield of carbon nanotube Y-junctions (scale bar=1 μm). (**b**) High-magnification SEM image showing neat smooth surface of a Y-junction nanotube with arms of 200 nm in diameter (scale bar=200 nm). (**a**) and (**b**) reprinted with permission from [223]. Copyright 2001, American Institute of Physics. (**c**) Schematic architecture of a CNT-diode and a CNT-transistor based on Y-junction nanotubes. (**d**) I-V characteristic of a Y-junction CNT. Reprinted with permission from [219]. Copyright 2000 by American Physical Society

a nanotube from bulk material, subsequent approaches have used CVD to bridge the electrodes with a nanotube [225,226]. In this regard, the successful synthesis of Y-junctions CNT is a big step forward [219–221]. Recent advances in the solubilization of CNTs have made possible the application of liquid-phase chemistry and chemical spectroscopy in the modification and characterization of the structure of CNTs. Appropriate functionalization of the nanotubes can help achieve the self-assembly of CNT circuits and selectivity. It has been shown that single-strained DNA forms stable complexes with CNTs which effectively disperse the CNTs in acqueous solution [227]. Furthermore, a particular sequence of single-strained DNA wraps around CNTs in such a way that the electrostatic properties of the DNA–CNT system depend on tube type, enabling CNT separation and selectivity. The separation is bimodal: the obtained fractions are richer of small-diameter and metallic tubes or larger-diameter and semiconducting tubes [228].

5.6.2 Chemical Sensors

The carbon nanotube electronic and transport properties are predicted to be extremely sensible to atomic structure, mechanical deformations or chemical doping [174–177]. When chemicals in the surrounding environment bond to the tube, the absorbed molecules may act as dopants, shifting the Fermi energy of the nanotube, or may change the band structure of the tube due to the orbital hybridizations for bond formation, possibly influencing the conducting properties. Since such changes can be easily detected when the nanotube are embedded in electronic devices, the large surface area and hollow geometry of CNTs make these nanostructures extremely promising candidates for miniaturized sensors having high sensitivity and fast response time to the chemical environment. These are important advantages for sensing applications in fields like pollution monitoring, space exploration, agricultural and medical diagnostics and hazardous gas detection. Therefore, CNT-based chemical, biological and physical sensors are the objectives of a significant research endeavor. These efforts can be roughly divided into two classes: one that exploits certain property of the pure nanotube (such as change in conductivity and/or electronic properties with gas adsorption or tube deformation) and the second that relies on the ability to functionalize the nanotube (tip and/or side wall) with molecular groups or metal atoms that serve as sensing elements. In the second case the challenge is to find a way to reproducibly and reliably alter the chemical reactivity of the nanotubes that, like graphite, are expected to be relatively unreactive.

However, Kong et al. [176] and Goldoni et al. [229] have found that pure carbon nanotubes can be used for miniature chemical sensors to detect small concentrations of toxic gas molecules with high sensitivity. Chemical sensors based on individual or ensembles of pure single-wall nanotubes can detect hazardous chemicals such as NO_2 [176,229], ammonia (NH_3) [176,229], SO_2 [229] and NO [229]. For a semiconducting single-wall nanotube exposed to 200 ppm of NO_2, it was found that the electrical conductance increases by up to three orders of magnitude in a few seconds [176]. On the other hand, exposure to 2% NH_3 caused the conductance to decrease by up to two orders of magnitude [176]. Gas sensors dedicated to toxic gases typically use semiconducting metal oxides (SnO_2 for example) and conducting polymers. As a general comparison, conventional solid-state sensors for NO_2 and NH_3 operate at temperatures over 400°C, while conducting polymers provide only limited sensitivity. SWCNT sensors may operate at room temperature and have response times of about 10 s. This is an advantage over metal-oxide sensors, which require high temperatures, and polymer sensors which have poor sensitivities and slow (10 min.) response times. Difficulties in manufacturing, however, are the main disadvantage of nanotube sensors. Another drawback for deploying CNT-sensors is the slow recovery of the CNTs to the initial state. In the experiment of Kong et al. [176], it took up to 12 hours for the conductivity of the CNTs to return to the original value after the

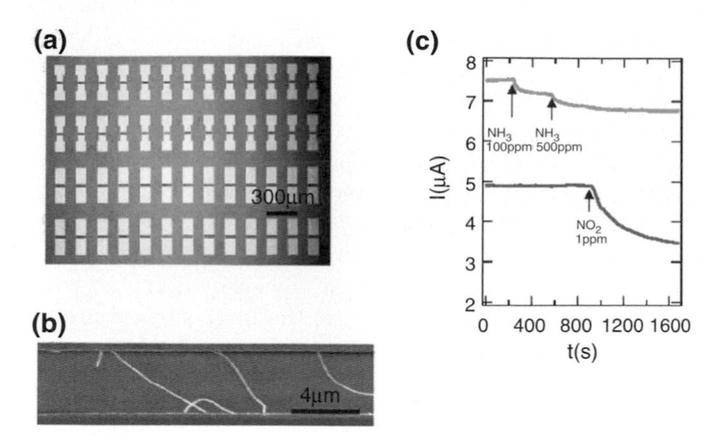

Fig. 5.39. (**a**) Optical image of an array of multiple Mo electrodes for SWCNT devices. (**b**) Scanning electron microscopy (SEM) image of several nanotubes bridging two opposing Mo electrodes in a device. The nanotubes appear bright and the apparent widths do not correspond to the true diameter of the nanotubes (average diameter=2 nm as measured by AFM and TEM). (**c**) Top curve: a device coated with Nafion exhibits response to 100 and 500 ppm of NH_3 in air, and no response when 1 ppm of NO_2 was introduced to the environment. Bottom curve: a PEI-coated device exhibiting no response to 100 and 500 ppm of NH_3 and large conductance decrease to 1 ppm of NO_2. Reprinted in part with permission from [230]. Copyright 2003 American Chemical Society

source gas is withdrawn. Heating the sensor to 200°C reduced this period to about one hour. Arrays of electrical devices, each made of multiple SWCNTs bridging metal electrodes have been obtained by chemical vapor deposition of nanotubes across prefabricated electrode arrays [230]. The ensemble of nanotubes in such a device collectively exhibits large electrical conductance changes under electrostatic gating, owing to the high percentage of semiconducting nanotubes. This leads to the fabrication of large arrays of low-noise electrical nanotube sensors with 100% yield for detecting gas molecules. Polymer functionalization is used to impart high sensitivity and selectivity to the sensors. Polyethyleneimine coating affords n-type nanotube devices capable of detecting NO_2 at less than 1 ppb (parts-per-billion) concentrations while being insensitive to NH_3. Coating Nafion (a polymeric perfluorinated sulfonic acid ionomer) on nanotubes blocks NO_2 and allows for selective sensing of NH_3. Multiplex functionalization of a nanotube sensor array is carried out by microspotting. Detection of molecules in a gas mixture is demonstrated with the multiplexed nanotube sensors [230].

A gas and organic vapor sensor, fabricated by the simple casting of SWCNTs on an interdigitated electrode, with detection limits of a few hundred of ppb at room-temperature, has been proposed by J. Li et al. [231]. Gao et al. [232] demonstrated that the use of multi-wall aligned carbon nanotubes coated with inherently conducting polymers provides a novel electrode platform for biosensors. These three-dimensional electrodes offer advantages in terms of large enzyme loadings within an ultra-thin layer, sensitivity and selective sensing. Electronic detection of a specific protein (streptavidin) has been obtained using a nanotube FET device [233]. An electrochemical DNA biosensor with a favorable performance for rapid detection of specific hybridization has been obtained by functionalizing CNTs with a carboxylic acid group [234].

A major role in understanding and optimizing the gas sensing behavior of nanotubes is played by computational investigations, which should precisely define the chemical reactivity, the bond geometry and predict the effect of gas-phase adsorption on the electronic characteristics nanotubes [177, 235–238]. First-principles calculations investigating the interaction of several representative molecules (NO_2, CO, NH_3, O_2, H_2 and H_2O) with SCWNTs have been performed using density functional theory. Among all the investigated molecules, there are at least three clear cases, which agree with experimental results.

The first case is the interaction of NO_2 with SWCNTs. All of calculations suggest that NO_2 binds to SWCNTs with an adsorption energy in the range 0.2–0.3 eV. Electron density analysis, according to the strong oxidizing nature of NO_2, shows that charge transfer is induced from C atom to the molecule leading to hole (or p-type) doping of nanotube. For example, in semiconducting (10,0) nanotubes the increase in hole carriers is responsible for the increase of conductance shown in Fig. 5.40.

The second one is CO. The calculations for the interaction of CO with SWCNTs show no binding, indicating that a perfect bare nanotube may not work as a CO sensor, and this is in agreement with the experimental results [229]. Based on the concept of high binding sites on SWCNT surface defects, it has been proposed the design of a new type of nanoscale sensor device developed by substitutional doping of impurity atoms (such as Boron or Nitrogen atoms) into intrinsic single-wall carbon nanotubes, or by using composite $B_xC_yN_z$ nanotubes. First-principle calculations suggest that these sensor devices can not only detect the presence of CO, but also the sensitivity of these devices can be controlled by the doping level of impurity atoms in a nanotube [237].

The third clear case is H_2O. The simulated molecular configurations show repulsive interaction, and no charge transfer is observed when the water molecule is placed close to the nanotube surface. This result is consistent with the experimental observation that the nanotube conductance and the electronic states are not significantly changed even when the nanotube is fully immersed in water or covered by ice. These findings foresee the important possibility of using nanotubes as biochemical sensors operating in water at physiological temperature. However, nanotube humidity sensors could be explored with

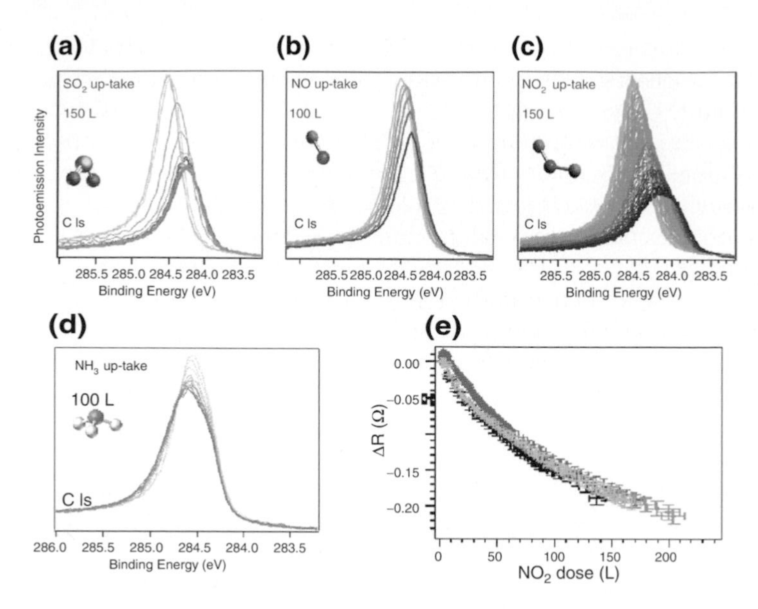

Fig. 5.40. Effect of the exposure to SO_2 (**a**), NO (**b**), NO_2 (**c**) and NH_3 (**d**) on the C 1s photoemission spectra of a clean SWCNT bucky-paper at 150 K. (**e**) SWCNT bucky-paper resistance variation as a function of NO_2 dose at 150 K in four different experiments after recovering the initial conditions of the clean bucky-paper (by annealing in ultra-high vacuum at 800 K). Adapted from [229]

suitable water-sensitive coatings as suggested by Kim et al. [207] or by using composite $B_xC_yN_z$ nanotubes [237].

Another way to exploit the high sensitivity of the electronic properties of SWCNTs to detect molecules that have a weak or repulsive interaction with them is the creation of nanotube-metal clusters (atoms) assemblies. Theoretical works [239] and experimental investigation [229,240] indicate that metallic clusters and atoms strongly interact with the tubes producing significant charge reorganization. Nardelli et al. [239], using NH_3 as a prototypical gas species and Al clusters, have performed first-principles quantum transport calculations predicting that gas adsorption onto the metal cluster drastically changes the electronic properties, for both metallic and semiconducting nanotubes. The electrical resistance of a metallic CNT–Al cluster assembly increases dramatically after the adsorption of NH_3 on the cluster, while the opposite occurs for a semiconducting assembly. The observed behavior can be understood in terms of interactions between the molecular species and the nanotube-cluster system, where successive charge transfers between components tailor the electronic and transport properties. Since the chemistry of adsorption for certain metal clusters is sufficiently well known, they can be easily engineered as receptors for specific molecules in nanotube based devices. This can be particularly useful for H_2O and oxygen detection. In fact, apart some studies that found stable absorption configurations of O_2 on pure SWCNTs, with a sizable charge transfer from the tube to O_2 in agreement with earlier experiments reporting large conductance changes upon exposure to oxygen, there is much experimental [229,241,242] and theoretical [238,243] evidence that O_2 does not bind to perfect SWCNTs above 50 K unless metal particles or alkali metals are already bonded to the tubes.

Finally, CN_x nanotubes have been demonstrated to sensing rapidly (0.1–0.5 s response time) relatively low concentrations of toxic gases, and organic vapors. It has been suggested that CN_x nanotubes could be even more efficient than carbon nanotubes for monitoring toxic species due to the presence of highly reactive pyridine-like sites on the tube surface [244].

5.6.3 Field Emission

The geometric properties of nanotubes such as the high aspect ratio and the small radius of curvature of the tip, coupled with the extraordinary mechanical strength and chemical stability, make them an ideal candidate for electron field emitters. CNT field emitters have several industrial and research applications: flat panel displays, outdoor displays, traffic signals and electron microscopy. de Heer et al. [170] demonstrated the first high intensity electron gun based on field emission from a film of nanotubes. A current density of $0.1\,mA/cm^2$ was observed for voltages as low as 200 V. For comparison most conventional field emitter displays operate at 300–5000 V whereas cathode ray tubes use 30000 V. Following this early result, several groups have studied the emission characteristics of SWCNTs and MWCNTs [170, 245–250].

A typical field emission test apparatus consists of a cathode and anode enclosed in an evacuated cell at a vacuum of $10^{-9} - 10^{-8}$ Torr. The cathode could consist of a glass or PTFE (polytetrafluoroethylene) substrate with metal-patterned lines where a film of nanotubes can be transplanted after the arc-grown or laser oven material is purified. Nanotubes can be also directly grown on the cathode using CVD or plasma CVD methods, or using SiC as cathode by direct annealing of the SiC substrate at high temperature [251]. The anode operating at positive potentials is placed at a distance of 20–500 μm from the cathode. The turn-on field, arbitrarily defined as the electric field required for generating 1 nA, can be as small as 1.5 V/μm. The threshold field (the electric field needed to yield a current density of $10\,mA/cm^2$) is in the range of 5–8 V/μm. At low emission levels, the emission behavior follows the Fowler–Nordheim relation, i.e., the plot of $\ln\,(I/V^2)$ vs. $\ln\,(1/V)$ is linear. The emission current significantly deviates from the F–N behavior in the high field region and indeed, emission current typically saturates. While most works report current densities of $0.1–100\,mA/cm^2$, very high current densities up to $4\,A/cm^2$ have been reported by Zhu et al. [248].

The next major commercial use of nanotubes probably will be that of SWCNTs in field-emission flat-panel displays. Working full color flat panel

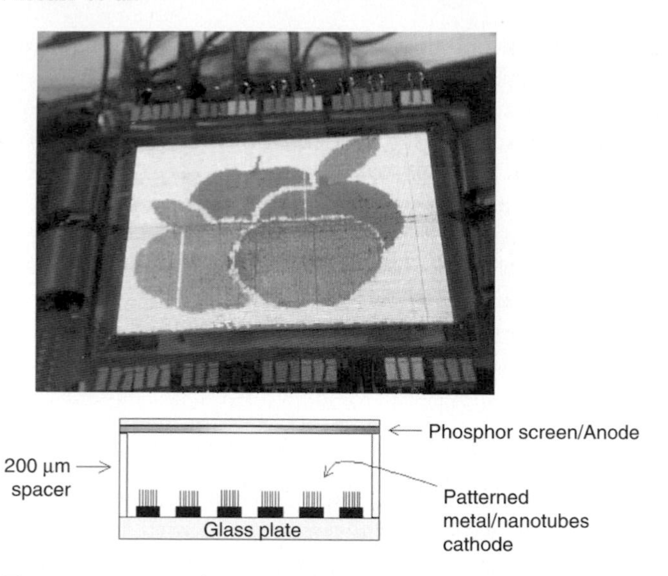

Fig. 5.41. Emission image of a 9 inch CNT field-emission-display from Samsung (top) and schematic drawing of an addressable CNT field-emission-display (bottom). Reprinted from [172], copyright 2001, with permission from Elsevier

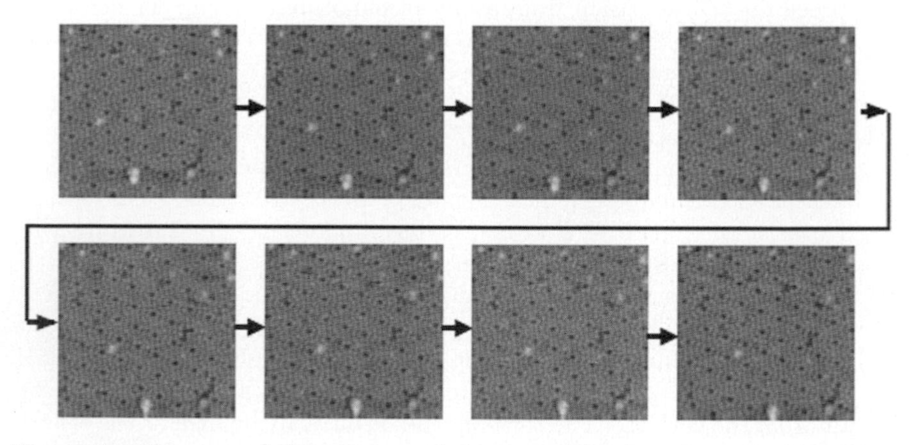

Fig. 5.42. Sequence of STM images of the 7×7-Si(111) surface obtained with a CNT tip. Reprinted from [265], copyright 2001, with permission from Elsevier

displays (FPD) and CRT-lighting elements have been demonstrated by research groups from Japan and Korea in collaboration with major electronic brands [246,250]. In the case of FPD, the anode structure consists of a glass substrate with phosphor coated indium tin oxide stripes. The anode and cathode are positioned perpendicular to each other to form pixels at their intersections. Appropriate phosphors such as Y_2O_2S: Eu, ZnS: Cu, Al, and

ZnS: Ag, Cl are used at the anode for red, green and blue colors respectively. A 4.5 inch display showed a uniform and stable image over the entire 4.5 inch panel. For lighting elements, phosphor screen is printed on the inner surface of the glass and backed by a thin Al film (about 100 nm) to give electrical conductivity. A lifetime test of the lighting element suggests a lifetime of over 10000 hours. The advantages over standard liquid-crystal displays include lower power consumption, wider viewing angles, higher brightness and resolution. The full-color display prototyped by Samsung is expected to reach the market in the late 2006.

5.6.4 CNTs in Microscopy

Scanning probe microscopes, such as Atomic Force Microscopes (AFM) and Scanning Tunneling Microscopes (STM), are widely used now by the research community to image, characterize and manipulate surfaces, atoms and molecules and also by the semiconductor industry as a metrology tool. A serious problem in scanning probe microscopy is the availability of tips having well-known structural and electrical properties as well as good mechanical and chemical stability. For example, a typical AFM probe consists of a silicon or silicon nitride cantilever with a pyramidal-shaped tip. This tip can be made as small as 10–20 nm offering reasonable resolution. However, the large cone angle of this tip (30–35 deg.) makes it difficult for probing narrow and deep features. Another serious drawback is that the tip is brittle thus limiting its use in applications; either the tip breaks after only a limited use or becomes blunt.

Carbon nanotubes have nanometer scale diameter and high aspect ratio, are robust and stable under a wide range of applied physical conditions and can be functionalized. For these reasons, CNT tips have become an attractive alternative in scanning probe microscopy.

CNTs probes can be manually glued to the apex tip of an AFM cantilever using acrylic adhesives [252, 253] or attached via electron beam deposition of amorphous carbon [254,255]. Alternatively, carbon nanotubes can be directly grown on the tip via CVD methods [256–258]. In a well-characterized CVD process, it may be possible to control the length of the probe by selecting the growth time. On the other hand, it is possible to shorten a long tip to the desired length by the application of an electric field to etch away the carbon atoms.

AFM carbon nanotube tips prepared as above described, have demonstraded high resolution and superb stability with no damages for both tip and sample and consequent long lifetime, in imaging biological material [257,259], carbon nanotubes [260] and deep sharp grooves [252, 259, 261], as well as in nanolithography patterning [262]. CNT atomic force microscopy tips have been also successfully applied as chemical sensitive probes of O_2, N_2 and H_2 via functionalization of the CNT by -COOH [263, 264].

The usefulness of carbon nanotube as STM tips with atomic resolution, has been demonstrated in air on chemically inactive surfaces, as the layered 1T-TaS2 compound [252], and in ultra-high-vacuum on the 7x7-Si(111) surface [265].

References

1. S. Iijima: Nature **354**, 56 (1991)
2. H.W. Kroto, J.R. Heath, S.C. O'Brien, R.F. Curl, R. Smalley: Nature **318**, 162 (1985)
3. Z.Y. Kosakovskaya, L.A. Chernozatonskim, E.A. Fedorov: JETP Lett. **56**, 26 (1992)
4. S. Iijima, T. Ichihashi: Nature **363**, 603 (1993)
5. D.S. Bethune, C.H. Kiang, M.S. de Vries, G. Gorman, R. Savoy, J. Vazquez, R. Beyers: Nature **363**, 605 (1993)
6. M.S. Dresselhaus, G. Dresselhaus, R. Saito: Phys. Rev. B **45**, 6234 (1992)
7. R. Saito, M. Fujita, G. Dresselhaus, M.S. Dresselhaus: Appl. Phys. Lett. **60**, 2204 (1992)
8. N. Hamada, S. Sawada, A. Oshiyama: Phys. Rev. Lett. **68**, 1579 (1992)
9. J.W.G. Wildoer, L.C. Venema, A.G. Rinzler, R.E. Smalley, C. Dekker: Nature **391**, 59 (1998)
10. T.W. Odom, J.L. Huang, P. Kim, C.M. Lieber: Nature **391**, 62 (1998)
11. R.A. Jishi, M.S. Dresselhaus, G. Dresselhaus: Phys. Rev. B **47**, 16671 (1993)
12. M.S. Dresselhaus, G. Dresselhaus, R. Saito: Carbon **33**, 883 (1995)
13. P.J.F. Harris: *Carbon Nanotubes and Related Structures: New Materials for the Twenty-First Century*, 1st edn (Cambridge University Press, Cambridge 1999) pp. 68–70
14. M. Fujita, R.Saito, G. Dresselhaus, M.S. Dresselhaus: Phys. Rev. B **45**, 13834 (1992)
15. M.S. Dresselhaus, G. Dresselhaus, P.C. Eklund: J. Mater. Res. **8**, 2054 (1992)
16. P.J.F. Harris: *Carbon Nanotubes and Related Structures: New Materials for the Twenty-First Century*, 1st edn (Cambridge University Press, Cambridge 1999) pp. 92–95
17. P.M. Ajayan, S. Iijima: Nature **358**, 23 (1992)
18. P.C. Eklund, J.M. Holden, R.A. Jishi: Carbon **33**, 959 (1995)
19. R. Saito, G. Dresselhaus, M.S. Dresselhaus: J. Appl. Phys **73**, 494 (1993)
20. L.F. Sun, S.S. Xie, W. Liu, W.Y. Zhou, Z.Q. Liu, D.S. Tang, G. Wang, L.X. Qian: Nature **403**, 384 (2000)
21. L.C. Qin, X. Zhao, K. Hirahara, Y. Miyamoto, Y. Ando, S. Iijima: Nature **408**, 50 (2000)
22. H.Y. Peng, N. Wang, Y.F. Zheng, Y. Lifshitz, J. Kulik, R.Q. Zhang, C.S. Lee, S.T. Lee: Appl. Phys. Lett. **77**, 2831 (2000)
23. N. Wang, Z.K. Tang, G.D. Li, J.S. Chen: Nature **408**, 50 (2000)
24. P. Launois, R. Moret, D. Le Bolloc'h, P.A. Albouy, Z.K. Tang, G.L. Chen: Solid State Commun. **116**, 99 (2000)
25. S. Iijima: Mater. Sci. Eng. B **19**, 172 (1993)
26. P.M. Ajayan, T.W. Ebbesen: Rep. Prog. Phys. **60**, 1025 (1997)

27. S. Iijima, T. Ichihashi, Y. Ando: Nature **356**, 776 (1992)
28. S. Iijima, P.M. Ajayan, T. Ichihashi: Phys. Rev. Lett. **69**, 3100 (1992)
29. J.W. Mintmire, C.T. White: Carbon **33**, 893 (1995)
30. R. Saito, G. Dresselhaus, M.S. Dresselhaus: *Physical Properties of Carbon Nanotubes*, 1st edn (Imperial College Press, London 1998) p. 25
31. P.R. Wallace: Phys. Rev. **71**, 622 (1947)
32. J.C. Slater, G.F. Koster: Phys. Rev. **94**, 1498 (1954)
33. M.S. Dresselhaus, G. Dresselhaus: Adv. Phys. **30**, 139 (1981)
34. M. Ouyang, J.-L. Huang, C.M. Lieber: Acc. Chem. Res. **35**, 1018 (2002)
35. R. Saito, M. Fujita, G. Dresselhaus, M.S. Dresselhaus: Phys. Rev. B **46**, 1804 (1992)
36. R. Saito, G. Dresselhaus, M.S. Dresselhaus: *Physical Properties of Carbon Nanotubes*, 1st edn (Imperial College Press, London 1998) p. 62–63
37. X. Blase, L.X. Benedict, E.L. Shirley, S.G. Louie: Phys. Rev. Lett. **72**, 1878 (1994)
38. J.W. Mintmire, C.T. White: Appl. Phys. A **67**, 65 (1998)
39. C.L. Kane, E.J. Mele: Phys. Rev. Lett. **78**, 1932 (1997)
40. S. Reich, C. Thomsen, P. Ordejn: Phys. Rev. B **65**, 155441 (2002)
41. M. Ouyang, J.-L. Huang, C.L. Cheung, C.M. Lieber: Science **292**, 702 (2001)
42. T. Wang, J.-L. Huang, P. Kim, C.M. Lieber: J. Phys. Chem. B **104**, 2794, (2000)
43. R. Saito, G. Dresselhaus, M. Dresselhaus: J. Appl. Phys. **73**, 494 (1993)
44. S. Roache, F. Triozon, A. Rubio, D. Mayour: Phys. Rev. Lett. B **64**, 121401 (2001)
45. S.M. Bachilo, L. Balzano, J.E. Herrera, F. Pompeo, D.E. Resasco, R.B. Weisman: J. Am. Chem. Soc. **125**, 11186 (2003)
46. C.N.R. Rao, B.C. Satishkumar, A. Govindaraj, M. Nath: ChemPhysChem **2**, 78 (2001)
47. J. Kong, H.T Soh, A.M. Cassell, C.F. Quate, H. Dai: Nature **395**, 878 (1998)
48. L. Delzeit, B. Chen, A.M. Cassel, R. Stevens, C. Nguyen, M. Meyyappan: Chem. Phys. Lett. **348**, 368 (2001)
49. J. Li, C. Papadopoulos, J.M. Xu, M. Moskovits: Appl. Phys. Lett. **75**, 367 (1999)
50. M. Meyyappan, L. Delzeit, A. Cassell, D. Hash: Plasma Sources Sci. Technol. **12**, 205 (2003)
51. J. Liu, A.G. Rinzler, H. Dai, J.H. Hafner, R.K. Bradley, P.J. Boul, A. Lu, T. Iverson, K. Shelimov, C.B. Huffman, F. Rodriguez-Macias, Y.-S. Shon, T.R. Lee, D.T. Colbert, R.E. Smalley: Science **280**, 1253 (1998)
52. J.L.Zimmerman, R.K. Bradley, C.B. Huffman, R.H. Hauge, J.L. Margrave: Chem. Mater. **12**, 1361 (2000)
53. I.W. Chiang, B.E. Brinson, R.E. Smalley, J.L. Margrave, R.H. Hauge: J. Phys. Chem. B **105**, 1157 (2001)
54. I.W. Chiang, B.E. Brinson, A.Y. Huang, P.A. Willis, M.J. Bronikowski, J.L. Margrave, R.E. Smalley, R.H. Hauge: J. Phys. Chem. B **105**, 8297 (2001)
55. D. Chattopadhyay, I. Galeska, F. Papadimitrakopoulos: Carbon **40**, 985 (2002)
56. R. Sen, S.M. Rickard, M.E. Itkis, R.C. Haddon: Chem. Mater. **15**, 4273 (2003)
57. S. Banerjee, S.S. Wong: J. Phys. Chem. B **106**, 12144 (2002)
58. M. Prato, D.M. Guldi, A. Kukovecz, H. Kuzmany: J. Am. Chem. Soc. **124**, 14318 (2002)

59. A. Graham, F. Kreupl, M. Liebau, W. Hoenlein: Curr. Appl. Phys. **2**, 107 (2002)
60. J.N. Coleman, A.B. Dalton, S. Curran, A. Rubio, A.P. Davey, A. Drury, B. McCarthy, B. Lahr, P.M. Ajayan, S. Roth, R.C. Barklie, W.J. Blau: Adv. Mater. **12**, 213 (2000)
61. A.B. Dalton, C. Stephan, J.N. Coleman, B. McCarthy, P.M. Ajayan, S. Lefrant, P. Bernier, W.J. Blau, H.J. Byrne: J. Phys. Chem. B **104**, 10012 (2000)
62. R. Murphy, J.N. Coleman, M. Cadek, B. McCarthy, M. Bent, A. Drury, R.C. Barklie, W.J. Blau: J. Phys. Chem. B **106**, 3087 (2002)
63. R. Bandyopadhyaya, E. Nativ-Roth, O. Regev, R. Yerushalmi-Rozen: Nano Lett. **2**, 25 (2002)
64. M.J. O'Connell, P. Boul, L.M. Ericson, C. Huffman, Y. Wang, E. Haroz, C. Kuper, J. Tour, K.D. Ausman, R.E. Smalley: Chem. Phys. Lett. **342**, 265 (2001)
65. S. Bandow, A.M. Rao, K.A. Williams, A. Thess, R.E. Smalley, P.C. Eklund: J. Phys. Chem. B **101**, 8839 (1997)
66. K.B. Shelimov, R.O. Esenaliev, A.G. Rinzler, C.B. Huffman, R.E. Smalley: Chem. Phys. Lett. **282**, 429 (1998)
67. T. Abatemarco, J. Stickel, J. Belfort, B.P. Frank, P.M. Ajayan, G. Belfort: J. Phys. Chem. B **103**, 3534 (1999)
68. G.S. Duesberg, M. Burghard, J. Muster, G. Philipp, S. Roth: Chem. Commun. 435 (1998)
69. S. Niyogi, H. Hu, M.A. Hamon, P. Bhowmik, B. Zhao, S.M. Rozenzhak, J.V Chen, M.E. Itkis, M.S. Meier, R.C. Haddon: J. Am. Chem. Soc. **123**, 733 (2001)
70. B. Zhao, H. Hu, S. Niyogi, M.E. Itkis, M.A. Hamon, P. Bhowmik, M.S. Meier, R.C. Haddon: J. Am. Chem. Soc. **123**, 11673 (2001)
71. S.K. Doorn, R.E. Fields, H. Hu, M.A. Hamon, R.C. Haddon, J.P. Selegue, V. Majidi: J. Am. Chem. Soc. **124**, 3169 (2002)
72. B. Chen, J.P. Selegue: Anal. Chem. **74**, 4774 (2002)
73. M. Callejas, A.M. Benito, W.K. Maser, M. Cochet, J.M. Andres, J. Schreiber, O. Chauvet, J.L.G. Fierro: Chem. Commun. 1000 (2002)
74. E.Vazquez, V. Georgakilas, M. Prato: Chem. Commun. 2308 (2002)
75. A.R. Harutyunyan, B.K. Pradhan, J. Chang, G. Chen, P.C. Eklund: J. Phys. Chem. B **106**, 8671 (2002)
76. L. Thien-Nga, K. Hernadi, E. Ljubovic, S. Garaj, L. Forro: Nano Lett. **2**, 1349 (2002)
77. L. Langer, V. Bayont, E. Grivei, J.-P. Issi, J.P. Heremans, C.H. Olk, L. Stockman, C. Van Heasendonck, Y. Bruynseraede: Phys. Rev. Lett. **76**, 479 (1996)
78. T.W. Ebbesen, H.J. Lezec, H. Hiura, J.W. Bennett, H.F. Ghaemi, T. Thio: Nature **382**, 54 (1996)
79. H. Dai, E.W. Wong, C.M. Lieber: Science **272**, 523 (1996)
80. A.Y. Kasumov, H. Bouchiat, B. Reulet, O. Stephan, I.I. Khodos, Y.B. Gorbatov, C. Colliex: Europhys. Lett. **43**, 89 (1998)
81. M. Ahlskog, E. Seynaeve, R.J.M. Vullers, C. Van Heaesendonk: J. Appl. Phys. **85**, 8432 (1999)

82. A. Thess, R. Lee, P. Nikolaev, H. Dai, P. Petit, J. Robert, C. Xu, Y.H. Lee, S.G. Kim, A.G. Rinzler, D.T. Colbert, G.E. Scuseria, D. Tomnek, J.E. Fischer, R.E. Smalley: Science **273**, 483 (1996)
83. M. Bockrath, D.H. Cobden, P.L. McEuen, N.G. Chopra, A. Zettl, A. Thess, R.E. Smalley: Science **275**, 1922 (1997)
84. S.J. Tans, M.H. Devoret, H. Dai, A. Thess, R.E. Smalley, L.J. Geerligs, C. Dekker: Nature **386**, 474 (1997)
85. S.J. Tans, A.R.M. Verschueren, C. Dekker: Nature **393**, 49 (1998)
86. R.D. Antonov, A.T. Johnson: Phys. Rev. Lett. **83**, 3274 (1999)
87. Z. Yao, H.W.C. Postma, L. Balents, C. Dekker: Nature **402**, 273 (1999)
88. M.S. Fuhrer, J. Nygrd, L. Shih, M. Forego, Y.-G. Yoon, M-S.C. Mazzoni, H.J. Choi, J. Ihm, S.G. Louie, A. Zettl, P.L. McEuen: Science **288**, 494 (2000)
89. M.R. Falvo, R.M. Taylor II, A. Helser, V. Chi, F.P. Brooks Jr, S. Washburn, R. Superfine: Nature **397**, 326 (1999)
90. T. Hertel, R. Martel, P. Avouris: J. Phys. Chem. B **102**, 910 (1998)
91. P. Avouris, T. Hertel, R. Martel, T. Schmidt, H.R. Shea, R.E. Walkup: Appl. Surf. Sci. **141**, 201 (1999)
92. T.W. Tombler, C. Zhou, L. Alexseyev, J. Kong, H. Dai, L. Liu, C. S. Jayanthi, M. Tang, S.-Y. Wu: Nature **405**, 769 (2000)
93. L. Roschier, J. Penttil, M. Martin, P. Hakonen, M. Paalanen, U. Tapper, E.I. Kauppinen, C. Journet, P. Bernier: Appl. Phys. Lett. **75**, 728 (1999)
94. P.G. Collins, H. Hersam, M. Arnold, R. Martel, P. Avouris: Phys. Rev. Lett. **86**, 3128 (2001)
95. P.G. Collins, M.S. Arnold, P. Avouris: Science **292**, 706 (2001)
96. Y. Huang, X. Duan, Q. Wei, C.M. Lieber: Science **291**, 630 (2001)
97. Herbert A. Pohl: *Dielectrophoresis* (Cambridge University Press, Cambridge, 1978)
98. K. Yamomoto, S. Akita, Y. Nakayama: J. Phys. D: Appl. Phys. **31**, L34 (1998)
99. V.T.S. Wong, W.J. Li: In: *The Sixteenth IEEE International Conference on Micro Electro Mechanical Systems (MEMS 2003) at Kyoto, Japan, January 19-January 23, 2003* ed. by IEEE Robotics and Automation Society, pp. 41–44
100. J. Chung, J. Lee: Sens. Actuator A-Phys. **104**, 229 (2003)
101. X.Q. Chen, T. Saito, H. Yamada, K. Matsushige: Appl. Phys. Lett. **78**, 3714 (2001)
102. L.A. Nagahara, I. Amiani, J. Lewenstein, R.K. Tsui: Appl. Phys. Lett. **80**, 3826 (2002)
103. M.R. Diehl, S.N. Yaliraki, R. A. Beckman, M. Barahona, J. R. Heath: Angew. Chem. Int. Ed. Engl. **41**, 353 (2002)
104. R. Krupke, F. Hennrich, H.V. Lhneysen, M.M. Kappes: Science **301**, 344 (2003)
105. M.J. O'Connell, S.M. Bachilo, C.B. Huffman, V.C. Moore, M.S. Strano, E.H. Haroz, K.L. Rialon, P.J. Boul, W.H. Noon, C. Kittrell, J. Ma, R.H. Hauge, R.B. Weisman, R.E. Smalley: Science **297**, 593 (2002)
106. R.B. Weisman: Nat. Mater. **2**, 569 (2003)
107. M. Zheng, A. Jagota, E.D. Semke, B.A. Diner, R.S. McLean, S.R. Lustig, R.E. Richardson, N.G. Tassi: Nat. Mater. **2**, 338 (2003)
108. D. Chattopadhyay, I. Galeska, F. Papadimitrakopoulos: J. Am. Chem. Soc. **125**, 3370 (2003)
109. J. Kong, H. Dai: J. Phys. Chem B **105**, 2890 (2001)

110. Z. Chen, D. Rankin, A.G. Rinzler: APS Meeting Abstr. 26013 (2003)
111. M. Burghard, G. Duesberg, G. Philipp, J. Muster, S. Roth: Adv. Mater. **10**, 584 (1998)
112. J. Muster. M. Burghard, S. Roth, G.S. Duesberg, E. Hernndez, A. Rubio: J. Vac. Technol. B **16**, 2796 (1998)
113. G.S. Duesberg, J. Muster, V, Krstic, M. Burghard, S. Roth: Appl. Phys A **67**, 117 (1998)
114. K.H. Choi, J.P. Bourgoin, S. Auvray, D. Esteve, G.S. Duesberg, S. Royh, M. Burghard: Surf. Sci. **426**, 195 (2000)
115. J.C. Lewenstein, T.P. Burgin, A. Ribayrol, L.A. Nagahara, R.K. Tsui: Nano Lett. **2**, 443 (2002)
116. J. Liu, M.J. Casavant, M. Cox, D.A. Walters, P. Boul, W. Lu, A.J. Rimberg. K.A. Smith, D.T. Colbert, R.E. Smalley: Chem. Phys. Lett. **303**, 125 (1999)
117. S. G. Rao, L. Huang, W. Setyawan, S. Hong: Nature **425**, 36 (2003)
118. D. Chattopadhyay, I. Galeska, F. Papadimitrakopoulos: J. Am. Chem. Soc. **123**, 9451 (2001)
119. P. Diao, Z. Liu, B. Wu, X. Nan, J. Zhang, Z. Wei: ChemPhysChem **10**, 898 (2002)
120. J.J. Gooding, R. Wibowo, J. Liu, W. Yang, D. Losic, S. Orbons, F.J. Mearns, J.G. Shapter, D.B. Hibbert: J. Am. Chem. Soc. **125**, 9006 (2003)
121. J. Zhu, M. Yudasaka, M. Zhang, D. Kasuya, S. Iijima: Nano Lett. **3**, 1239 (2003)
122. R.J. Chen, Y. Zhang, D. Wang, H. Dai: J.Am. Chem. Soc **123**, 3838 (2001)
123. H. Xin, A.T. Woolley: J. Am. Chem. Soc. **125**, 8710 (2003)
124. A. Hirsch: Angew. Chem. Int. Ed. Engl. **41**, 1853 (2002)
125. J.L. Bahr, J.M. Tour: J. Mater. Chem. **12**, 1952 (2002)
126. Y.-P. Sun, K. Fu, Y. Lin, W. Huang: Acc. Chem. Res. **12**, 1096 (2002)
127. M. A. Hamon, H. Hu, B. Zhao, P. Bhowmik, R. Sen, M.E. Itkis, R.C. Haddon, S. Niyogi: Acc. Chem. Res. **12**, 1105 (2002)
128. D. Tasis, N. Tagmatarchis, V. Georgakilas, M. Prato: Chem. Eur. J. **9**, 4000 (2003)
129. J.E. Riggs, Z. Guo, D.L. Carroll, Y.-P. Sun: J. Am. Chem. Soc. **122**, 5879 (2000)
130. Y.-P. Sun, W. Huang, Y. Lin, K. Fu, A. Kitaygorodskiy, L. Riddle, Y.J. Yu, D.L. Carroll: Chem. Mater. **13**, 2864 (2001)
131. M.G.C. Kahn, S. Banerjee, S.S. Wong: Nano Lett. **2**, 1215 (2002)
132. F. Pompeo, D.E. Resasco: Nano Lett. **2**, 369 (2002)
133. S. Niyogi, M.A. Hamon, H. Hu, B. Zhao, P. Browmik, R. Sen, M.E. Itkis, R.C. Haddon: Acc. Chem. Res. **35**, 1105 (2002)
134. J. Chen, M.A. Hamon, H. Hu, Y. Chen, A.M. Rao, P.C. Eklund, R.C. Haddon: Science **282**, 95 (1998)
135. M.A. Hamon, J. Chen, H. Hu, Y. Chen, A.M. Rao, P.C. Eklund, R.C. Haddon: Adv. Mater. **11**, 834 (1999)
136. J. Chen, A.M. Rao, S. Lyuksyutov, M.E. Itkis, M.A. Hamon, H. Hu, R.W. Cohn, P.C. Eklund, D.T. Colbert, R.E. Smalley, R.C. Haddon: J. Phys. Chem. B **105**, 2525 (2001)
137. Z. Liu, Z. Shen, T. Zhu, S. Hou, L. Ying, Z. Shi, Z. Gu: Langmuir **16**, 3569 (2000)
138. B. Wu, J. Zhang, Z. Wei, S. Cai, Z. Liu: J. Phys. Chem. B **105**, 5075 (2001)

139. Y. Chen, R.C. Haddon, S. Fang, A.M. Rao, P.C. Eklund, W.H. Lee, E.C. Dickey, E.A. Grulke, J.C. Pendergrass, A. Chavan, B.E. Haley, R.E. Smalley: J. Mater. Res. **13**, 2423 (1998)
140. E.T. Mickelson, C.B. Huffman, A.G. Rinzler, R.E. Smalley, R.H. Hauge, J.L. Margrave: Chem. Phys. Lett. **296**, 188 (1998)
141. E.T. Mickelson, I.W. Chiang, J.L. Zimmerman, P. Boul, J. Lozano, J. Liu, R.E. Smalley, R.H. Hauge, J.L. Margrave: J. Phys. Chem. B **103**, 4318 (1999)
142. Z. Gu, H. Peng, R.H. Hauge, R.E. Smalley, J.L. Margrave: Nano Lett. **2**, 1009 (2002)
143. M. Holzinger, O. Vostrovsky, A. Hirsch, F. Hennrich, M. Kappes, R. Weiss, F. Jellen: Angew. Chem. Int. Ed. Engl. **40**, 4002 (2001)
144. P.J. Boul, J. Liu, E.T. Mickelson, C.B. Huffman, L.M. Ericson, I.W. Chiang, K.A. Smith, D.T. Colbert, R.H. Hauge, J.L. Margrave, R.E. Smalley: Chem. Phys. Lett. **310**, 367 (1999)
145. J.L. Bahr, J. Yang, D.V. Kosynkin, M.J. Bronikowski, R.E. Smalley, J.M. Tour: J. Am. Chem. Soc. **123**, 6536 (2001)
146. J.L. Bahr, L.M. Tour: Chem. Mater. **13**, 3823 (2001)
147. C.A. Mitchell, J.L. Bahr, S. Arepalli, J.M. Tour, R. Krishnamoorti: Macromolecules **35**, 8825 (2002)
148. S.E. Kooi, U. Schlecht, M. Burghard, K. Kern: Angew. Chem. Int. Ed. Engl. **41**, 1353 (2002)
149. N. Tagmatarchis, V. Georgakilas, M. Prato, H. Shinohara: Chem. Commun. 2010 (2002)
150. M. Maggini, G. Scorrano, M. Prato: J. Am. Chem. Soc. **115**, 9798 (1993)
151. M. Prato, M. Maggini: Acc. Chem. Res. **31**, 519 (1998)
152. V. Georgakilas, K. Kordatos, M. Prato, D.M. Guldi, M. Holzinger, A. Hirsch: J. Am. Chem. Soc. **124**, 760 (2002)
153. V. Georgakilas, N. Tagmatarchis, D. Pantarotto, A. Bianco, J.-P. Briand, M. Prato: Chem. Commun. 3050 (2002)
154. K. Kordatos, T. Da Ros, S. Bosi, E. Vazquez, M. Bergamin, C. Cusan, F. Pellarini, V. Tomberli, B. Baiti, D. Pantarotto, V. Georgakilas, G. Spalluto, M Prato: J. Org. Chem. **66**, 4915 (2001)
155. B.F. Erlanger, B.-X. Chen, M. Zhu, L. Brus: Nano Lett. **1**, 465 (2001)
156. M. Shim, N.W. Shi Kam, R.J. Chen, Y. Li, H. Dai: Nano Lett. **2**, 285 (2002)
157. W. Huang, S. Taylor, K. Fu, Y. Lin, D. Zhang, T.W. Hanks, A.M. Rao, Y.-P. Sun: Nano Lett. **2**, 311 (2002)
158. C.V. Nguyen, L. Delzeit, A.M. Cassell, J. Li, J. Han, M. Meyyappan: Nano Lett. **2**, 1079 (2002)
159. B.R. Azamian, J.J. Davis, K.S. Coleman, C.B. Bagshaw, M.L.H. Green: J. Am. Chem. Soc. **124**, 12664 (2002)
160. S.E. Baker, W. Cai, T.L. Lasseter, K.P. Weidkamp, R.J. Hamers: Nano Lett. **2**, 1413 (2002)
161. N. Nakashima, Y. Tomonari, H. Murakami: Chem. Lett., 638 (2002)
162. A. Star, J.F. Stoddart, D. Steuerman, M. Diehl, A. Boukai, E.W. Wong, X. Yang, S.-W. Chung, H. Choi, J.R. Heath: Angew. Chem. Int. Ed. Engl. **40**, 1721 (2001)
163. D.W. Steuerman, A. Star, R. Narizzano, H. Choi, R.S. Ries, C. Nicolini, J.F. Stoddart, J.R. Heath: J. Phys. Chem. B **106**, 3124 (2002)
164. A. Star, J.F. Stoddart: Macromolecules **35**, 7516 (2002)

165. B.Z. Tang, H. Xu: Macromolecules **32**, 2569 (1999)
166. J. Chen, H. Liu, W.A. Weimer, M.D. Halls, D.H. Waldeck, G.C. Walker: J. Am. Chem. Soc. **124**, 9034 (2002)
167. A. Star, D.W. Stewerman, J.R. Heath, J.F. Stoddart: Angew. Chem. Int. Ed. Engl. **41**, 2508 (2002)
168. P.M. Ajayan, O. Z. Zhou: In: *Carbon Nanotubes–Synthesis, Structure, Properties, and Applications*, ed. by M.S. Dresselhaus, G. Dresselhaus, P. Avouris, p. 401 (Springer-Verlag Berlin Heildelberg New York 2000).
169. R. Martel, T. Schmidt, H.R. Shea, T. Hertel, P. Avouris: Appl. Phys. Lett. **73**, 2447 (1998)
170. W.A. de Heer, A. Chaterlain, D. Ugarte: Science **270**, 1179 (1995)
171. A.G. Rinzler, J.H. Hafner, P. Nikolaev, L. Lou, S.G. Kim, D. Tomànek, P. Nordlander, D.T. Colbert, R.E. Smalley: Science **269**, 1550 (1995)
172. N.S. Lee, D.S. Chung, I.T. Han, J.H. Kang, Y.S. Choi, H.Y. Kim, S.H. Park, Y.W. Jin, W.K. Yi: Diamond Relat. Mater. **10**, 265 (2001)
173. A. Modi, N. Koratkar, E. Lass, B. Wei, P.M. Ajayan: Nature **424**, 171 (2003)
174. L. Yang, J. Han: Phys. Rev. Lett. **85**, 154 (2000)
175. D. Srivastava et al: J. Phys. Chem. B **103**, 4330 (1999)
176. J. Kong, N.R. Franklin, C. Zhou, M.G. Chapline, S. Peng, K.J. Cho, H. Dai: Science **287**, 622 (2000)
177. S. Peng, K.J. Cho: Nanotechnology **11**, 57 (2000)
178. A.C. Dillon, K.M. Jones, T.A. Bekkedahl, C.H. Kiang, D.S. Bethune, M.J. Heben: Nature **386**, 377 (1997)
179. P. Chen, X. Wu, J. Lin, K. Tan: Science **285**, 91 (1999)
180. C. Liu, Y.Y. Fan, M. Liu, H.T. Cong, H.M. Cheng, M.S. Dresselhaus: Science **286**, 1127 (1999)
181. C. Nutenadel, A. Zuttel, D. Chartouni, L. Schlapbach: Solid State Lett. **2**, 30 (1999)
182. A. Chambers, C. Park, R.T.K. Baker, N.M. Rodriguez: J. Phys. Chem. B **102**, 4253 (1998)
183. M.S. Dresselhaus, K.A. Williams, P.C. Eklund: MRS Bull. **24**, 45 (1999)
184. N. Rajalakshmi, K.S. Dhathathreyan, A. Govindaraj, B.C. Satiskumar: Electrochim. Acta **45**, 4511 (2000)
185. C. Guerret-Piecourt, Y. Le Bouar, A. Loiseau, H. Pascard: Nature **372**, 761 (1994)
186. A. Loiseau, H. Pascard: Chem. Phys. Lett. **256**, 246 (1996)
187. S.T. Lee, N. Wang, Y.F. Zhang, Y.H. Tang: MRS Bull. **24**, 36 (1999)
188. M. Terrones, N. Grobert, W.K. Hsu, Y.Q. Zhu, W.B. Hu, H. Terrones, J.P. Hare, H.W. Kroto, D.R.M. Walton: MRS Bull. **24**, 43 (1999)
189. B. Gao, C. Bower, J.D. Lorentzen, L. Fleming, A. Kleinhammes, X.P. Tang, L.E. McNeil, Y. Wu, O. Zhou: Chem. Phys. Lett. **327**, 69 (2000)
190. Y. H. Lee, K.H. An, S.C. Lim, W.S. Kim, H.J. Jeong, C.-H. Doh, S.-I. Moon: New Diamond Frontier Carbon Technol. **12**, 209 (2002)
191. Y.H. Gao, Y. Bando: Nature **415**, 599 (2002)
192. S. Berber, Y.K. Kwon, D. Tomanek: Phys. Rev. Lett. **84**, 4613 (2000)
193. M. Karlsson, M. Davidson, R. Karlsson, A. Karlsson, J. Bergenholtz, Z. Konkoli, A. Jesorka, T. Labovkina, J. Urtigh, M. Voinova, O. Orwar: Annu. Rev. Phys. Chem. **55**, 613 (2004)
194. R.H. Baughman, A.A. Zakhidov, W.A. de Heer: Science **297**, 787 (2002)

195. E.D. Steinle, D.T. Mitchell, M. Wirtz, S.B. Lee, V.Y. Young, C.R. Martin: Anal. Chem. **74**, 2416 (2002)
196. S. Joseph, R.J. Mashl, E. Yakobsson, N.R. Aluru: Nano Lett. **3**, 1399 (2003)
197. C.R. Martin, P. Kohli: Nat. Rev. Drug Discov. **2**, 29 (2003)
198. A. Bianco, M. Prato: Adv. Mater. **15**, 1765 (2003)
199. R. Gasparac, P. Kholi, M.O. Mota, L. Trofin, C.R. Martin: Nano Lett. **4**, 513 (2004)
200. J.E. Fischer, H. Dai, A. Thess, R. Lee, N.M. Hanjani, D.L. Dehaas, R.E. Smalley: Phys. Rev. B **55**, R4921 (1997).
201. A. Bachtold, M.S. Fuhrer, S. Plyasunov, M. Forero, E.H.Z. Anderson, Z.A. Zettl, P.L. McEuen: Phys. Rev. Lett. **84**, 6082 (2000)
202. J. Kong, E. Yenilmez, T.W. Tombler, W. Kim, H. Dai, R.B. Laughlin, L. Liu, C.S. Jayanthi, S.Y. Wu: Phys. Rev. Lett. **87**, 106801 (2001)
203. A. Javey, J. Guo, M. Paulsson, Q. Wang, D. Mann, M. Lundstrom, H. Dai: Phys. Rev. Lett. **92**, 106804 (2004)
204. D. Mann, A. Javey, J. Kong, Q. Wang, H. Dai: Nano Lett. **3**, 1541 (2003)
205. A. Jevey, J. Guo, D.B. Farmer, Q. Wang, V.W. Wang, R.G. Gordon, M. Lundstrom, H.J. Dai: Nano Lett. **4**, 447 (2204)
206. P. Avouris: Acc. Chem. Res. **35**, 1026 (2002)
207. W. Kim, A. Javei, O. Vermesh, Q. Wang, Y. Li, H. Dai: Nano Lett. **3**, 193 (2003)
208. A. Javey, J. Guo, Q. Wang, M. Lundstrom, H. Dai: Nature **424**, 654 (2003)
209. A. Javey, H. Kim, M. Brink, Q. Wang, A. Ural, J. Guo, P. McIntyre, P. McEuen, M. Lundstrom, H. Dai: Nat. Mater. **1**, 241 (2002)
210. T. Durkop, S.A. Getty, E. Cobas, M. S. Fuhrer: Nano Lett. **4**, 35 (2004)
211. M. Menon, D. Srivastava: Phys. Rev. Lett. **79**, 4453 (1997)
212. T. Rueckes, K. Kim, E. Joselevich, G.Y. Tseng, C.L. Cheung, C.M. Lieber: Science 289, **94** (2000)
213. G.U. Sumanasekera, C.K.W. Adu, S. Fang, P.C. Eklund: Phys. Rev. Lett. **85**, 1096 (2000)
214. P.G. Collins, K. Bradley, M. Ishigami, A. Zettl: Science **287**, 1801 (2000)
215. S.-H. Jhi, S.G. Louie, M.L. Cohen: Phys. Rev. Lett. **85**, 1710 (2000)
216. V. Derycke, R. Martel, J. Appenzeller, P. Avouris: Appl. Phys. Lett. **80**, 2773 (2002)
217. L. Chico, V. H. Crespi, L. X. Benedict, S. G. Louie, M. L. Cohen: Phys. Rev. Lett. **76**, 971 (1996)
218. M. Menon, D. Srivastava: J. Mater. Res. **13**, 2357 (1998)
219. C. Papadopoulos, A. Rakitin, J. Li, A.S. Vedeneev, J.M. Xu: Phys. Rev. Lett. **85**, 3476 (2000)
220. C. Satishkumar, P.J. Thomas, A. Govindraj, C.N.R. Rao: Appl. Phys. Lett. **77**, 2530 (2000)
221. N. Gothard, C. Daraio, J. Gaillard, R. Zidan, S. Jin, A.M. Rao: Nano Lett. **4**, 213 (2004)
222. A. Antonis, M. Menon, D. Srivastava, L. A. Chernozatonskii: Appl. Phys. Lett. **79**, 266 (2001)
223. W.Z. Li, J.G. Wen, Z.F. Ren: Appl. Phys. Lett. **79**, 1879 (2001)
224. A. Bachtold, P. Hadley, T. Nakanishi, C. Dekker: Science **294**, 1317 (2001)
225. H. Dai et al: J. Phys. Chem. B **103**, 11246 (1999)
226. X. Liu, C. Lee, C. Zhou, J. Han: Appl. Phys. Lett. **79**, 3329 (2001)

227. M. Zheng, A. Jagota, E.D. Semke, B.A. Diner, R.S. Mclean, S.R. Lustig, R.E. Richardson, N.G. Tassi: Nat. Mater. **2**, 338 (2003)
228. M. Zheng, A. Jagota, M.S. Strano, A.P. Santos, P. Barone, S.C. Chou, B.A. Diner, M.S. Dresselhaus, R.S. McLean, G.B. Onoa, G.G. Samsonidze, E.D. Semke, M. Usrey, D.J. Walls: Science **302**, 1545 (2003)
229. A. Goldoni, R. Larciprete, L. Petaccia, S. Lizzit: J. Am. Chem. Soc. **125**, 11329 (2003)
230. Q.F. Pengfei, H. Vermesh, M. Grecu, A. Javey, Q. Wang, H. Dai, S. Peng, K.J. Cho: Nano Lett. **3**, 347 (2003)
231. J. Li, Y. J. Lu, Q. Ye, M. Cinke, J. Han, M. Meyyappan: Nano Lett. **3**, 929 (2003)
232. M. Gao, L.M. Dai, G.G. Wallace: Electroanalysis **15**, 1089 (2003)
233. A. Star, J.C.P. Gabriel, K. Bradley, G. Gruner: Nano Lett. **3**, 459 (2003)
234. H. Cai, X.N. Cao, Y. Jang, P.G. He, Y.Z. Fang: Anal. Bioanal. Chem. **375**, 287 (2003)
235. H. Chang, J.D. Lee, S.M. Lee, Y.H. Lee: Appl. Phys. Lett. **79**, 3863 (2001)
236. J.J. Zhao, A. Buldum, J. Han, J.P. Lu: Nanotechnology **13**, 195 (2002)
237. S. Peng, K. Cho: Nano Lett. **3**, 513 (2003)
238. P. Giannozzi, R. Car, G. Scoles: J. Chem Phys. **118**, 1003 (2003)
239. Q. Zhao, M.B. Nardelli, J. Bernholc: In: *International Conference on the Science and Applications of Nanotubes NT02, Boston, Massachusetts, 6–11 July 2002*
240. H. Dai: Surf. Sci. **500**, 218 (2002)
241. H. Ulbricht, G. Moos, T. Hertel: Phys. Rev. B **66**, 075404 (2002)
242. R. Larciprete, A. Goldoni, S. Lizzit: Nucl. Instrum. Methods Phys. Res. Sect. B **200**, 5 (2003)
243. A. Ricca, C.W. Bauschlicher, A. Maiti: Phys. Rev. B **68**, 035433 (2003)
244. F. Villalpando-Paez, A.H. Romero, E. Munoz-Sandoval, L.M. Martinez, H. Terrones, M. Terrones: Chem. Phys. Lett. **386**, 137 (2004)
245. P.G. Collins, A. Zettl: Phys. Rev. B **55**, 9391 (1997)
246. Y. Saito, S. Uemura, K. Hamaguchi: Jpn. J. Appl. Phys. **37**, L 346 (1998)
247. X. Xu, G.R. Brandes: Appl. Phys. Lett. **74**, 2549 (1999)
248. W. Zhu, C. Bower, O. Zhou, G. Kochanski, S. Jin: Appl. Phys. Lett. **75**, 873 (1999)
249. K.A. Dean, B.R. Chalamala: J. Appl. Phys. **85**, 3832 (1999)
250. W.B. Choi, Y.H. Lee, D.S. Chung, N.S. Lee, J.M. Kim: In: *Science and Application of Nanotubes*, ed by D. Tomanek, R. Enbody (Kluwer/Plenum, New York 2000) pp. 355–356
251. M. Kusunoki, T. Suzuki, C. Honjo, T. Hirayama, N. Shibata: Chem. Phys. Lett. **366**, 458 (2002)
252. H. Dai, J.H. Hafner, A.G. Rinzler, D.T. Colbert, R.E. Smalley: Nature **384**, 147 (1996)
253. S.S. Wang, J.D. Harper, P.T. Lansbury, C.M. Lieber: J. Am. Chem. Soc. **120**, 603 (1998)
254. M.F. Yu, M.J. Dyer, G.D. Skidmore, H.W. Rohrs, X.K. Lu, K.D. Ausman, J.R.V. Ehr, R.S. Ruoff: Nanotechnology **10**, 244 (1999)
255. R.M.D. Stevens, N.A. Frederick, B.L. Smith, D.E. Morse, G.D. Stucky, P.K. Hansma: Nanotechnology **11**, 1 (2000)
256. C.L. Cheung, J.H. Hafner, C.M. Lieber: Proc. Natl. Acad. Sci. USA **97**, 3809 (2000)

257. J.H. Hafner, C.L. Cheung, C.M. Lieber: Nature **398**, 761 (1999)
258. C.V. Nguyen, K.J. Cho, R.M.D. Stevens, L. Delzeit, A. Cassell, J. Han, M. Meyyappan: Nanotechnology **12**, 363 (2001)
259. Y. Nakayama, H. Nishijima, S. Akita, K.I. Hohmura, S.H. Yoshimura, K Takeyasu: J. Vacuum Sci. Technol. B **18**, 661 (2000)
260. N. Choi, T. Uchihashi, H. Nishijima, T. Ishida, W. Mizutani, S. Akita, Y. Nakayama, H. Tokumoto: Jpn. J. Appl. Phys. **39**, 3707 (2000)
261. A. Rothschild, S.R. Cohen, R. Tenne: Appl. Phys. Lett. **75**, 4025 (1999)
262. H. Dai, N. Franklin and J. Han: Appl. Phys. Lett. **73**, 1508 (1998)
263. S.S. Wong, E. Joselevich, A.T. Wolley, C.L. Cheung, C.M. Lieber: Nature **394**, 52 (1998)
264. S.S. Wong, A.T. Wolley, E. Joselevich, C.M. Lieber: Chem. Phys. Lett. **306**, 219 (1999)
265. T. Shimizu, H. Tokumoto, S. Akita, Y. Nakayama: Surf. Sci. **486**, L455 (2001)

6 Calculating Transport Properties of Nanometer-Scale Systems: Nanodevice Applications of Carbon Nanotubes and Organic Molecules

Amir A. Farajian, Rodion V. Belosludov, Olga V. Pupysheva,
Hiroshi Mizuseki, and Yoshiyuki Kawazoe

6.1 Introduction

There is no doubt that in the rapidly growing fields of nanotechnology, theoretical calculations and experimental techniques should support each other. In particular, in nanoelectronics and molecular electronics, calculating transport properties of nanometer-scale systems can provide the explanations necessary for a clear understanding of the experimental results. In addition, transport calculations can suggest new assemblies of nanostructures for nanodevice applications. In this chapter, we focus on transport calculations at nanometer scale. We show, using a nonequilibrium Green function approach, how the transport properties, for example, conductance and current–voltage (I–V) curves, of various nanostructures can be calculated. The importance of the contact phenomena, i.e., the attachment of nanometer-scale objects to macroscopic electrodes, is investigated using atomistic models and ab initio calculations. It is shown how exact, relative positions and orientations of the nanometer-scale objects and macroscopic electrodes can affect the transport properties. As specific nanometer-scale systems, we consider carbon nanotubes and some organic molecules. Transport properties of these systems attached to, for example, gold electrodes are calculated, with emphasis on nanodevice applications.

Nanoelectronics and molecular electronics basically make use of functional components that are at the molecular length scale. The essential requirement of a theoretical model that concerns electronic transport through molecular-scale devices is the full quantum-mechanical description of the transport phenomena. Only an accurate calculation approach based on a quantum-mechanical description of the system can reveal the peculiarities of electronic transport at the molecular scale. As an example of these peculiarities, we may mention the issue of contact resistance of a molecular-scale ballistic quasi-1D (Q1D) system [1, 2]: As the number of conducting channels in the Q1D conductor is limited, the injection of carriers to/from the infinite number of channels of a "macroscopic" contact limits the conduction of the ballistic conductor. For an ideal molecular conductor, i.e., a ballistic conductor without scattering, this mechanism results in quantization of the conduction in units of $2e^2/h$. The number of quantum units that determine the quantized conduction equals the number of available conducting channels of the ideal Q1D conductor. Although the quantized conductance is assigned to the ideal conductor, the resistance is in fact "contact" resistance, as there is no potential drop along the ballistic Q1D conductor and the whole potential drop occurs at the contacts.

The main functional parts in nanoelectronics and molecular electronics applications are a few nanometers across. Two examples are depicted in Fig. 6.1. Figure 6.1a shows a polythiophene molecule and parts of the gold nanocontacts to which the molecule is attached. The nanocontacts are semi-infinite in extent, i.e., are bounded only by the molecule sandwiched between them. In Fig. 6.1b part of an infinitely long carbon nanotube is shown. The nanotube is deposited on a double-crystal substrate; hence, its "left" and "right" parts are subject to different doping effects. Applying bias will cause current to flow through these systems.

When there are extra scattering centers available within the Q1D conductor, like in the cases depicted in Fig. 6.1, for each channel the transmission probability across the Q1D conductor will be less than 1, and part of the potential drop will occur around the scattering centers. In this case, the formalism that is usually used in calculating conductance and, subsequently, I–V characteristics, is Landauer's formalism, which we explain in detail in Sect. 6.2. Making use of this formalism, we can calculate the quantum conductance of an open contact–molecule–contact system, in which the contacts are semi-infinite. If we consider semi-infinite Q1D nanocontacts, then the band structures of the nanocontacts will also play an important role in calculating conduction. This issue has practical importance because with the advancement of nanolithographic techniques, contacts whose cross sections lie in the nanometer scale are experimentally realized nowadays. When two such nanocontacts sandwich a molecule as in Fig. 6.1a, the conduction of the whole system can be calculated by deriving the probability for an

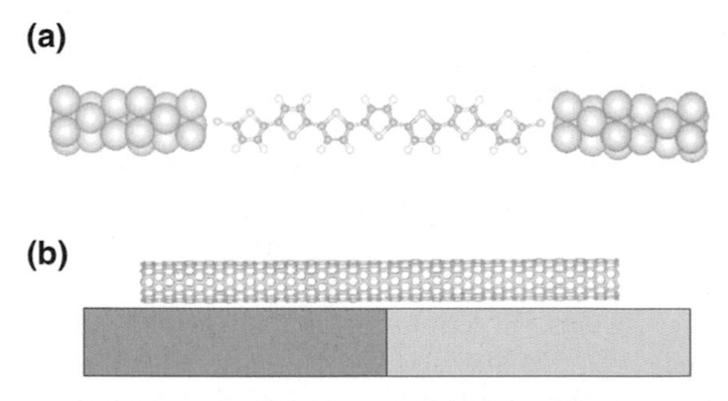

Fig. 6.1. Two examples of nanometer-scale systems whose transport properties can be calculated using the methods of this chapter. (**a**) A polythiophene molecule attached to gold nanocontacts via sulfur "clips." (**b**) A carbon nanotube deposited on a double-crystal substrate

excitation well inside the "left" nanocontact to travel first to the left contact's "surface," then across the middle junction onto the "surface" of the "right" contact, and finally well within the right contact. The problem of calculating the transition probability is conveniently solved by calculating the "surface Green functions" of the left and right contacts, and by "attaching" them to the Green function of the middle junction [3].

For the system of Fig. 6.1b, the left and right contacts are the left and right parts of the nanotube located "far enough" from the interface of the substrate crystals, and the middle junction part is the middle part of the nanotube that joins the left side to the right side. By mentioning "far enough" we mean that the effects of the disturbance caused by the interface of the substrate crystals are negligible for the left and right parts of the nanotube. These effects, however, are not negligible for the middle junction part. In other words, the middle part is long enough to effectively screen the disturbance [4].

This chapter is devoted to transport calculations for nanoscale systems. First, the Landauer formalism is explained, with a general survey and derivation of the formalism for the two- and four-probe cases. Next, the mathematical formulation of nonequilibrium surface Green function matching (NSGFM) is shown to provide a natural framework for nanoscale transport calculations. The implementation of the formalism is then employed for calculating transport characteristics of some actual systems of interest, such as carbon nanotubes and organic molecules. Within our formalism, we are able to assess the localization pattern of the density of states. This in turn enables us to make a correspondence between discrete molecular orbitals and conductance characteristics of the device. These results pave the way for insightful design of nanoelectronics and molecular electronics components via proper engineering of the electronic structure of the functional molecule, and its attachment to the contacts.

6.2 Landauer Formalism

Landauer's formalism defines the conductance of the system in terms of its transmission and reflection probabilities. In this way, the conductance of the system for any (carrier) energy is obtained. The conductance functions thus obtained are then integrated over all energies, taking into account the Fermi–Dirac distribution of the electrons of the two sides of the junction. This is carried out according to the Landauer–Büttiker formula, which gives the current passing through a system, once its conductance is given as a function of energy. The Landauer–Büttiker formula was originally proposed within the linear-response formalism, but later it was used to include nonlinear transport properties to some extent. In the formalism presented in this chapter, as we shall see, the nonlinearity of transport stems from the fact that the transmission probability is a functional of the potential drop profile at the junction, which itself depends on the external potential applied to the two sides of the junction.

As the Landauer formula is used throughout this chapter for calculating transport characteristics of various nanostructures, here we present a general survey, followed by derivation of the formalism for the two- and four-probe measurements.

6.2.1 Landauer Formula: an Overview

Interest in the tunneling transparency of 1D and Q1D systems, as well as their behavior in an electric field, was significantly stimulated by the famous Landauer formula [5]. It establishes a relation between the electron transmission and reflection coefficients of a 1D disordered structure (T and R, respectively), i.e., its quantum-mechanical characteristics, with a macroscopic value of electrical conductance Γ:

$$\frac{1}{\Gamma} = \frac{2\pi\hbar}{e^2}\frac{R}{T} . \qquad (6.1)$$

In [5] this formula was derived from the following considerations. Let us consider a 1D structure with a random potential, which transmits carriers in a quantum-mechanically coherent and elastic way. Let the charge carriers approach its left-hand side and get reflected from it with probability R. Then the relative electron density to the left of the structure equals $1 + R$, and that to the right of it equals $T = (1 - R)$. Consequently, its gradient equals $-2R/L$, where L is the length of the structure. The current of the electrons through the structure can be expressed, from one point, through the diffusion equation, and from another point, as a product of the transmission coefficient with the carrier velocity v. Thus, one can find the diffusion coefficient

$$D = \frac{vL}{2}\frac{(1 - R)}{R}$$

and use it in the Einstein relation. As a result, the resistance is proportional to the ratio R/T.

Let us note that Landauer's paper [6] of 1957, which is often cited in connection with (6.1), actually did not contain this result, as it used a semi-classical approach and the scatterers were assumed to act incoherently (see Landauer's comment [7]).

An alternative derivation of (6.1) was suggested by Anderson et al. [8]. On the basis of the scattering theory, a more strict formalism was developed and generalized for the case of a few scattering channels. In particular, Anderson et al. [8] proposed a simple scaling method for the problems of localization in 1D disordered systems, which was further developed by Andereck and Abrahams [9].

However, Abrikosov [10] argued that the scaling method of [8] is not strict enough, and its results are valid only at the qualitative, but not quantitative, level.

Economou and Soukoulis [11] discussed the applicability of the Landauer formula when the leads in contact with the structure under consideration are ideal 1D conductors. This distinguishes this work from [5, 8], where the electric field was considered as a result of the charge build-up at the sample boundaries, and the leads thus played, in a sense, the role of a current source. The Kubo formula [12] for alternating current with the frequency approaching zero was used in [11]. The obtained conductivity was proportional to the transmission coefficient T, but not to the ratio T/R, as follows from (6.1). Obviously, these two results are practically the same for small T values, but differ significantly for a highly transparent sample.

In his comment entitled "Why Landauer's formula for resistance is right," Thouless [13] pointed out that the usage of the Kubo formula in [11] was incorrect, as the alternating current electric field must lead to the current oscillations in the ideal conductor. In their reply, Economou and Soukoulis [14] discussed the conditions when the relation $\Gamma \propto T$ can be satisfied.

The approach of [11] was further developed by Fisher and Lee [15] for any given number of scattering channels. The connection between the conductivity of the finite system with a static disorder and the scattering $\mathbf{T}$ matrix was found in the form

$$\Gamma = \frac{e^2}{2\pi\hbar}\mathrm{Tr}(\mathbf{T}^\dagger\mathbf{T})\,. \tag{6.2}$$

However, the resistance of an ideal conductor (or a sample of an infinitely small length $L \to 0$) must approach zero, while its transmission coefficient would be close to 1. This obvious consideration is in agreement with the Landauer formula (6.1), but contradicts (6.2). This contradiction was explained [15] by the impossibility to create a finite electric field in an ideal conductor, if the leads outside the sample are in equilibrium.

At the same time, Langreth and Abrahams [16] obtained the same result as Landauer, also starting from the Kubo formula. It was noticed [16] that the correct modeling of the electrodes is extremely important for solving this problem, and a non-self-consistent electric field considered in [11, 15] can result in fluctuations of charge in the system, which contradict the condition of stationary current. In addition, a detailed derivation of the expression for a 1D system with few scattering channels was given [16].

In Landauer's comment [17] regarding this discussion, entitled "Can a length of perfect conductor have a resistance?," [10, 11, 15] were criticized for ignoring the screening effects. It was shown by means of a simple semiclassical argument that the origin of the expressions other than (6.1) is the complete ignoring of self-consistency and violation of the global charge conservation.

Azbel [18] generalized the Landauer formula (6.1) for systems of arbitrary dimensionality, and discussed the relation between the resistance of a thin wire and the reflection coefficient averaged over all the scattering channels.

If the number of channels is large enough, then $< R >$ is expressed through the scattering section of the impurity and the size-quantized momentum.

A many-dimensional problem was considered in detail by Büttiker et al. [19]. A sample with a few leads attached to it was considered later [20–22].

Other variants of the Landauer formula were also suggested for some particular structures; for example, for systems with Kronig–Penney potential [23]. A review of such works was presented by Landauer [24]. In [24] various contact geometries were considered, and the possible influence of the measurement on the R value obtained was discussed.

Landauer–Büttiker theory has numerous applications. For example, this formalism was used for the study of the electron transport through a chain of scatterers by Burmeister et al. [25], Maschke and Schreiber [26], Berthod et al. [27] and Hey et al. [28], including the case of energy dissipation (i.e., inelastic scattering). In [29] asymmetric double quantum wells were studied in a similar way.

6.2.2 Derivation of the Landauer Formula for Two- and Four-Probe Measurements

Here we discuss the Landauer formula [8] for conductance, as well as the Landauer–Büttiker formula [19] for current, in a Q1D system. Consider a sample connected to perfect and identical 1D conductors which in turn connect to electron reservoirs. The reservoir to the left injects carriers into the perfect wire up to a Fermi energy μ_1 and the reservoir to the right emits carriers up to a Fermi energy μ_2. Without loss of generality we assume $\mu_1 > \mu_2$. The reservoirs are incoherent; the waves emerging from separate reservoirs do not have any phase relationship. The current, for two spin directions, emitted by the left reservoir in the energy range between μ_2 and μ_1 is

$$I = ev\frac{\partial n}{\partial E}(\mu_1 - \mu_2). \tag{6.3}$$

Here v is the Fermi velocity, and $\partial n/\partial E$ is the density of states for two spin directions and for carriers with positive velocities. In one dimension $\partial n/\partial k = 1/\pi$ and $\partial n/\partial E = 1/\pi\hbar v$. Thus, the total current emitted by the left reservoir owing to the difference in the Fermi levels is $I = (e/\pi\hbar)(\mu_1 - \mu_2)$. These carriers have a probability T for traversal of the sample and a probability R of being reflected; therefore, the net current flow is given by

$$I = \frac{e}{\pi\hbar}T(\mu_1 - \mu_2). \tag{6.4}$$

The difference between the Fermi energies μ_1 and μ_2 is chosen to be small enough so that the energy dependence of T (and R) within this range can be neglected. Next, we have to determine the voltage across the sample. This potential difference is determined by the piled-up charges to the left and right of the sample, and the screening of these charges. The carrier densities can be characterized by the chemical potentials μ_A to the left and μ_B to the right of the sample. (The assignment of a chemical potential to a particular class of electrons can be considered simply as a mathematical alternative to the use of carrier densities.) The levels corresponding to μ_A and μ_B, between μ_1 and μ_2, are determined such that the number of occupied states (electrons) above μ_A (μ_B) is equal to the number of empty states (holes) below μ_A (μ_B). Below the energy μ_2 all the states are fully occupied and we need to consider the energy range from μ_2 to μ_1 only. The total number of states in this range is $2(\partial n/\partial E)(\mu_1 - \mu_2)$. The factor 2 arises because we have a state with positive velocity and a state with negative velocity at each energy, and $\partial n/\partial E$ is the density of states for carriers moving to the right (or to the left) only. Consider now the perfect wire to the right of the sample. Since carriers have a transmission probability T, the number of occupied states is $T(\partial n/\partial E)(\mu_1 - \mu_B)$—which is the number of electrons tunneling from left to right—and the number of unoccupied states is $(2 - T)(\partial n/\partial E)(\mu_B - \mu_2)$—which is the number of holes less those tunneling from right to left. Thus, the chemical potential μ_B to the right of the sample is determined by

$$T\frac{\partial n}{\partial E}(\mu_1 - \mu_B) = (2 - T)\frac{\partial n}{\partial E}(\mu_B - \mu_2). \tag{6.5}$$

To the left of the barrier we have both incident carriers and reflected carriers. The number of occupied states is $(1 + R)(\partial n/\partial E)(\mu_1 - \mu_A)$ and the number of unoccupied states is $[2-(1+R)](\partial n/\partial E)(\mu_A - \mu_2)$; therefore, the chemical potential μ_A to the left of the sample is determined by

$$(1 + R)\frac{\partial n}{\partial E}(\mu_1 - \mu_A) = [2 - (1 + R)]\frac{\partial n}{\partial E}(\mu_A - \mu_2). \tag{6.6}$$

Charge neutrality does not allow different densities to the left and right of the sample over distances large compared to a screening length. This requires that the separation between the chemical potential μ_A (or μ_B) and the band bottom must be the same as in equilibrium. Thus, the conduction-band bottoms of the perfect wires are displaced against each other by a potential difference

$$eV = \mu_A - \mu_B. \tag{6.7}$$

Therefore, (6.5) and (6.6) can be used to determine the voltage across the sample. The result of this calculation yields $eV = R(\mu_1 - \mu_2)$, and therefore (6.4) leads to

$$\Gamma = \frac{I}{V} = \frac{e^2}{\pi\hbar}\frac{T}{R} \tag{6.8}$$

for conductance. This result is in fact for a "four-terminal" or a "four-probe" measurement [30, 31], in which the pair of contacts providing current is different from the pair of contacts through which the voltage across the sample is measured. For a two-probe measurement, on the other hand, we obtain

$$eV = \mu_1 - \mu_2. \tag{6.9}$$

The reason is that in a two-probe measurement, the potential is measured inside the reservoirs which provide the current. Therefore, we obtain

$$\Gamma = \frac{I}{V} = \frac{e^2}{\pi\hbar} T \tag{6.10}$$

for the conductance in a two-probe measurement.

To arrive at (6.8) we have assumed that the reservoirs feed all states up to their Fermi energies equally. This is strictly correct only at zero temperature. At higher temperatures, we assume that the reservoirs fill the states according to the Fermi distribution. The left reservoir fills the states with probability

$$f(E - \mu_1) = \frac{1}{\mathrm{e}^{\beta(E-\mu_1)} + 1}, \tag{6.11}$$

and the right reservoir fills the states with probability

$$f(E - \mu_2) = \frac{1}{\mathrm{e}^{\beta(E-\mu_2)} + 1}. \tag{6.12}$$

As the transmission probability T is now energy-dependent, and the distribution of the state energies is given by the Fermi distribution, the net current flowing from the left reservoir to the right reservoir is now given by

$$I = \frac{e}{\pi\hbar} \left[\int \mathrm{d}E \frac{\mathrm{d}f(E - E_{\mathrm{F}})}{\mathrm{d}E} T(E) \right] (\mu_1 - \mu_2) \tag{6.13}$$

instead of (6.4). Here

$$\frac{\mathrm{d}f(E - E_{\mathrm{F}})}{\mathrm{d}E} \equiv \frac{-\mathrm{d}f}{\mathrm{d}E} \simeq \frac{f(E - \mu_1) - f(E - \mu_2)}{\mu_1 - \mu_2} \tag{6.14}$$

is the derivative of the equilibrium Fermi distribution $f(E - E_{\mathrm{F}})$. Similarly, we find

$$eV = \frac{\int \mathrm{d}E(-\mathrm{d}f/\mathrm{d}E)R(E)(\partial n/\partial E)}{\int \mathrm{d}E(-\mathrm{d}f/\mathrm{d}E)(\partial n/\partial E)} (\mu_1 - \mu_2). \tag{6.15}$$

Therefore

$$\Gamma = \frac{e^2}{\pi\hbar} \left[\int \mathrm{d}E \frac{-\mathrm{d}f}{\mathrm{d}E} T(E) \right] \frac{\int \mathrm{d}E(-\mathrm{d}f/\mathrm{d}E)v^{-1}(E)}{\int \mathrm{d}E(-\mathrm{d}f/\mathrm{d}E)R(E)v^{-1}(E)}, \tag{6.16}$$

where we have used $\partial n/\partial E = 1/\pi\hbar v(E)$.

Up to now only the one-channel case was considered. In other words, there was a single incoming and a single outgoing channel on each side of the system. The multichannel counterparts of (6.8) and (6.10) can be obtained as follows [8]. Since the voltage difference is the same for every channel while the current adds, the conductance is just the sum of all channel conductances. Now, we indicate the transmission and reflection matrices by **t** and **r**, respectively, with matrix elements $t_{\alpha\beta}$ and $r_{\alpha\beta}$ between conduction channels α and β. If it were possible to go to a basis in which $\mathbf{t}$ and $\mathbf{r}$ are both diagonal, then (6.8) would lead to

$$\Gamma = \frac{e^2}{\pi\hbar} \sum_{\alpha} \frac{|t_{\alpha\alpha}|^2}{|r_{\alpha\alpha}|^2}. \tag{6.17}$$

As in general it is not possible to diagonalize $\mathbf{t}$ and $\mathbf{r}$ using the same representation, (6.17) is written as

$$\Gamma = \frac{e^2}{\pi\hbar} \sum_{\alpha} \frac{1}{|r_{\alpha\alpha}|^2} \sum_{\beta} |t_{\alpha\beta}|^2 = \frac{e^2}{\pi\hbar} \sum_{\alpha\beta} \frac{|t_{\alpha\beta}|^2}{1 - \sum_{\beta} |t_{\alpha\beta}|^2}. \tag{6.18}$$

This is the multichannel four-terminal conductance. Similarly, for the multichannel two-terminal conductance we obtain

$$\Gamma = \frac{e^2}{\pi\hbar} \sum_{\alpha\beta} |t_{\alpha\beta}|^2 = \frac{e^2}{\pi\hbar} \mathrm{Tr}\{\mathbf{t}\mathbf{t}^{\dagger}\}. \tag{6.19}$$

It is shown [15] that (6.19), which was derived within the Landauer formalism here, can be derived from the Kubo approach as well. Further comparison of these two approaches has been reported recently [32].

6.3 Calculation of Transport for Contact–Junction–Contact Systems

The NSGFM that is employed here for transport calculations is a generalization of the surface Green function matching approach [3, 33], which was previously applied to some nanotube junctions under equilibrium conditions, i.e., without external bias [34]. Both the equilibrium surface Green function matching and the NSGFM formalisms are independent of the particular description of the electronic structure of the system. In particular, NSGFM can be applied to ab initio or semiempirical model Hamiltonians. Using a tight-binding description, we have previously applied this method to various carbon nanotube systems [4,35–38]. Generalization to an ab initio description with nonorthogonal basis, which we discuss in this chapter, makes it possible to calculate transport characteristics of any desired contact–molecule–contact system with high accuracy. Specifically, we use ab initio modeling in order to calculate transport characteristics of benzene-based and polythiophene-based molecular devices, as examples of the application of our general approach.

6.3.1 General Assumptions

The NSGFM method is schematically shown in Fig. 6.2. As is seen from Fig. 6.2a, an excitation far inside the left contact travels to its surface after

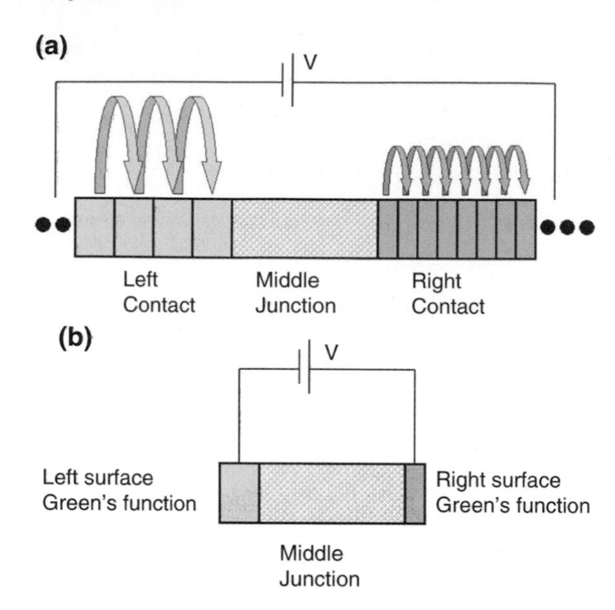

Fig. 6.2. The nonequilibrium surface Green function matching (NSGFM) method. Applying the NSGFM method to an open system (**a**) mathematically closes it (**b**) by introducing the surface Green functions of the contacts

passing through successive layers. After tunneling across the middle junction, the excitation travels from the surface of the right contact to layers far inside. We make use of the NSGFM method in order to calculate the conductance and I–V characteristics of an open system that consists of a general finite system (e.g., the functional molecule) attached on its left-hand side and right-hand side to two semi-infinite contacts or electrodes. The method applies to the general case of nonorthogonal basis. In Fig. 6.2a, the left and right contacts are divided into successive "principal" layers. A principal layer includes the minimum number of unit cells such that each principal layer can be assumed to interact only with its nearest neighboring principal layers. Moreover, the middle molecular junction is assumed to have direct interactions only with the surface principal layers of the left and right contacts. These will be clarified later in terms of Hamiltonian matrices. Throughout this chapter, whenever we use layer we mean principal layer, unless otherwise stated.

6.3.2 Closing the Open System: Surface Green Functions

As is clear from Fig. 6.2a, the general system that we would like to consider for transport calculations is "open." This means that the contacts, which are responsible for applying bias to the functional middle part and letting the current pass through the system, are semi-infinite. The mathematically open system corresponds to the physical case of macroscopic contacts attached to a nanometer-scale functional part. The starting point of the transport calculation is obtaining the necessary Hamiltonian and overlap matrices. As mentioned before, the transport calculation is independent of the particular electronic structure calculation (ESC) procedure employed for obtaining the Hamiltonian and overlap matrices. The only requirement of the ESC is that the Hamiltonian and overlap matrices should be available in some spatially localized basis. Examples include linear combination of atomic orbitals ab initio and tight-binding descriptions. Within the localized basis, the infinite-dimensional Green functions $\mathbf{G}_{\mathrm{L,R}}$ corresponding to the semi-infinite left and right contacts satisfy [39–41]

$$(z\mathbf{S}_{\mathrm{L,R}} - \mathbf{H}_{\mathrm{L,R}})\mathbf{G}_{\mathrm{L,R}} = \mathbf{I}\,. \tag{6.20}$$

Here, $\mathbf{H}_{\mathrm{L,R}}$ and $\mathbf{S}_{\mathrm{L,R}}$ are the infinite-dimensional Hamiltonian and overlap matrices corresponding to the left and right contacts, and z is the complex energy. Projecting (6.20) onto the spaces defined by individual layers, we obtain a series of coupled equations [39, 40] for the layer Green functions $\mathbf{G}^{n,0}_{\mathrm{L,R}}$, where n is the layer index and 0 indicates the surface layer. It should be noted that if the numbers of localized basis functions corresponding to each layer of the left and right contacts are m_{L} and m_{R}, then the matrices $\mathbf{G}^{n,0}_{\mathrm{L}}$ and $\mathbf{G}^{n,0}_{\mathrm{R}}$ will be $m_{\mathrm{L}} \times m_{\mathrm{L}}$ and $m_{\mathrm{R}} \times m_{\mathrm{R}}$, respectively. The same is true for other layer matrices of the left and right contacts, such as Hamiltonian and overlap matrices.

In order to solve the coupled equations for the layer Green functions, the following layer matrices are required: the Hamiltonians of one layer of the left and right contacts, $\mathbf{H}_{\mathrm{L}}$ and $\mathbf{H}_{\mathrm{R}}$; the Hamiltonians coupling one layer to its right neighboring layer for the left and right contacts, $\mathbf{H}^{-1,0}_{\mathrm{L}}$ and $\mathbf{H}^{0,1}_{\mathrm{R}}$; the overlap matrices of one layer of the left and right contacts, $\mathbf{S}_{\mathrm{L}}$ and $\mathbf{S}_{\mathrm{R}}$; and the overlap matrices coupling one layer to its right neighboring layer for the left and right contacts, $\mathbf{S}^{-1,0}_{\mathrm{L}}$ and $\mathbf{S}^{0,1}_{\mathrm{R}}$. These matrices are in fact defined by the particular contact geometries, the method used for ESC and the basis functions used for ESC. Having obtained these matrices from the ESC of the left and right contacts, we employ a powerful algorithm [39,40] that provides the transfer matrices $\mathbf{T}_{\mathrm{L}}$ and $\bar{\mathbf{T}}_{\mathrm{R}}$ responsible for decreasing and increasing the layer indices of $\mathbf{G}^{n,0}_{\mathrm{L}}$ and $\mathbf{G}^{n,0}_{\mathrm{R}}$, respectively;

$$\mathbf{T}_{\mathrm{L}}\, \mathbf{G}^{n,0}_{\mathrm{L}} = \mathbf{G}^{n+1,0}_{\mathrm{L}}; n \to -\infty\,, \tag{6.21}$$

$$\bar{\mathbf{T}}_{\mathrm{R}}\, \mathbf{G}^{n,0}_{\mathrm{R}} = \mathbf{G}^{n-1,0}_{\mathrm{R}}; n \to +\infty\,. \tag{6.22}$$

It should be mentioned that although the original algorithm of López Sancho et al. [39, 40] was proposed for orthogonal basis, the generalization to the case of nonzero overlap is straightforward. It is only necessary to make the substitutions [41, 42]

$$z - \mathbf{H}^{0,0}_{\mathrm{L,R}} \to z\mathbf{S}^{0,0}_{\mathrm{L,R}} - \mathbf{H}^{0,0}_{\mathrm{L,R}}\,, \tag{6.23}$$

$$\mathbf{H}^{0,1(\dagger)}_{\mathrm{L,R}} \to \mathbf{H}^{0,1(\dagger)}_{\mathrm{L,R}} - z\mathbf{S}^{0,1(\dagger)}_{\mathrm{L,R}}\,. \tag{6.24}$$

The transfer matrices $\mathbf{T}_{\mathrm{L}}$ and $\bar{\mathbf{T}}_{\mathrm{R}}$ allow us to mathematically "close" the open contact–junction–contact system by calculating the "surface Green functions" of the left and right contacts, as depicted in Fig. 6.2. The surface Green functions are finite matrices with the dimensionality of one layer of the corresponding contacts, and include the effect of all the layers of the semi-infinite left and right systems. This is carried out by projecting the Green functions $\mathbf{G}_{\mathrm{L}}$ and $\mathbf{G}_{\mathrm{R}}$ of (6.20) onto the spaces of the surface layers of the left and right contacts [3, 33, 34]. The surface Green functions $\mathbf{G}_{\mathrm{s;L}}$ and $\mathbf{G}_{\mathrm{s;R}}$ of the left and right contacts are defined as [3, 33, 34, 39, 40, 43, 44]

$$\mathbf{G}_{\mathrm{s;L}} = [z\mathbf{S}_{\mathrm{L}}^{0,0} - \mathbf{H}_{\mathrm{L}}^{0,0} - (\mathbf{H}_{\mathrm{L}}^{-1,0} - z\mathbf{S}_{\mathrm{L}}^{-1,0})\mathbf{T}_{\mathrm{L}}]^{-1}\,, \tag{6.25}$$

$$\mathbf{G}_{\mathrm{s;R}} = [z\mathbf{S}_{\mathrm{R}}^{0,0} - \mathbf{H}_{\mathrm{R}}^{0,0} - (\mathbf{H}_{\mathrm{R}}^{0,1} - z\mathbf{S}_{\mathrm{R}}^{0,1})^{\dagger}\bar{\mathbf{T}}_{\mathrm{R}}]^{-1}\,. \tag{6.26}$$

Using $\mathbf{T}_{\mathrm{L}}$ and $\bar{\mathbf{T}}_{\mathrm{R}}$, we can calculate the propagation of an excitation from far within the left contact to its surface layer and from the surface layer of the right contact to far within the right contact [34]. It remains to calculate the transition across the middle junction. This is accomplished by matching the surface Green functions of the left and right contacts to the Green function of the middle junction.

6.3.3 Matching at the Junction

The middle junction (e.g., the functional molecule) and its connections to the left and right contacts are characterized by the following matrices: the Hamiltonian of the middle junction, $\mathbf{H}_{\mathrm{M}}$; the Hamiltonians coupling the middle system to the first layers of the left and right contacts, $\mathbf{H}_{\mathrm{LM}}$ and $\mathbf{H}_{\mathrm{MR}}$; the overlap matrix of the basis functions of the middle junction, $\mathbf{S}_{\mathrm{M}}$; and the overlap matrices coupling the middle junction to the first layers of the left and right contacts, $\mathbf{S}_{\mathrm{LM}}$ and $\mathbf{S}_{\mathrm{MR}}$. From the definition of principal layers, it is known that the elements of the $\mathbf{H}$ matrices are negligible unless we consider neighboring layers in the contacts or the interactions of the middle junction and the surface principal layers of the contacts. The aforementioned $\mathbf{H}$ and $\mathbf{S}$ matrices are obtained from ESC, and are considered as input data for the transport calculation. Self-consistency can be achieved by feeding the transport results, in particular the charge distribution and potential drop within the middle junction, back into the ESC, as was carried out for nanotube junctions [4, 35].

In general, the dimensions of the middle junction matrices are different from the dimensions of the matrices of each layer of the left and right contacts; therefore, from the point of view of the dimensions of the matrices involved alone, it is necessary to "match" the surface Green functions of the left and right contacts to the Green function of the middle junction. By doing so, we obtain the projection of the *total* Green function of the system onto the junction region [3, 33]. Matching is necessary from a physical point of view too, as the middle system is coupled to the left and right contacts. The total

Green function of the system, $\mathbf{G}$, is represented by an infinite-dimensional matrix in the localized basis describing the whole system, and satisfies

$$(z\mathbf{S} - \mathbf{H})\mathbf{G} = \mathbf{I}\,. \tag{6.27}$$

After projecting $\mathbf{G}$ onto the space defined by the middle junction, $\mathbf{G}$ turns into $\mathbf{G}_{\mathrm{t;M}}$, i.e., an $m_{\mathrm{M}} \times m_{\mathrm{M}}$ matrix, where we assume that there are m_{M} localized basis functions corresponding to the middle junction. It should be noted that the matrix $\mathbf{G}_{\mathrm{t;M}}$ includes the effects of the semi-infinite contacts coupled to the middle junction system.

In order to match the surface Green functions of the left and right contacts to the Green function of the middle junction, we define the self-energies [2] of the left and right contacts as

$$\mathbf{\Sigma}_{\mathrm{L}} = (\mathbf{H}_{\mathrm{LM}} - z\mathbf{S}_{\mathrm{LM}})^{\dagger}\mathbf{G}_{\mathrm{s;L}}(\mathbf{H}_{\mathrm{LM}} - z\mathbf{S}_{\mathrm{LM}})\,, \tag{6.28}$$

$$\mathbf{\Sigma}_{\mathrm{R}} = (\mathbf{H}_{\mathrm{MR}} - z\mathbf{S}_{\mathrm{MR}})\mathbf{G}_{\mathrm{s;R}}(\mathbf{H}_{\mathrm{MR}} - z\mathbf{S}_{\mathrm{MR}})^{\dagger}\,. \tag{6.29}$$

The total Green function of the system projected onto the junction region, $\mathbf{G}_{\mathrm{t;M}}$, is then obtained by matching at the junction [3, 33, 34, 43, 44]:

$$\mathbf{G}_{\mathrm{t;M}} = (z\mathbf{S}_{\mathrm{M}} - \mathbf{H}_{\mathrm{M}} - \mathbf{\Sigma}_{\mathrm{L}} - \mathbf{\Sigma}_{\mathrm{R}})^{-1}\,. \tag{6.30}$$

It is clear that the surface Green function matching method [3,33] is a powerful approach for obtaining the Green function of a system which generally consists of several subsystems connected together. The Green function of the whole system can be derived from its projection on the interface regions, together with the transfer matrices of the subsystems. The formalism that we present is therefore applicable to the general case of several functional junctions attached by metallic contacts.

6.3.4 Determination of the Fermi Energy

Having obtained the total Green function $\mathbf{G}_{\mathrm{t;M}}$, it is straightforward to calculate the local density of states (LDOS) at any position within the molecular junction. This is simply carried out by summing over the relevant diagonal elements of $-\mathrm{Im}(\mathbf{G}_{\mathrm{t;M}})/\pi$, with Im indicating the imaginary part. One should notice, however, that calculating the normalized LDOS for a general nonorthogonal basis needs some care [45, 46]. From a strict mathematical point of view, the matrix $\mathbf{G}_{\mathrm{t;M}}$, whose derivation we discussed earlier, is the contravariant representation [45, 47] of the total Green function. However, in order to calculate the normalized LDOS, we need the mixed representation, $\mathbf{G}_{\mathrm{m;t;M}}$, that is obtained through

$$\mathbf{G}_{\mathrm{m;t;M}} = \mathbf{G}_{\mathrm{t;M}}(\mathbf{S}_{\mathrm{M}} + \mathbf{D}_{\mathrm{M}})\,, \tag{6.31}$$

in which

$$\mathbf{D}_{\mathrm{M}} = (\mathbf{H}_{\mathrm{LM}} - z\mathbf{S}_{\mathrm{LM}})^{\dagger}\mathbf{G}_{\mathrm{s;L}}\mathbf{S}_{\mathrm{LM}} + (\mathbf{H}_{\mathrm{MR}} - z\mathbf{S}_{\mathrm{MR}})\mathbf{G}_{\mathrm{s;R}}\mathbf{S}^{\dagger}_{\mathrm{MR}} \,. \tag{6.32}$$

Using $\mathbf{G}_{\mathrm{m;t;M}}$, we can obtain the normalized LDOS, from which the charge distribution can be obtained by integration. We perform the LDOS integration in the complex plane [48, 49] to avoid LDOS singularities for real energies. If the total charge of the molecular junction attached to the contacts is known, integrating the normalized LDOS will determine the Fermi energy.

6.3.5 Conductance and Current

The conductance $\Gamma(E,V)$ of the device is related to the transmission probability $T(E,V)$ through the Landauer–Büttiker formula [19, 22, 50]

$$\Gamma(E,V) = \frac{2e^2}{h} T(E,V). \tag{6.33}$$

By expressing the elements of the scattering matrix for different conducting channels in terms of the Green function [15], the transmission probability $T(E,V)$ can be written as [2]

$$T(E,V) = \mathrm{Tr}[\mathbf{\Gamma}_{\mathrm{R}}\ \mathbf{G}_{\mathrm{t;M}}\mathbf{\Gamma}_{\mathrm{L}}\ \mathbf{G}^{\dagger}_{\mathrm{t;M}}], \tag{6.34}$$

in which [2, 42, 47, 51]

$$\mathbf{\Gamma}_{\mathrm{L,R}} = i(\mathbf{\Sigma}_{\mathrm{L,R}} - \mathbf{\Sigma}^{\dagger}_{\mathrm{L,R}}). \tag{6.35}$$

Finally, the I–V relation is obtained as [19]

$$I = \frac{2e}{h}\int_{-\infty}^{+\infty} \mathrm{d}E\, T(E,V)[f_{\mathrm{L}}(E,V) - f_{\mathrm{R}}(E,V)], \tag{6.36}$$

where E is the carrier energy, V is the bias potential and $f_{\mathrm{L,R}}(E,V)$ are the Fermi–Dirac distributions of the contacts.

6.4 The Contact Issue

The molecules that are to be used as the main functional parts of nanoelectronic devices are typically a few angstroms/nanometers across. This gives rise to a natural question; namely, when these small building blocks are attached to the electrodes, how will their main characters of interest depend on the actual arrangement of the atoms in the electrode. This is an issue of concern for both experimentalists and theoreticians. In experiments, it should be known which surface in general, and which part of the surface in particular, is more suited to be used as the electrode. As for the theoretical calculations, it should be known what model should be used in the description of the electrodes. A particular adsorption site, for example, may result in specific transport characteristics that differ from those of the other adsorption sites. The nature of the bonding between the molecule and the electrodes can have a similar effect.

In this subsection, we investigate the aforementioned issues for a typical case of a simple organic molecule, i.e., benzene, attached to two gold electrodes via sulfur clips.

6.4.1 Attachment to Contact as Viewed at Nanoscale

In a pioneering experiment, Reed et al. [52] measured charge transport through single benzene-1,4-dithiol molecules, using a mechanically controllable break junction. Subsequent experimental [47] and theoretical [47, 53] works attempted to explain the observed transport characteristics. Although the explanations were based on adjustable parameters [47] or quantitatively differed [53] from the experimental results, they provided valuable insight into the essential mechanisms behind the transport at the molecular scale.

When the benzene-1,4-dithiol molecules are self-assembled onto the gold surface, even for a clean, defect-free surface, there are several sites where the terminal sulfur atoms can be adsorbed. These include the "hollow" and "on-top" sites depicted in Fig. 6.3. Each of these different adsorption sites can in principle result in different transport properties.

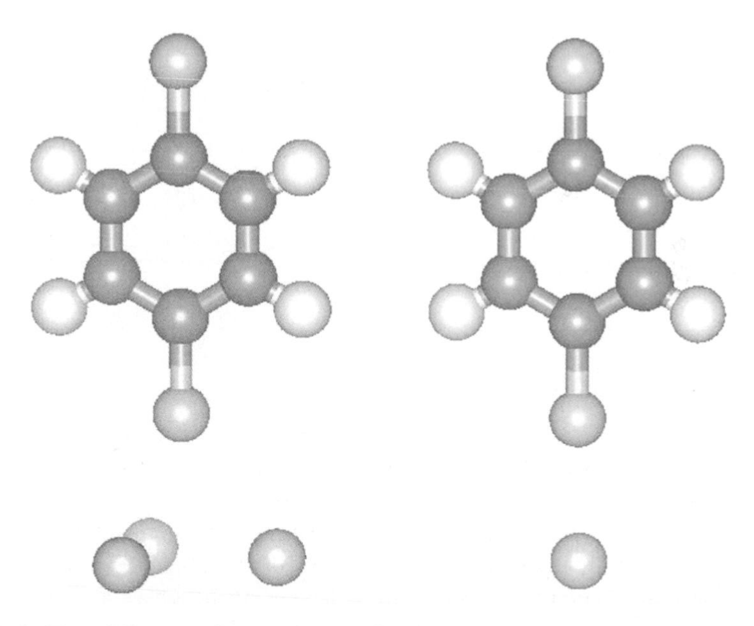

Fig. 6.3. Two different adsorption sites for the benzene-1,4-dithiol molecule on the Au(111) surface: the hollow site (*left*) and the on-top site (*right*)

What makes the situation even more complex is the fact that in the actual experiments, the electrode surface is usually neither clean nor defect-free. Any undesired adatom on the surface, as well as any undesired defect, will cause the molecule to be adsorbed in an unexpected way. This may include strong chemical bonds that normally do not form for the molecule's adsorption on the clean, defect-free surface; therefore, we see that the adsorption geometry can modify the electrode–molecule coupling: For one attachment geometry the coupling may be strong, while for another it may be weak. Moreover, depending the attachment geometry, the electrode may couple more strongly to some particular atoms of the molecule than to the others. Obviously, the contact issue can affect the transport characteristics. We show this in the following subsection by calculating the transport properties of benzene-1,4-dithiol sandwiched between two gold electrodes. As we will see, the conductance and I–V curves strongly depend on the particular model used for the electrodes and the adsorption geometry.

6.4.2 Contact Modeling Effects on Transport Properties

Consider a single benzene-1,4-dithiol attached to two gold electrodes via its two sulfur atoms. There are several ways to model the contacts in the corresponding open system and apply the formalism of Sect. 6.3 in order to calculate conductance and I–V curves. In Fig. 6.4 we show three different models for the contacts.

The continuum contact model [47, 54, 55], depicted in the upper panel of Fig. 6.4, is based on the assumption that the density of states of the contacts is constant (i.e., independent of energy) at and near the Fermi energy. The

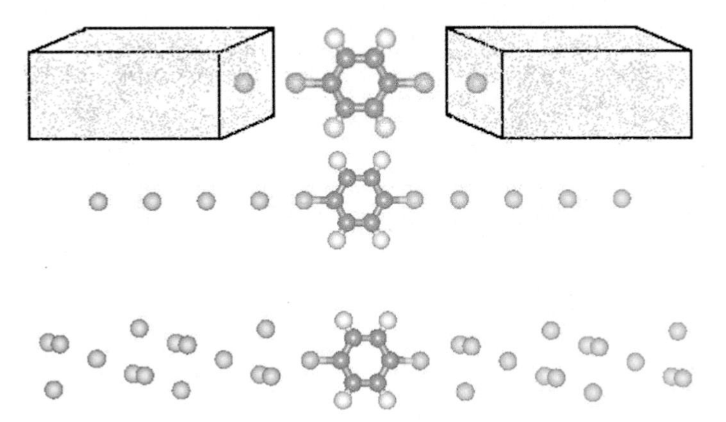

Fig. 6.4. Three different models for the contacts for calculating the transport properties. Continuum contacts with the bulk density of states at Fermi energy (*top*) is the easiest, as the surface Green functions of the contacts are independent of energy and applied bias. The chain (*middle*) and partial Au(111) surface (*bottom*) models are relatively more sophisticated

corresponding density of states and its distribution among states with different symmetries (s, p and d) are taken from separate calculations for bulk gold, as explained, for example, by Papaconstantopoulos [56]. Within this model, the carriers are supposed to be injected from the source into the molecule at the Fermi energy of the source, and to be collected from the molecule in the drain at the Fermi energy of the drain. The surface Green functions $\mathbf{G}_{\mathrm{s;L}}$ and $\mathbf{G}_{\mathrm{s;R}}$ of the left and right contacts, introduced in Sect. 6.3, are assumed to be independent of energy and applied bias. If we assume that the coupling Hamiltonians $\mathbf{H}_{\mathrm{LM}}$ and $\mathbf{H}_{\mathrm{MR}}$ are independent of the applied bias, then the self-energies $\mathbf{\Sigma}_{\mathrm{L}}$ and $\mathbf{\Sigma}_{\mathrm{R}}$ of Sect. 6.3 will also be independent of bias. Therefore, once the conduction Γ has been calculated from (6.33) and (6.34) for zero bias and different carrier energies, the I–V curve is readily obtained via (6.36). Therefore, within the continuum contact model the bias voltage only enters the formalism through the Fermi–Dirac distributions of the contacts; $f_{\mathrm{L,R}}(E, V)$.

Figure 6.5 represents the relaxed structure of a benzene-1,4-dithiol molecule attached to two gold atoms. The structure optimization was achieved using Gaussian 03 software [57] with method/basis b3pw91/lanl2mb. This structure is used to extract the Hamiltonian and overlap matrices describing the molecule, as well as those describing the coupling of the molecule to the contacts, within the continuum approximation. In fact the two gold atoms serve as representing the whole bulk electrodes: The matrices coupling the molecule to these two gold atoms are taken as the matrices coupling the molecule to the bulk electrodes. It should be mentioned that geometry optimization is crucial in this kind of modeling. Using a configuration that is not fully optimized within the desired method/basis for extracting the coupling matrices may cause unphysical results such as negative density of states.

The other two models of the contacts depicted in Fig. 6.4, however, represent multichannel contact configurations [58], with different numbers of channels. For the chain configuration, the distances between the sulfur atoms

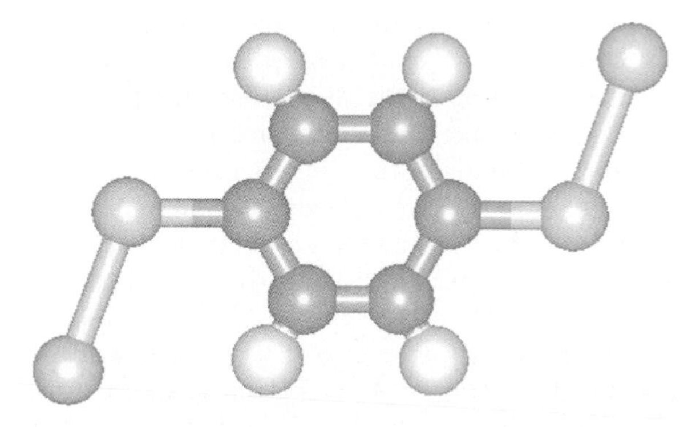

Fig. 6.5. The relaxed structure of a benzene-1,4-dithiol molecule attached to two gold atoms

and the first gold atoms of the chain, as well as the distance between neighboring gold atoms, were taken to be 3.0 Å. For the partial Au(111) configuration, the distances between the sulfur atoms and their nearest gold neighbors were 3.0 Å. This distance was obtained by geometry optimization, while fixing the partial Au(111) configuration, and was used for the chain too. The electronic structures of these two configurations were calculated using Gaussian 98 [59] software with the same method/basis as that used for the bulk modeling. The middle molecules in these configurations are attached to two unit cells of gold nanocontacts. The gold unit cells closest to the molecule were used to extract the coupling matrices, while the outer unit cells served as buffers preventing the unphysical effects of an immediate open boundary. Similar calculations, using only four unit cells of gold atoms, were performed to extract the necessary matrices representing the gold nanocontacts.

Figure 6.6a depicts the calculated density of states projected onto the middle molecular junction, while Fig. 6.6b shows the conductances at different carrier energies. In this figure the middles of the highest occupied molecular orbital (HOMO)–lowest unoccupied molecular orbital (LUMO) gaps, obtained from the cluster ECS calculations, are shifted to zero. Considering these results, and, for example, the results of a previous calculation of ours using different interatomic distances [60], it is obvious that contact modeling crucially affects the transport calculation results. As a result, many theoretical calculations use adjusting parameters in order to reproduce experimental transport observations. One of these assumptions concerns the way the potential drops along the molecule. A self-consistent calculation of the potential-drop pattern, although possible in principle, becomes forbiddingly difficult, especially if the ECS is to be carried out with ab initio accuracy.

For calculating the I–V characteristics of the benzene-1,4-dithiol systems with different model contacts, here we assume that the potential drops evenly at the junctions between the molecule and the nanocontacts. In other words, we take the molecule's potential to be fixed at zero, while shifting the potentials of the Fermi-Dirac distributions of the contacts by $\pm V/2$ for a total potential drop of V. The results are depicted in Fig. 6.7. From this figure, although an overall qualitative agreement with the results in the literature can be seen especially for the Au(111) modeling, the quantitative results are different. The reason, as mentioned earlier, might be the lack of knowledge about the exact junction configuration in experiments, and an ad hoc choice of the contact modeling in theoretical calculation. A model based on more accurate experimental knowledge of the atomistic contact configuration will undoubtedly reconcile the experimental and theoretical transport results. "It is worth mentioning that the choice of the imaginary part of the complex energy, taken to be 0.0001 Hartree in our benzene-1,4-dithiol calculations, also affects the transport results. A larger imaginary part, e.g., results in more smooth conductance curves and a better agreement between the continuum model and experimental current-voltage results."

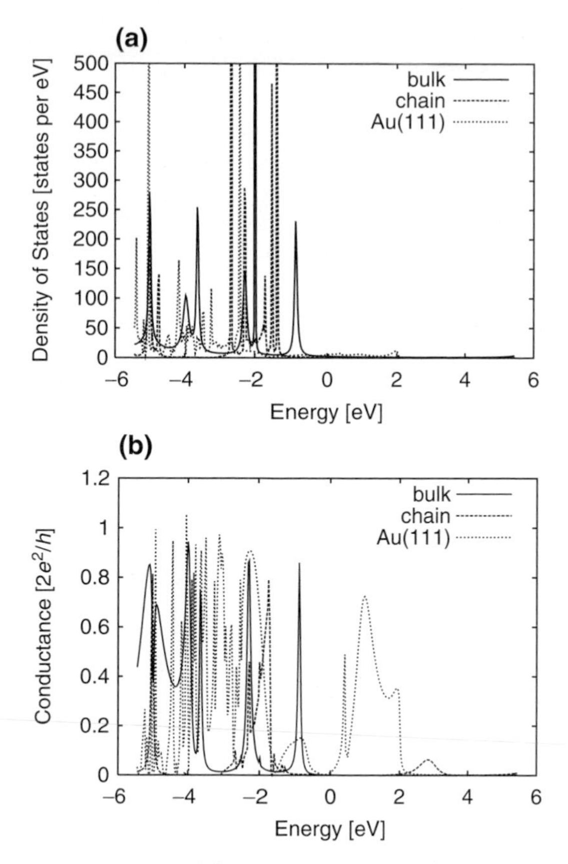

Fig. 6.6. The density of states (**a**) and conductance (**b**) corresponding to different contact models of Fig. 6.4, with the optimized configuration of Fig. 6.5 for the continuum bulk approximation

6.5 Applications of the Method

The general formalism for calculating transport properties of nanoscale systems was introduced in previous subsections. We also showed an application of the method to the case of transport through benzene-1,4-dithiol sandwiched between two gold electrodes, where the significance of different contact models was discussed. Here we present other applications of the method, namely, screening at nanotube junctions, nanoelectromechanical sensors and switches, and a polythiophene-based nanodevice. As we shall see, the transport formalism of Sect. 6.3 is applied to a one orbital per atom tight-binding description of screening at nanotube junctions (Sect. 6.5.1), a four orbital per atom tight-binding description of transport through bent nanotubes (Sect. 6.5.2) and an ab initio description of transport through a polythiophene-based nanodevice (Sect. 6.5.3). The transport formalism of

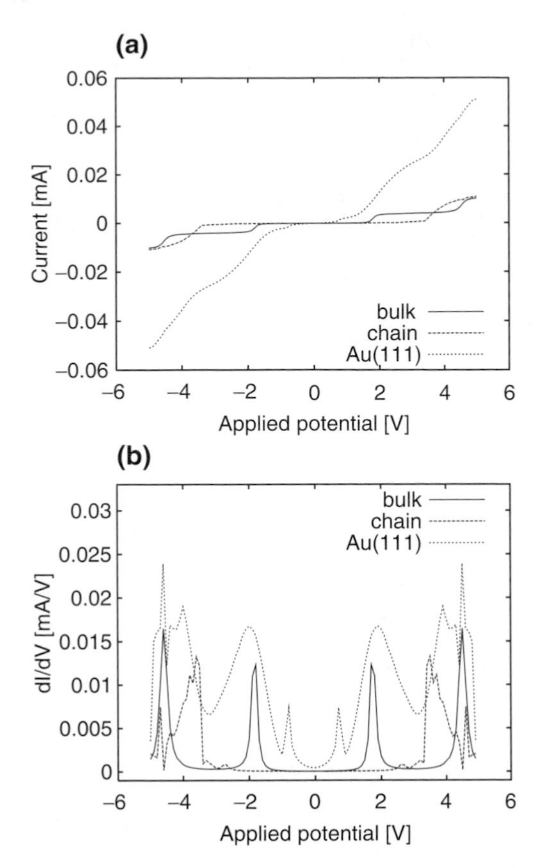

Fig. 6.7. The current–voltage characteristics (**a**) and differential conductance (**b**) corresponding to different contact models of Fig. 6.4, with the optimized configuration of Fig. 6.5 for the continuum bulk approximation

Sec. 6.3 is therefore shown to be quite general and capable of treating systems at the desired level of accuracy. While the ab initio description may be the best choice for accurate calculations of small systems (whose number of atoms is at most a few hundred), various tight-binding schemes can be used to study transport properties of systems with a few thousand atoms.

6.5.1 Screening at Carbon Nanotube Junctions

Consider the nanotube system shown in Fig. 6.1b. A long carbon nanotube is depicted deposited on a double-crystal substrate. One may consider the system shown in this figure to part of an experimental setup in which a nanotube is deposited on a substrate containing two electrodes and an insulating spacer between them, as used, for example, in [61–65]. In general, the work functions of the two surfaces of Fig. 6.1b on which the nanotube is deposited would be different. This results in different amounts of charge transferred to/from the left and right sides of the nanotube. In this subsection, we would like to examine the nanotube junction of Fig. 6.1b when there are different charge transfers and/or biases on the left- and right-hand sides of the junction. In particular, we are interested in the screening of the disturbance caused by the left–right asymmetry of work functions and bias. The same asymmetry can also be introduced by inserting two different types of dopant atoms within the left and right sides [66]. Figure 6.8 schematically represents a (5,5) armchair carbon nanotube with different dopings on the left and right of the junction. We consider an infinite ideal nanotube doped with two different dopings on its left and right sides [4, 35, 66, 67]. Different dopings result in different electronic properties owing to the addition or subtraction of charge and the eventual shifts in the left and right Fermi levels. It is assumed that there are no charged impurities present in the proximity of the junction, and therefore the effect of disorder is neglected in this study.

In order to apply the formalism of Sect. 6.3 self-consistently, we consider a one orbital per atom tight-binding model that is generalized by including long-range Coulomb interactions [68]. Within this model, the Hamiltonian of a perfect, infinite nanotube is written as

$$H = \sum_i \Big(\varepsilon_i + \sum_j \frac{\delta n_j}{\sqrt{r_{ij}^2 + U_{\mathrm{H}}^{-2}}}\Big)\, c_i^\dagger c_i + \sum_{\langle i,j\rangle} V_\pi\, c_i^\dagger c_j, \tag{6.37}$$

where U_{H} is the Hubbard parameter used in describing on-site interactions [69, 70]. Using the experimental estimation of U_{H} for carbon [71], we set $U_{\mathrm{H}} = 4V_\pi$. In (6.37), $\delta n_i = n_i - n_i^0$ is the change in the self-consistent and screened occupation number n_i at site i in the system under bias, as compared with the occupation number imposed by the substrates/dopants, n_i^0. As we mentioned before, the effect of the substrate or dopants is to

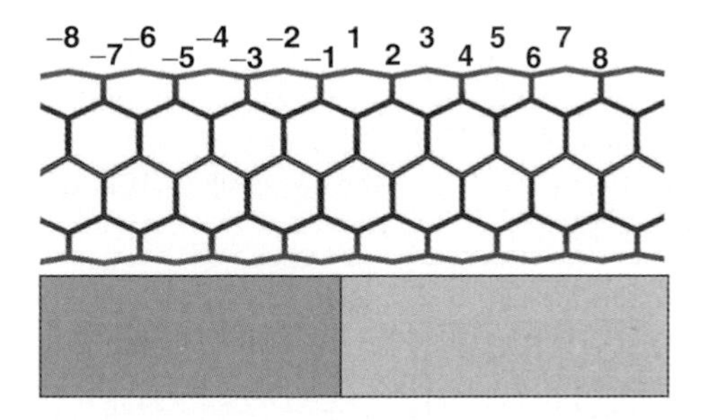

Fig. 6.8. A (5,5) carbon nanotube deposited on a double-crystal substrate. The border between the two bulk electrodes indicates the junction border of the nanotube where the nanotube parts to the left and right would have different charge transfers to/from the substrates. Successive nanotube rings to the left and right of the junction are indicated by *negative and positive integers*, respectively

transfer some charge to the tube. This modifies the on-site energies which we assume to be known for the two semi-infinite tubes. From this, the additional charge can be easily deduced without the need to perform a self-consistent calculation. At the junction, however, a self-consistent calculation needs to be performed in order to determine the charge and potential profiles, even if the two separated half-tubes are treated non-self-consistently. Therefore, δn_i would be the induced charge at site i, in response to the applied bias, taking into account the initial shifts of the chemical potentials, due to, for example, substrate dopings. In our results, the initial shifts of the chemical potentials are indicated by the equivalent doping levels of the on-site energies. We consider two different doping levels; namely, $U_{\mathrm{L,R}} = \pm 0.1 V_\pi$ and $\pm 0.5 V_\pi$. The on-site energies ε_i are therefore the sum of the initial dopings and the applied bias [4,67].

We distinguish three different parts in the infinite nanotube: an unperturbed semi-infinite part to the left, a perturbed finite part in the middle and an unperturbed semi-infinite part to the right. In the unperturbed parts, $\delta n_i = 0$ and n_i^0 is determined solely by the initial shifts of the chemical potentials. This implies that the length of the perturbed finite part, for which the values of δn_i are assumed to be nonzero, is chosen to be sufficiently long such that the applied bias is fully screened within this part of the nanotube. For concreteness, we assume that the external potential drop only modifies the occupations of the sites belonging to 16 carbon rings of the nanotube. The rings are depicted in Fig. 6.8, and are indexed by negative and positive integers to the left and right of the junction, respectively. This number of carbon rings for the perturbed part has proven [4] to be sufficiently long to enable the derivation of the converged induced charges, as increasing the number of rings has had no effect on the self-consistent occupations.

The self-consistent procedure of calculating the induced charge δn_i proceeds as follows [4,35]. Starting with an initial guess of zero δn_i in the Hamiltonian (6.37), we calculate the surface Green functions, $\mathbf{G}_{\mathrm{s;L}}$ and $\mathbf{G}_{\mathrm{s;R}}$, of the two semi-infinite unperturbed parts to the left and right of the perturbed region [39,40]. These surface Green functions are then attached to the Green function of the perturbed middle part, $\mathbf{G}_{\mathrm{M}} = (z - \mathbf{H}_{\mathrm{M}})^{-1}$, where z is the complex energy and $\mathbf{H}_{\mathrm{M}}$ is the Hamiltonian of the perturbed region, in order to obtain the total Green function of the system, $\mathbf{G}_{\mathrm{t;M}}$, projected onto the perturbed region [2,3,33,34,43,44]:

$$\mathbf{G}_{\mathrm{t;M}} = (\mathbf{G}_{\mathrm{M}}^{-1} - \mathbf{H}_{\mathrm{ML}}\mathbf{G}_{\mathrm{s;L}}\mathbf{H}_{\mathrm{LM}} - \mathbf{H}_{\mathrm{MR}}\mathbf{G}_{\mathrm{s;R}}\mathbf{H}_{\mathrm{RM}})^{-1} . \tag{6.38}$$

Here, $\mathbf{H}_{\mathrm{ML}} = \mathbf{H}_{\mathrm{LM}}^{\dagger}$ and $\mathbf{H}_{\mathrm{RM}} = \mathbf{H}_{\mathrm{MR}}^{\dagger}$ matrices indicate the coupling between the perturbed region and the first unit cells of the unperturbed parts to the left and right. Notice that here we use orthogonal basis, and therefore all the overlap matrices of Sect. 6.3 are zero. Using 6.38, we derive the LDOS at site j in the perturbed region, $g_j(E) = [\mathbf{G}_{\mathrm{t;M}}(j,j) - \mathbf{G}_{\mathrm{t;M}}^{*}(j,j)]/2\pi\mathrm{i}$. The induced charges can be obtained through integrating $g_j(E)$. Specifically, we assume that the externally applied potential drops sharply right in the middle of the perturbed part of the system. In other words, the left semi-infinite unperturbed part plus the left eight rings of the perturbed part are assumed to be subject to the external bias $V/2$, while the right eight rings of the perturbed part together with the right semi-infinite unperturbed part are subject to the external bias $-V/2$. (An alternative assumption regarding the external potential drop is explained in [4].) The values of n_i of the left eight rings of the perturbed part are obtained by integrating the LDOS up to $+V/2$, while for the right eight rings the integrations are carried out up to $-V/2$. The values of δn_i are then used to initialize the Hamiltonian for another round of δn_i calculations. This procedure is continued until the induced charges are determined self-consistently. Throughout the self-consistent calculation, the chemical potentials of the left and right parts of the system are assumed to be pinned to $\pm V/2$ as imposed by the external source applying the bias. This together with the specific length of the perturbed region, beyond which the external potential drop is assumed to be screened, constitute the boundary conditions of the system. It should be noted that although the chemical potentials of the left and right parts of the perturbed region are fixed, the potential-drop pattern in this region depends on the values of n_i, and is thus self-consistently calculated. The junction problem is therefore solved in real space within the one-π-orbital tight-binding formalism.

For the (5,5) armchair metallic nanotube depicted in Fig. 6.8, with the initial shifts of chemical potentials $U_{\mathrm{L,R}} = \pm 0.1 V_{\pi}$ and $\pm 0.5 V_{\pi}$, the results of self-consistent charge calculations are shown in Fig. 6.9. In this figure, the results are presented for two values of the external applied bias V. It is clearly seen that the perturbation caused at the junction is effectively screened within a distance of 19.4 Å that includes eight unit cells (i.e., 16 rings) of the carbon nanotube. In other words, the occupations converge to their corresponding values for the semi-infinite left and right nanotube parts, when one moves eight rings away from the junction in each direction. Similar results were obtained for semiconducting nanotubes, and for nanotubes with different diameters and external potential drops [4].

The screening length being of the order of a few carbon-ring separations is an important conclusion with practical consequences: The relatively short screening length makes it possible for different functional parts of nanoelectronic devices based on nanotubes, for example, those of [61–65], to be designed independently. That is, the effects of one junction on another could be neglected if they are separated by a distance safely longer than the relevant screening length.

6.5.2 Transport Through Bent Carbon Nanotubes

In the preceding subsection, we saw how the general formalism of Sect. 6.3 was applied to carbon nanotube junctions within a one-π-orbital tight-binding approximation. As the tight-binding formulation was orthogonal, the overlap

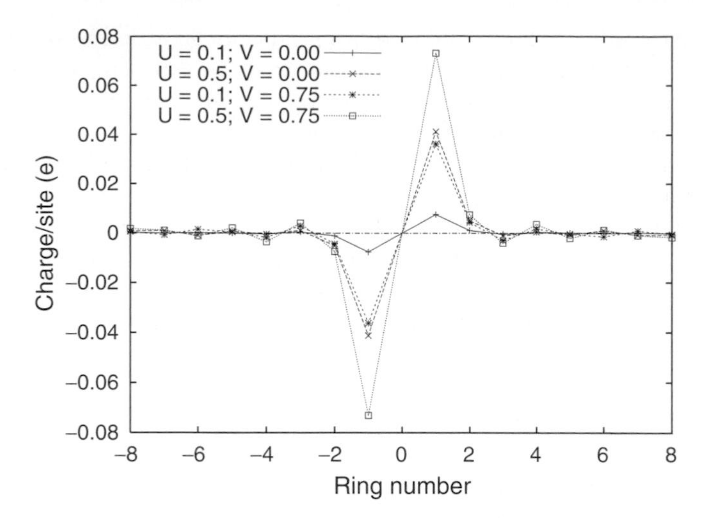

Fig. 6.9. The induced occupations as a result of screening external perturbation in a (5,5) armchair nanotube [67]. U and V indicate the initial shifts of the chemical potentials, and the external bias, respectively. The ring indices correspond to those defined in Fig. 6.8. The unit for energy or voltage is the hopping matrix element $V_\pi = 2.7\,\text{eV}$

matrices were zero. In the present subsection, we again use an orthogonal tight-binding formulation, but make use of a four orbital per atom parameterization introduced by Xu et al. [72] for carbon. This parameterization has proven to provide a transferable potential in tight-binding studies of carbon systems, and has recently been applied to the study of C_{60}-doped nanotubes [73] and nanotubes under large strain [74]. Our aim is to investigate the interrelationship of mechanical response and transport properties of typical metallic and semiconducting carbon nanotubes. We use the general formalism of Sect. 6.3 in order to derive the I–V characteristics of the bent nanotubes. This enables us to show the applicability of the transport of the bent nanotubes in nanoelectromechanical sensors and switches.

Several experiments [75–78] have indicated that, under severe bendings, buckling is the usual way for the nanotubes to reduce strain. The usual approach to the theoretical modeling of the buckling phenomenon has so far made use of classical potentials [76, 79, 80]. Even for calculations of electronic properties of the bent tubes, which make use of relatively more sophisticated quantum tight-binding Hamiltonians, the starting geometry of the bent nanotubes have been obtained using classical potentials [81–83]. The relaxation procedure based on classical potentials does not take into account the increased σ–π hybridization as a result of bending, specially at large bending angles. However, the increased hybridization is expected to have a decisive effect not only on electronic structure, but also on the forces and relaxed atomic configurations themselves.

The interrelationship of the electronic and mechanical properties of carbon nanotubes [81–87] gives rise to natural speculations for possible applications. It has been shown both experimentally [88] and theoretically [89, 90] that the reversible bending of nanotubes can be used to alter their conduction, which, in turn, may be used in nanoelectromechanical switch/sensor applications.

In order to obtain the relaxed structures under bending, and to calculate transport properties, we consider parts of (6,6) armchair and (10,0) zigzag nanotubes that contain 972 and 940 carbon atoms, respectively. The diameters of the (6,6) armchair and the (10,0) zigzag nanotubes are 8.1 and 7.8 Å, respectively. The lengths of these nanotube portions are 98 Å. Considering the large number of atoms in the systems that makes ab initio geometry optimization formidable, and taking into account the disadvantage of using classical potentials mentioned earlier, we chose the four orbital per atom tight-binding approach of Xu et al. [72] for carbon, both to obtain the optimized geometries and to calculate the electronic/transport properties. Geometry optimizations were performed via the *O(N)* density-matrix ESC method of Li et al. [91], combined with the Broyden minimization scheme [92], within the aforementioned tight-binding approach. The optimization of the bent structures proceeds as follows. For successive bending angles, while fixing eight carbon rings (96 atoms for the armchair case, and 80 atoms for the zigzag case) at each end of the nanotube, the structure is optimized such that the maximum force acting on the unconstrained atoms becomes less than 0.05 eV/Å. The results of the geometry optimizations are depicted in Fig. 6.10.

It should be mentioned that the (6,6) and (10,0) nanotubes can be joined to construct a nanotube kink [63, 93, 94] structure, containing a pentagon–heptagon defect. It is shown that such a structure is quite stable even under severe bendings [36, 37].

The transport characteristics of the bent structures shown in Fig. 6.10 were calculated using the general formalism of Sect. 6.3. For calculating conductance, two semi-infinite perfect nanotubes, i.e., "leads," are assumed to be

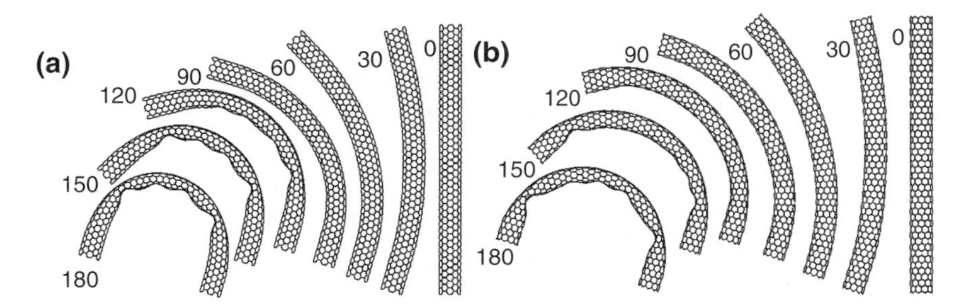

Fig. 6.10. The optimized geometries of the (6,6) armchair (**a**) and the (10,0) zigzag (**b**) nanotubes at different bending angles [36]. The optimized geometries are used to obtain the current that passes through the bent nanotubes

attached to the two ends of the bent region. When a bias voltage V is applied to the bent region, the chemical potentials of the two leads, μ_L and μ_R, are assumed to be pinned at $V/2$ and $-V/2$. This determines the shifts of the band structures, and the density of states, of the leads. This is achieved by shifting the on-site elements of the tight-binding Hamiltonians of the two leads by $V/2$ and $-V/2$. Within the bent region, on the other hand, the on-site shifts are determined by the potential-drop pattern. As for the functional form of the potential drop, which is necessary in calculating I–V characteristics, we assume a linear drop across the bent region. Although, as we showed in the previous subsection, self-consistent calculation of the accurate pattern of the potential drop is possible in principle, applying this approach to the bent nanotubes of this study is currently formidable, owing to the large number of carbon atoms involved. Our assumption of a linear drop is justified by the observation that, as we shall see shortly, the deformations within the bent nanotubes are distributed more or less uniformly.

Within the formalism of Sect. 6.3, we set all overlap matrices equal to zero owing to the orthogonal tight-binding formulation used here. Using the tight-binding Hamiltonians and coupling matrices for the bent systems under bias, we obtain the conductance curves from (6.33). Using the corresponding chemical potentials of the left and right leads, and integrating the conductance curves as formulated in (6.36), we derive the I–V characteristics of the bent nanotubes. The I–V curves are depicted in Fig. 6.11. From this figure, it is evidently seen that as the bending angle increases, the current passing through the bent armchair (6,6) tube decreases, while that of the zigzag (10,0) tube *increases*.

We notice that the conduction of the bent tube under bias is determined by the number of conducting channels available within the two semi-infinite leads and the corresponding transmission coefficient $T(E,V)$ between any pair of them, for the energies belonging to the integration window $f_L(E - \mu_L) - f_R(E - \mu_R)$ in (6.36). The number of conducting channels of the leads at

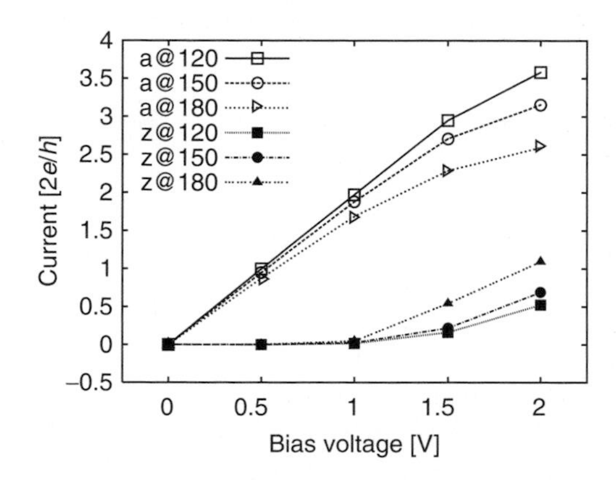

Fig. 6.11. The current–voltage characteristics for the three most bent structures

any energy is independent of the bending angle. The transmission coefficient, on the other hand, depends on the deformation of the LDOS within the bent region, as one moves from the two ends of this region toward its center. Upon a detailed examination of the LDOS within the bent region under nonzero bias [38], we observe that the LDOS in the bent region includes oscillations whose amplitude increases with increasing bending angle. The increase of the amplitude, however, is not symmetric: Taking the LDOS of the straight tube as a reference, on the average, the "positive" amplitudes are dominant over the "negative" ones. This is attributed to the creation of additional localized electron/hole states for higher bendings.

We first consider the case of the metallic (6,6) tube, with a pseudogap of 2.0 eV. For this tube, the number of available conducting channels in one of the leads at any energy in the integration window is equal to the number of bands within the pseudogap, up to bias 2.0 V. Whenever the oscillations of the LDOS of the bent region result in a smaller number of states as compared with that of straight tube at a certain energy, the transmission coefficient $T(E, V)$ reduces—owing to the reduction of the states available for tunneling across the bent region—which results in reduction of conductance. Whenever the LDOS oscillations of the bent region increase the number of states compared with that of the straight tube, however, the conductance cannot increase owing to the limited number of channels within the pseudogap of one of the leads. Therefore, the net effect of LDOS oscillations within the bent region is *reduction* of conductance and current for higher bendings. Next, consider the case of the semiconducting (10,0) tube. In this case, the number of available channels of the leads, within the integration window, is not as restricted as that of the metallic (6,6) tube. In particular, some of the van Hove singularities corresponding to the density of states of one of the leads become aligned with those of the other lead, as a result of a shift of the density of states owing to the applied bias. Therefore, the LDOS oscillations in the bent region can result in both increase and decrease of conductance. However, as the "positive" amplitudes in the LDOS oscillations are on average dominant over the "negative" ones, the conductance and current *increase* at higher bending angles.

From the results depicted in Fig. 6.11 it is evident that there is a correspondence between the bending angle and current of the metallic and semiconducting nanotubes considered here. This correspondence may be employed in order to design nanoelectromechanical sensors and switches: The amount of current that passes through the nanotube depends on its bending, and can be used to measure the amount of bending. For example, if the nanotube is deposited on a double-crystal substrate, like that of Fig. 6.1b, where the two crystals may tilt with respect to each other, then the current passing through the nanotube can be used to measure the tilt angle. On the other hand, the bending of the semiconducting nanotube can cause the current to switch from a negligible to a finite value, as seen in Fig. 6.11 at a fixed bias of, for example, 1.5 V. Taking the example of the nanotube deposited on the double-crystal substrate of Fig. 6.1b and maintained at the desired fixed bias, the turning on of the current would indicate that the tilt angle has exceeded a certain amount.

6.5.3 Transport Characteristics of a Polythiophene-Based Nanodevice

As a final application, a polythiophene molecule that is attached to two gold nanocontacts, is considered (Figs. 6.1, 6.12, top). For this calculation, we used ab initio modeling in order to obtain the necessary input matrices for the formalism of Sect. 6.3. The gold nanocontacts, resembling Q1D systems, are extracted from the bulk gold structure without any reconstruction. The

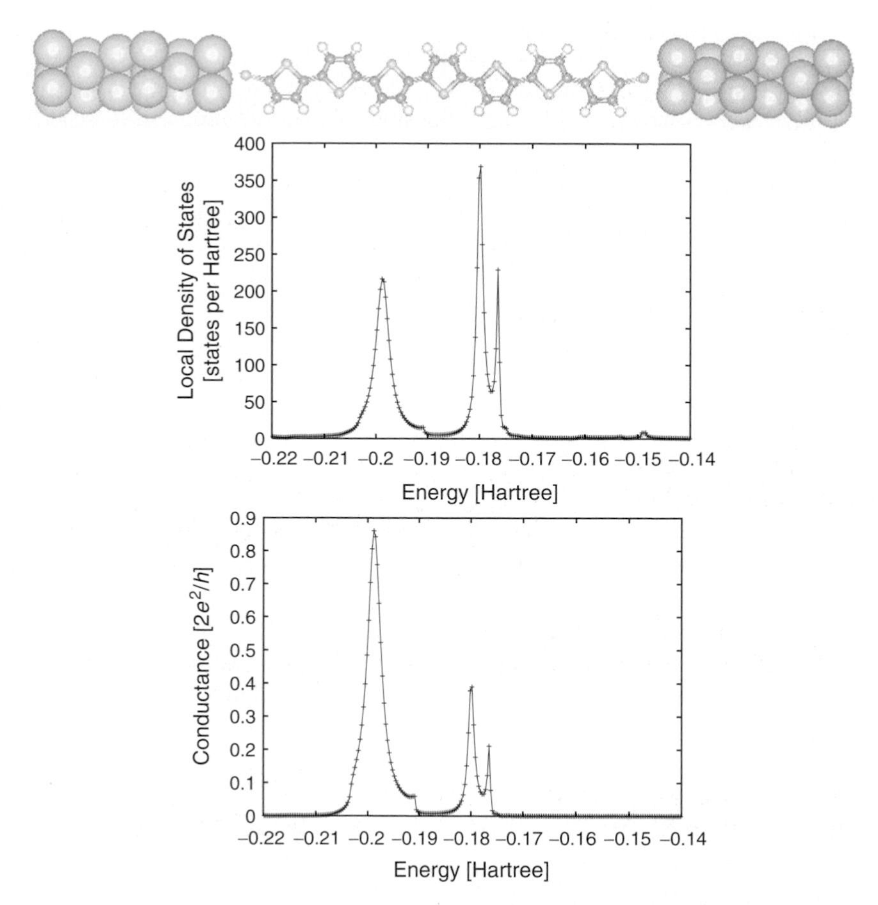

Fig. 6.12. A polythiophene molecule attached to two gold nanocontacts by sulfur clips (*top*). The local density of states projected onto the molecular junction region (*middle*) and the corresponding zero bias conductance (*bottom*)

polythiophene is assumed to be attached to the contacts by two sulfur atoms, located over the center of the surface gold triangles, that act like clips. The cluster structure shown at the top of Fig. 6.12 is used in the ESC by Gaussian 98 [59] with the method/basis b3pw91/lanl2mb, and describes the interaction of the molecule with the surface (principal) layers of the nanocontacts. Each surface gold (principal) layer contains three atomic layers. The extra terminal layers act as buffer layers in order to avoid an open boundary.

From the results of the ESC for the cluster structure, we obtain the Hamiltonian and overlap matrices $\mathbf{H}_{\mathrm{M}}$, $\mathbf{H}_{\mathrm{LM}}$, $\mathbf{H}_{\mathrm{MR}}$, $\mathbf{S}_{\mathrm{M}}$, $\mathbf{S}_{\mathrm{LM}}$ and $\mathbf{S}_{\mathrm{MR}}$. A separate ESC for a four principal layer pure gold structure provides us with $\mathbf{H}_{\mathrm{L,R}}^{0,0}$, $\mathbf{H}_{\mathrm{L,R}}^{0,1(\dagger)}$, $\mathbf{S}_{\mathrm{L,R}}^{0,0}$ and $\mathbf{S}_{\mathrm{L,R}}^{0,1(\dagger)}$. Using these matrices, we obtain the LDOS and zero-bias conductance of Fig. 6.12, through the surface Green function matching method, as described in Sect. 6.3. It is worth mentioning that the gap between the HOMO and the LUMO as derived by the ESC is 0.005 hartree. This indicates the metallic character of the gold clusters attached to the polythiophene.

The ESC of the polythiophene attached to the gold nanocontacts determines the HOMO, HOMO $-$ 2 and HOMO $-$ 6 to be located at -0.174, -0.179 and -0.198 hartree, respectively. From Fig. 6.12 we observe that there are two peaks for HOMO $-$ 2 and HOMO $-$ 6 in the LDOS and conductance characteristics, while no such peak exists for the HOMO. In order to understand the reason, we investigated the spatial distribution of these cluster eigenstates depicted in Fig. 6.13. It is observed that the HOMO is basically localized on the gold nanocontacts, while HOMO $-$ 2 and HOMO $-$ 6 are delocalized and extend over the whole length of the molecule. The delocalization provides

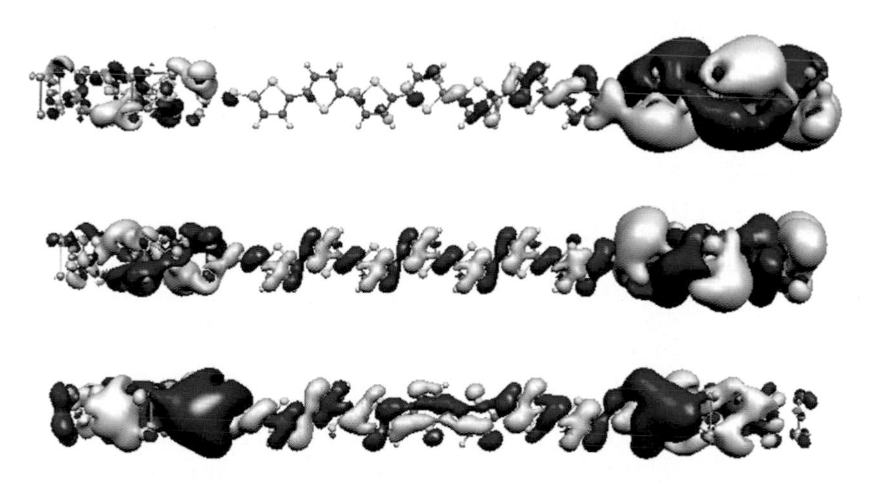

Fig. 6.13. Spatial distribution of the highest occupied molecular orbital (HOMO) (*top*), HOMO $-$ 2 (*middle*) and HOMO $-$ 6 (*bottom*), for the cluster configuration of Fig. 6.12

better tunneling routes, and causes the LDOS and conductance curves to peak at the corresponding energies.

6.6 Conclusions

In this chapter, we introduced and applied some of the essential theoretical and computational approaches that are used for calculating transport in nanoscale systems. We first introduced Landauer's formalism, followed by the nonequilibrium Green function approach that makes use of it within the context of nanoscale transport. The contact issue was next described in some detail, using a typical example of transport through benzene-1,4-dithiol sandwiched between two gold electrodes. Finally, some other applications of the method were discussed, namely, screening at the nanotube junctions, I–V characteristics of bent nanotubes and transport through a polythiophene-based nanodevice. The emphasis was on showing the generality of the method; it was applied to such divers nanodevice models as orthogonal one and four orbitals per atom tight binding as well as ab initio. The resulting transport data were analyzed within the nonequilibrium Green function approach, and were shown to be useful in making application-oriented conclusions.

The importance of the methods introduced in this chapter is evident considering the massive efforts to realize nanoelectronic and molecular electronic components. The huge amount of experimental data needs theoretical explanation, and new configurations of atoms and molecules need to be designed. The methods of this chapter, and similar other ones, will provide partial answers to these needs, and will serve as seeds for future advancements and more complete solutions.

References

1. Y. Imry: In: *Directions in Condensed Matter Physics*, ed by G. Grinstein and G. Mazenko (World Scientific, Singapore 1986), pp. 120–131
2. S. Datta: *Electronic Transport in Mesoscopic Systems* (Cambridge University Press, Cambridge 1995)
3. M. C. Munoz, V. R. Velasco, F. Garcia-Moliner: Prog. Surf. Sci. **26**, 117 (1987)
4. A. A. Farajian, K. Esfarjani, M. Mikami: Phys. Rev. B **65**, 165415 (2002)
5. R. Landauer: Philos. Mag. **21**, 863 (1970)
6. R. Landauer: IBM J. Res. Dev. **1**, 223 (1957). Reprinted in J. Math. Phys. **37**, 5260 (1996) and IBM J. Res. Dev. **44**, 251 (2000)
7. R. Landauer: J. Math. Phys. **37**, 5259 (1996)
8. P. W. Anderson, D. J. Thouless, E. Abrahams, D. S. Fisher: Phys. Rev. B **22**, 3519 (1980)
9. B. S. Andereck, E. Abrahams: J. Phys. C Solid State Phys. **13**, L383 (1980)
10. A. A. Abrikosov: Solid State Commun. **37**, 997 (1981)
11. E. N. Economou, C. M. Soukoulis: Phys. Rev. Lett. **46**, 618 (1981)

12. S. Doniach, E. H. Sondheimer: *Green's Functions for Solid State Physicists*, 2nd edn (Imperial College Press, London 1998)
13. D. J. Thouless: Phys. Rev. Lett. **47**, 972 (1981)
14. E. N. Economou, C. M. Soukoulis: Phys. Rev. Lett. **47**, 973 (1981)
15. D. S. Fisher, P. A. Lee: Phys. Rev. B **23**, 6851 (1981)
16. D. C. Langreth, E. Abrahams: Phys. Rev. B **24**, 2978 (1981)
17. R. Landauer: Phys. Lett. **85A**, 91 (1981)
18. M. Y. Azbel: J. Phys. C: Solid State Phys. **14**, L225 (1981)
19. M. Büttiker, Y. Imry, R. Landauer, S. Pinhas: Phys. Rev. B **31**, 6207 (1985)
20. M. Büttiker: Phys. Rev. B **32**, 1846 (1985)
21. M. Büttiker: Phys. Rev. B **33**, 3020 (1986)
22. M. Büttiker: IBM J. Res. Dev. **32**, 317 (1988)
23. D. Lenstra, R. T. M. Smokers: Phys. Rev. B **38**, 6452 (1988)
24. R. Landauer: J. Phys. Condens. Matter **1**, 8099 (1989)
25. G. Burmeister, K. Maschke, M. Schreiber: Phys. Rev. B **47**, 7095 (1993)
26. K. Maschke, M. Schreiber: Phys. Rev. B **49**, 2295 (1994)
27. C. Berthod, F. Gagel, K. Maschke: Phys. Rev. B **50**, 18299 (1994)
28. R. Hey, K. Maschke, M. Schreiber: Phys. Rev. B **52**, 8184 (1995)
29. H. Vaupel, P. Thomas, O. Kühn, V. May, K. Maschke, A. P. Heberle, W. W. Rühle, K. Köhler: Phys. Rev. B **53**, 16531 (1996)
30. R. Landauer: Z. Phys. B **68**, 217 (1987)
31. D. K. Ferry, S. M. Goodnick: *Transport in Nanostructures* (Cambridge University Press, Cambridge 1997)
32. I. Safi: Phys. Rev. B **55**, R7331 (1997)
33. F. Garcia-Moliner, V. R. Velasco: *Theory of Single and Multiple Interfaces* (World Scientific, Singapore 1992)
34. L. Chico, L. X. Benedict, S. G. Louie, M. L. Cohen: Phys. Rev. B **54**, 2600 (1996)
35. A. A. Farajian, K. Esfarjani, Y. Kawazoe: Phys. Rev. Lett. **82**, 5084 (1999)
36. A. A. Farajian, B. I. Yakobson, H. Mizuseki, Y. Kawazoe: Phys. Rev. B **67**, 205423 (2003)
37. A. A. Farajian, H. Mizuseki, Y. Kawazoe: Physica E **22**, 675 (2004)
38. A. A. Farajian, B. I. Yakobson, H. Mizuseki, Y. Kawazoe: Int. J. Nanoscience **3**, 131 (2004)
39. M. P. López Sancho, J. M. López Sancho, J. Rubio: J. Phys. F Met. Phys. **14**, 1205 (1984)
40. M. P. López Sancho, J. M. López Sancho, J. Rubio: J. Phys. F Met. Phys. **15**, 851 (1985)
41. E. Artacho, F. Ynduráin: Phys. Rev. B **44**, 6169 (1991)
42. M. Buongiorno Nardelli: Phys. Rev. B **60**, 7828 (1999)
43. P. A. Lee, D. S. Fisher: Phys. Rev. Lett. **47**, 882 (1981)
44. A. MacKinnon: Z. Phys. B Condens. Matter **59**, 385 (1985)
45. D. Lohez, M. Lannoo: Phys. Rev. B **27**, 5007 (1983)
46. K. W. Sulston, S. G. Davison: Phys. Rev. B **67**, 195326 (2003)
47. W. Tian, S. Datta, S. Hong, R. Reifenberger, J. I. Henderson, C. P. Kubiak: J. Chem. Phys. **109**, 2874 (1998)
48. A. R. Williams, P. J. Feibelman, N. D. Lang: Phys. Rev. B **26**, 5433 (1982)
49. J. Taylor, H. Guo, J. Wang: Phys. Rev. B **63**, 245407 (2001)
50. R. Landauer: IBM J. Res. Dev. **32**, 306 (1988)

51. A. Rochefort, P. Avouris, F. Lesage, D. R. Salahub: Phys. Rev. B **60**, 13824 (1999)
52. M. A. Reed, C. Zhou, C. J. Muller, T. P. Burgin, J. M. Tour: Science **278**, 252 (1997)
53. M. Di Ventra, S. T. Pantelides, N. D. Lang: Phys. Rev. Lett. **84**, 979 (2000)
54. L. E. Hall, J. R. Reimers, N. S. Hush, K. Silverbrook: J. Chem. Phys. **112**, 1510 (2000)
55. P. A. Derosa, J. M. Seminario: J. Phys. Chem. B **105**, 471 (2001)
56. D. A. Papaconstantopoulos: *Handbook of the Band Structure of Elemental Solids* (Plenum, New York 1986)
57. M. J. Frisch, G. W. Trucks, H. B. Schlegel, G. E. Scuseria, M. A. Robb, J. R. Cheeseman, J. A. Montgomery Jr, T. Vreven, K. N. Kudin, J. C. Burant, J. M. Millam, S. S. Iyengar, J. Tomasi, V. Barone, B. Mennucci, M. Cossi, G. Scalmani, N. Rega, G. A. Petersson, H. Nakatsuji, M. Hada, M. Ehara, K. Toyota, R. Fukuda, J. Hasegawa, M. Ishida, T. Nakajima, Y. Honda, O. Kitao, H. Nakai, M. Klene, X. Li, J. E. Knox, H. P. Hratchian, J. B. Cross, C. Adamo, J. Jaramillo, R. Gomperts, R. E. Stratmann, O. Yazyev, A. J. Austin, R. Cammi, C. Pomelli, J. W. Ochterski, P. Y. Ayala, K. Morokuma, G. A. Voth, P. Salvador, J. J. Dannenberg, V. G. Zakrzewski, S. Dapprich, A. D. Daniels, M. C. Strain, O. Farkas, D. K. Malick, A. D. Rabuck, K. Raghavachari, J. B. Foresman, J. V. Ortiz, Q. Cui, A. G. Baboul, S. Clifford, J. Cioslowski, B. B. Stefanov, G. Liu, A. Liashenko, P. Piskorz, I. Komaromi, R. L. Martin, D. J. Fox, T. Keith, M. A. Al-Laham, C. Y. Peng, A. Nanayakkara, M. Challacombe, P. M. W. Gill, B. Johnson, W. Chen, M. W. Wong, C. Gonzalez, J. A. Pople: Gaussian 03, revision B.04. Gaussian, Pittsburgh, 2003
58. E. G. Emberly, G. Kirczenow: Phys. Rev. B **58**, 10911 (1998)
59. M. J. Frisch, G. W. Trucks, H. B. Schlegel, G. E. Scuseria, M. A. Robb, J. R. Cheeseman, V. G. Zakrzewski, J. A. Montgomery Jr, R. E. Stratmann, J. C. Burant, S. Dapprich, J. M. Millam, A. D. Daniels, K. N. Kudin, M. C. Strain, O. Farkas, J. Tomasi, V. Barone, M. Cossi, R. Cammi, B. Mennucci, C. Pomelli, C. Adamo, S. Clifford, J. Ochterski, G. A. Petersson, P. Y. Ayala, Q. Cui, K. Morokuma, P. Salvador, J. J. Dannenberg, D. K. Malick, A. D. Rabuck, K. Raghavachari, J. B. Foresman, J. Cioslowski, J. V. Ortiz, A. G. Baboul, B. B. Stefanov, G. Liu, A. Liashenko, P. Piskorz, I. Komaromi, R. Gomperts, R. L. Martin, D. J. Fox, T. Keith, M. A. Al-Laham, C. Y. Peng, A. Nanayakkara, M. Challacombe, P. M. W. Gill, B. Johnson, W. Chen, M. W. Wong, J. L. Andres, C. Gonzalez, M. Head-Gordon, E. S. Replogle, J. A. Pople: Gaussian 98, revision A.11.1. Gaussian, Pittsburgh, 2001
60. A. A. Farajian, R. V. Belosludov, H. Mizuseki, Y. Kawazoe: Physica E **18**, 253 (2003)
61. S. J. Tans, M. H. Devoret, H. Dai, A. Thess, R. E. Smalley, L. J. Geerligs, C. Dekker: Nature **386**, 474 (1997)
62. S. J. Tans, A. R. M. Verschueren, C. Dekker: Nature **393**, 49 (1998)
63. Z. Yao, H. W. Ch. Postma, L. Balents, C. Dekker: Nature **402**, 273 (1999)
64. H. W. C. Postma, T. Teepen, Z. Yao, M. Grifoni, C. Dekker: Science **293**, 76 (2001)
65. A. Bachtold, P. Hadley, T. Nakanishi, C. Dekker: Science **294**, 1317 (2001)
66. K. Esfarjani, A. A. Farajian, Y. Hashi, Y. Kawazoe: Appl. Phys. Lett. **74**, 79 (1999)

67. K. Esfarjani, A. A. Farajian, Y. Kawazoe, S. T. Chui: J. Phys. Soc. Jpn. **74**, 515 (2005)
68. K. Harigaya, S. Abe: Phys. Rev. B **49**, 16746 (1994)
69. F. Liu: Phys. Rev. B **52**, 10677 (1995)
70. K. Esfarjani, Y. Kawazoe: J. Phys. Condens. Matter **10**, 8257 (1998)
71. R. G. Pearson: Inorg. Chem. **27**, 734 (1988)
72. C. H. Xu, C. Z. Wang, C. T. Chan, K. M. Ho: J. Phys. Condens. Matter **4**, 6047 (1992)
73. A. A. Farajian, M. Mikami: J. Phys. Condens. Matter **13**, 8049 (2001)
74. T. Ozaki, Y. Iwasa, T. Mitani: Phys. Rev. Lett. **84**, 1712 (2000)
75. J. F. Despres, E. Daguerre, K. Lafdi: Carbon **33**, 87 (1995)
76. S. Iijima, C. Barbec, A. Maiti, J. Bernholc: J. Chem. Phys. **104**, 2089 (1996)
77. O. Lourie, D. M. Cox, H. D. Wagner: Phys. Rev. Lett. **81**, 1638 (1998)
78. T. Kizuka: Phys. Rev. B **59**, 4646 (1999)
79. B. I. Yakobson, C. J. Brabec, J. Bernholc: Phys. Rev. Lett. **76**, 2511 (1996)
80. C. F. Cornwell, L. T. Wille: Solid State Commun. **101**, 555 (1997)
81. A. Rochefort, D. R. Salahub, P. Avouris: Chem. Phys. Lett. **297**, 45 (1998)
82. A. Rochefort, P. Avouris, F. Lesage, D. R. Salahub: Phys. Rev. B **60**, 13824 (1999)
83. M. B. Nardelli, J. Bernholc: Phys. Rev. B **60**, R16338 (1999)
84. S. Paulson, M. R. Falvo, N. Snider, A. Helser, T. Hudson, A. Seeger, R. M. Taylor, R. Superfine, S. Washburn: Appl. Phys. Lett. **75**, 2936 (1999)
85. P. E. Lammert, P. Zhang, V. H. Crespi: Phys. Rev. Lett. **84**, 2453 (2000)
86. D. Tekleab, D. L. Carroll, G. G. Samsonidze, B. I. Yakobson: Phys. Rev. B **64**, 035419 (2001)
87. D. Bozovic, M. Bockrath, J. H. Hafner, C. M. Lieber, H. Park, M. Tinkham: Appl. Phys. Lett. **78**, 3693 (2001)
88. T. W. Tombler, C. Zhou, L. Alexseyev, J. Kong, H. Dai, L. Liu, C. S. Jayanthi, M. Tang, S. Y. Wu: Nature **405**, 769 (2000)
89. L. Liu, C. S. Jayanthi, M. Tang, S. Y. Wu, T. W. Tombler, C. Zhou, L. Alexseyev, J. Kong, H. Dai: Phys. Rev. Lett. **84**, 4950 (2000)
90. A. Maiti, A. Svizhenko, M. P. Anantram: Phys. Rev. Lett. **88**, 126805 (2002)
91. X. -P. Li, R. W. Nunes, D. Vanderbilt: Phys. Rev. B **47**, 10891 (1993)
92. K. Ohno, K. Esfarjani, Y. Kawazoe: *Computational Materials Science from ab Initio to Monte Carlo Methods* (Springer, Berlin Heidelberg New York 1999)
93. P. Lambin, A. Fonseca, J. P. Vigneron, J. B. Nagy, A. A. Lucas: Chem. Phys. Lett. **245**, 85 (1995)
94. J. Han, M. P. Anantram, R. L. Jaffe, J. Kong, H. Dai: Phys. Rev. B **57**, 14983 (1998)

7 Molecular Wires

Hitoshi Nejo

7.1 Introduction

The molecular electronics age has started [1,3–6,224] using the ubiquitous underlying law of various types of nanosystem behavior. This law is incarnated as a huge variety of organic molecules and biomolecules. Future electronics may use molecules advantageously in the fabrication of nanometer-scale structures; molecular self-assembly may be used for this purpose. Molecular self-assembly is thought to be related to molecular interactions on molecular or even atomic scales. Even though the self-assembly phenomenon can be used for fabricating nanostructures, the mechanism of self-assembly itself is not well understood. Attempts to control self-assembly are still under way, although there has been much research in this direction to fabricate nanostructures. As a matter of course, great effort has been made to get molecular or even atomic resolution images using local probes, to reveal the role of organic functional groups in this self-assembly effect. Prediction was made from scanning tunneling microscope images of furan and pyrrole on Pd(111) [7]. In this chapter, not only fabrication of two-dimensional monolayer structures and molecular wires but also technologies to characterize electronic transport in them are reviewed.

Nowadays, it is possible to control intramolecular-level conformation and hence then the characteristics of materials for further engineering. Controllability of conformational and interpolymer effects in conjugated polymers has been shown [8]. Also, conductance switching in single molecules through conformational changes has been demonstrated [9,10]. The capability to control individual single atoms has been applied to individual organic molecules using local probes [11], and the molecular flexure and motion has been used for controlled positioning of individual molecules [12]. One drawback of positioning individual molecules using probes is it is time-consuming. Molecular self-assembly in a desired way has been sought to replace the individual manipulation technique previously described. Recently the possibility of tuning the epitaxy of large aromatic adsorbates by molecular design has been demonstrated [13]. Since the conformation change reflects the vibrational modes between atoms within molecules, it is important to try to detect vibrational modes which characterize each molecular species, and the individual vibrational modes within a single molecule can be actually probed [14–16].

The fullerene molecule, among many types of organic molecules, is one of the promising candidates for fabricating single-molecule-based structures and has been well studied. To clarifying the characteristics of fullerenes, they have to be supported on substrates and hence the effects of the underlying substrates have been studied, including C_{60} bonding and energy-level alignment on metal and semiconductor surfaces [17–23]. A lot of studies of intramolecular structures have been done [24–29], and they are thoroughly reviewed in other chapters in this book.

The cooperative self-assembly of Au atoms and C_{60} on Au(111) surfaces [30] and self-trapped polaron excitation in neutral fullerene C_{60} have been shown [31] from the viewpoint of manipulation of individual fullerene molecules. Lattice distortion and energy-level structures in doped C_{70} [32,33] and doping effects and electronic states in C_{60} polymer have been studied theoretically [34].

Although a lot of work has been done to characterize the fullerene molecule formed, the formation mechanisms of the fullerenes are still being clarified. Electron spin resonance and optical properties of the radical anion of C_{60} have been studied [35]. Remarkable technologies have been used to do this, such as the gas-phase production and photoelectron spectroscopy of the smallest fullerene, C_{20} [36], and controlled atomic doping of a single C_{60} molecule [37]. Nanomechanical oscillations in a single-C_{60} transistor have been shown [38] as a remarkable application of the fullerene molecule.

After it had been shown that fullerene molecules are a good basis for establishing nanostructures, carbon nanotubes (CNTs) were shown to be excellent candidates for fabricating nanostructures. The electronic structures of atomically resolved CNTs were determined in the 1990s [39]. The length control of individual CNTs has been shown by nanostructuring [40] for fabrication of desired structures. The reversible intercalation of charged iodine chains into CNT ropes has been shown [41] for engineering the electronic characteristics of CNT. There have been many works from the chemistry side on the handling CNTs. Chromatography of CNTs has developed for getting desired CNTs out of the pristine materials in order to process CNTs for technical applications [42, 43]. Controlled adsorption of CNTs on chemically modified electrode arrays has been shown for characterization of the conductivity of nanotubes [44–46]. There have been a lot of works from the physics side aimed at revealing the characteristics of CNTs. A lot of unique characteristics of CNTs have been investigated owing to their ideally perfect one-dimensionality and high aspect ratio. As one example, supercurrents through single-walled CNTs (SWNTs) have been shown recently [47]. They are also well reviewed in other chapters in this book.

Organic molecules show a wide variety of characteristics and noble features owing to their ability to be tuned at functional groups. Recently, single-molecule-level engineering has been shown, such as single photons on demand from a single molecule at room temperature [48], a soluble and air-stable organic semiconductor with high electron mobility [49], efficient organic photovoltaic diodes based on doped pentacene [50], and molecular-scale interface engineering for polymer light-emitting diodes [51]. Bright high-efficiency blue organic light-emitting diodes [52], the orientation-controlled organic electroluminescence of *p*-sexiphenyl films [53] and blue–green stimulated emission using a high-gain conjugated polymer [54] have been shown to be further applications. It was in 1980s that Coulomb blockade was first observed at room temperature [55]. A Coulomb staircase at room temperature in a self-assembled molecular nanostructure has been shown [56]. Even atomic-scale coupling of photons to single-molecule junctions has been successfully shown [57].

This chapter describes some possible techniques to characterize such a small object, a molecule. Near-field optical microscopy is a powerful tool for characterizing molecular systems and miniaturization of the probes has been successfully shown [58]. Also a micromachined aperture probe tip was fabricated for multifunctional scanning probe microscopy [59]. Illuminating single molecules in condensed matter has been shown to be possible without probes [60].

A low-energy electron projection microscope is a useful tool for investigating free-standing objects such as CNTs and is one of the useful techniques for evaluating molecules. The overall concepts and instrumentation can be found, for example, in [61, 62]. Quantum-mechanical theory of field electron emission under axially symmetric forces has been developed [63, 64].

This chapter is arranged as follows. Sections 7.2–7.5 review the molecular assembly techniques and molecular wire fabrication based on them. Sections 7.6–7.8 review the control of individual molecular wires which have been fabricated in advance, using probes. Sections 7.9 and 7.10 present typical examples of the application of CNTs and evaluation of technology, respectively.

7.2 1-Dodecanethiol Molecules on Graphite

The use of scanning tunneling microscopy (STM) to investigate the arrangement of organic molecules on surfaces has been of great interest in the last few years [65]. Many molecular arrangements have been successfully revealed, such as those of liquid crystals [66, 67], *n*-alkanes [68, 69], alkoxylated molecules [70], alkanols and alkanethiols [71]. Data obtained from STM images can help to understand the relative importance of molecule–molecule and molecule–substrate interactions as well as the types of forces responsible for a particular packing order on the surface. On the other hand, adsorption of alkanethiols onto graphite provides a good model system for the study of molecular adsorption and the structure of the adsorbed layer. Previous experiments [68–72] have suggested that the molecules are adsorbed with their long axis parallel to the graphite surface. The results have been conformed in STM studies of hydrocarbons on graphite, which reveal that the molecules lie on the graphite surface and exhibit a high degree of order.

Sheng et al. reported an STM investigation of an ordered 1-dodecanethiol layer which was formed on graphite substrates by a simple thermodynamic deposition method. The substrates were prepared by cleaving highly oriented pyrolytic graphite (HOPG) through tape rendering of surfaces. 1-Dodecanethiol, $CH_3(CH_2)_{11}SH$, has a molecular length of approximately 1.5 nm and a van der Waals diameter of approximately 0.45 nm [73], as shown in Fig. 7.1. A drop of the solution was deposited on the substrate. An STM image of 1-dodecanethiol on graphite is shown in Fig. 7.2, which was taken at sample bias $V = 1000\,\mathrm{mV}$ and tunneling current $I = 150\,\mathrm{pA}$. A formation of rows is obtained on the surface of the HOPG substrate with a periodicity of 5.5 nm across the rows. According to the width of the rows it appears that three molecules are arranged along the molecular axis (i.e., perpendicular to the rows). Figure 7.3 shows a higher-resolution STM image of 1-dodecanethiol taken at bias voltage $V = 407\,\mathrm{mV}$ and tunneling current $I = 850\,\mathrm{pA}$. The image further resolves the aforementioned 5.5-nm wide row into two subrows, 1.5 and 3.5 nm in width, respectively. Since approximately 1.5 nm is the molecular length, this observation suggests that the 1.5-nm wide subrow corresponds to a single molecule and the 3.5-nm wide one to two

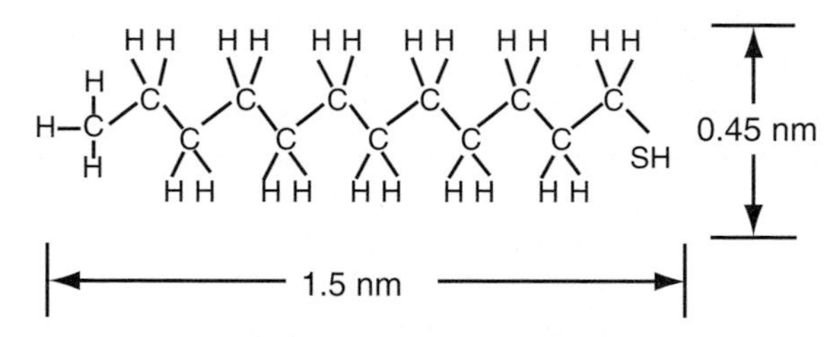

Fig. 7.1. Chemical structure of 1-dodecanethiol [65]

Fig. 7.2. A 32 × 32 nm^2 scanning tunneling microscopy (*STM*) image of 1-dodecanethiol adsorbed on highly oriented pyrolytic graphite (*HOPG*), taken at a bias voltage $V = 1000\,\mathrm{mV}$ and tunneling current $I = 150\,\mathrm{pA}$ [65]

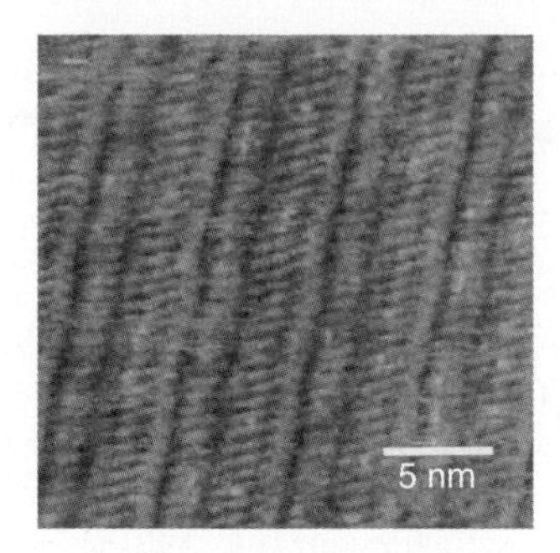

Fig. 7.3. Molecular-resolution STM image ($23 \times 23\,\mathrm{nm}^2$) of 1-dodecanethiol on HOPG. The image was taken at a bias voltage $V = 407\,\mathrm{mV}$ and tunneling current $I = 850\,\mathrm{pA}$ [65]

molecules in series. These dimension values also indicate that the molecules lie on the surface. The periodicity vertical to the molecular axis is approximately 0.55 nm, comparable with the van der Waals diameter of the molecule. The observation of the ordered self-assembled monolayers (SAMs) of 1-dodecanethiol on graphite with parallel orientation indicates a physisorption interaction between adsorbate molecules and the surface. Thermodynamics measurements suggest that the long-chain hydrocarbon molecules have a high affinity for graphite by physisorption [74]. In addition, the image in Fig. 7.3 shows discontinuities in the bright and dark bands. We attribute the bright bands to the thiol (SH) functional group at the end of the carbon chain since sulfur atoms in the long-chain molecule dihexadecyl disulfide have been reported to have enhanced conductivity compared with the alkyl chains when imaged on graphite [70]. This observation indicates that the current tunneling into the thiol functional group is larger than that into the hydrocarbon chain and the methyl (CH_3) functional group. This is similar to the STM results observed for liquid crystals, in which the aromatic rings appear much brighter than the alkyl chains attached to the rings [66, 67]. Hence, the dark subrows probably correspond to the methyl ends of the molecules. According to the locations and sizes of the bright and dark features in Fig. 7.3, it appears that, with a single row, the SH groups face head-to-head in the 3.5-nm wide subrow, while a head-to-tail arrangement exists between the two subrows. The sequence between rows is tail-to-tail. In other words the order sequence across rows is ||...S...C||C...S|C...SS...C||C...S...||. In order to explain the observed pattern, we propose a model for the arrangement of the molecules on the graphite surface, as shown in Fig. 7.4. The open and shaded circles in Fig. 7.4 represent SH and CH_3 terminal groups, respectively. The head-to-head and head-to-tail arrangements of the functional groups imply van der Waals interactions between the functional groups. The molecules forming a parallel structure can be explained assuming that the molecules are commensurate with the graphite lattice. The long hydrocarbon chains

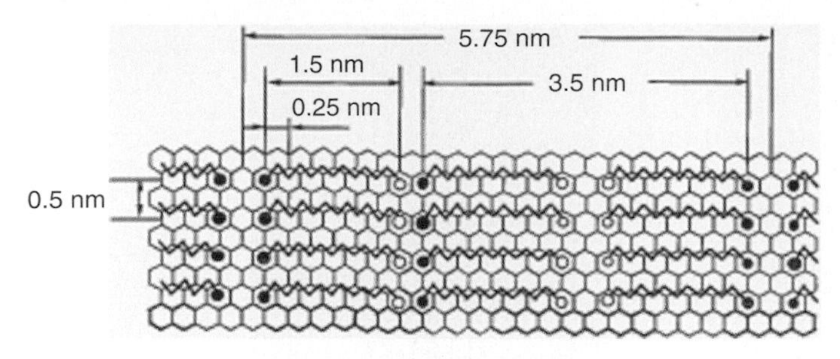

Fig. 7.4. The molecular arrangement for the packing of 1-dodecanethiol on graphite. The *open circles* and the *shaded circles* represent thiol (SH) and methyl (CH_3) terminal groups, respectively [65]

have an affinity for the graphite surface, and this affinity has been attributed to almost perfect match of the distance between the C–C–C zigzag group along the hydrocarbon chain (0.251 nm) and the hole in the hexagonal lattice of graphite (0.246 nm) [74]. This match promotes commensurate adsorption. Some hydrogen bonds may be formed if the hydrocarbon backbone of the CH_3 groups lies in the hollows of the graphite lattice. The stability of the system is contributed by the lattice match between the 1-dodecanethiol hydrocarbon chains and the graphite substrate as well as the hydrogen-bonding network.

7.3 Two-Dimensional Ordering of Octadecanethiol Molecules on Graphite

Molecular-resolution images of *n*-octadecanethiol monolayer molecules on a graphite substrate have been successfully obtained (Fig. 7.5) with a scanning tunneling microscope in air [75]. The images reveal that the molecules arrange in a hexagonal p(3×3) structure with a periodicity of 7.5 Å, a feature associated with the graphite substrate. X-ray photoelectron spectroscopy (XPS) studies suggest that the methyl terminal group is adsorbed on the graphite (Fig. 7.6). A model for the possible arrangement of the molecules on graphite surfaces is proposed (Fig. 7.7).

7.4 Silver Deposited Octanethiol Self-Assembled Monolayers

SAMs of thiol molecules on Au(111) have attracted much attention because of the simplicity of preparation and their applicability [90]. Especially, their applications as tunnel barriers and insulating layers are expected in the field of nanoscale electronics, because the film thickness can be precisely controlled on the atomic scale by changing the length of molecules. The interface structure

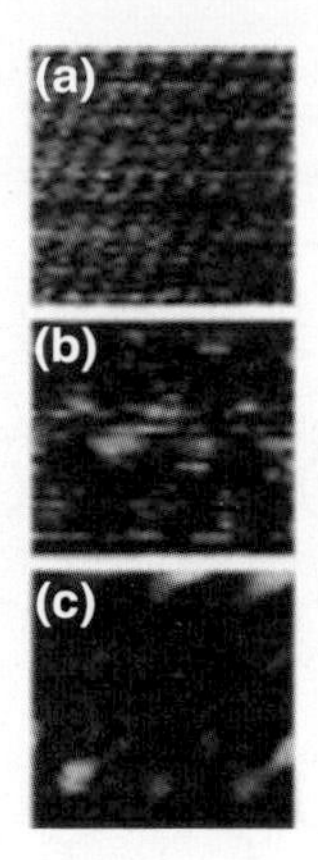

Fig. 7.5. **(a)** STM image of n-octadecanethiol adsorbed on a HOPG substrate. The image was taken in constant-current mode at $V = 200\,\text{mV}$ and $I = 1.52\,\text{nA}$ in an area $8\,\text{nm} \times 8\,\text{nm}$. **(b)** A small scanning area STM image of molecules; bias voltage $V = 200\,\text{mV}$, tunneling current $I = 1.52\,\text{nA}$, image area $2.7\,\text{nm} \times 2.7\,\text{nm}$. **(c)** A fast Fourier transform filtered image of **(b)** [75]

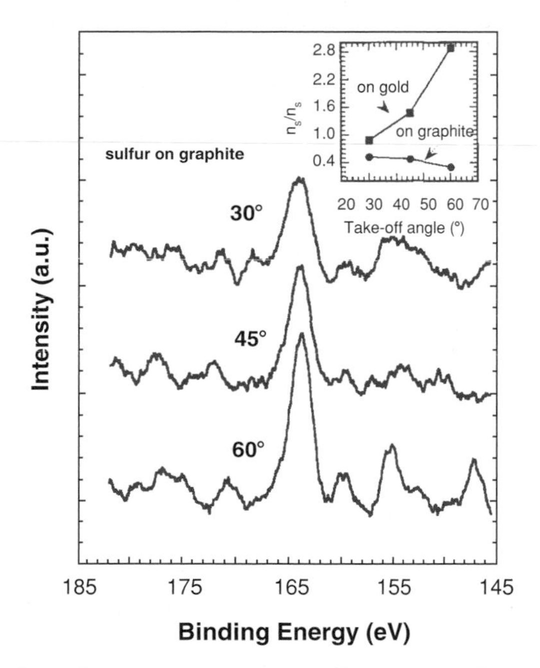

Fig. 7.6. X-ray photoelectron spectroscopy sulfur spectra of n-octadecanethiol on graphite, the take-off angles were 30°, 45° and 60°. *Insert*: The area density of n/n, as a function of take-off angle [75]

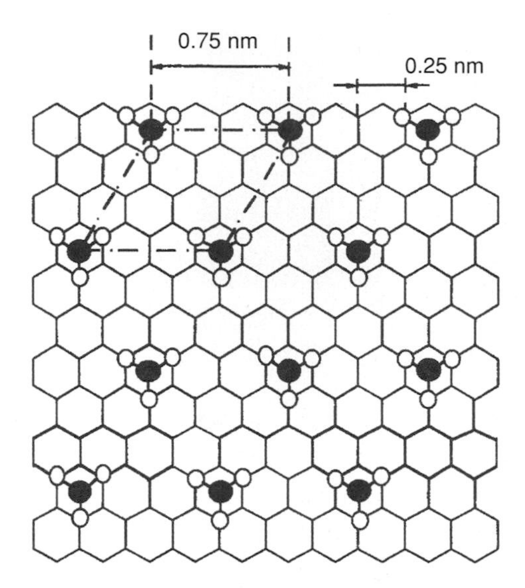

Fig. 7.7. A model for arrangement of the *n*-octadecanethiol molecules adsorbed on graphite constructed from the image of Fig. 7.5. The *open circles* and *shaded circles* represent the hydrogen and carbon atoms of the CH_3 terminal group, respectively, and *dot-dashed lines* indicate the unit cell [75]

between SAMs and substrates and the interaction between SAMs and deposited metal overlayers are therefore of great importance in this field and a lot of research on the structures [91,94] and the electrical transport properties of metal/SAMs/metal heterostructures has been reported [95, 98]. Moreover this could be a good model for fundamental studies of organic molecules and metal atoms in their interacting systems since the structure is very simple and the results obtained can be easily interpreted oeing to the well defined and very simple structure on the atomic scale.

Ohgi et al. studied gold deposited SAMs of octanethiol, $CH_3(CH_2)_7SH$, on Au(111) using STM. Deposited Au atoms cannot stay on the surface of the layers, but they penetrate through them and form monoatomic-height islands at the interface between SAMs and the Au(111) substrate after submonolayer deposition. In this system, the molecules act as surfactants and hinder the diffusion of metal atoms, resulting in the high nucleation density ($2 \times 10^{12}\,cm^2$) and different growth characteristics from the normal metal-on-metal epitaxial growth.

Ag deposited SAMs of octanethiol have been studied using STM and XPS. At the initial stage of Ag deposition, monoatomic-height islands, $7 \times 10^{11}\ cm^2$ in density, grow at the SAMs/Au(111) interface and become larger as more Ag atoms are deposited up to a full monolayer coverage of Ag (Fig. 7.8). The differences in the nucleation density and the growth property between Ag and Au islands can be attributed to the higher mobility of Ag atoms and the difference in the molecular packing on these islands (Fig. 7.9).

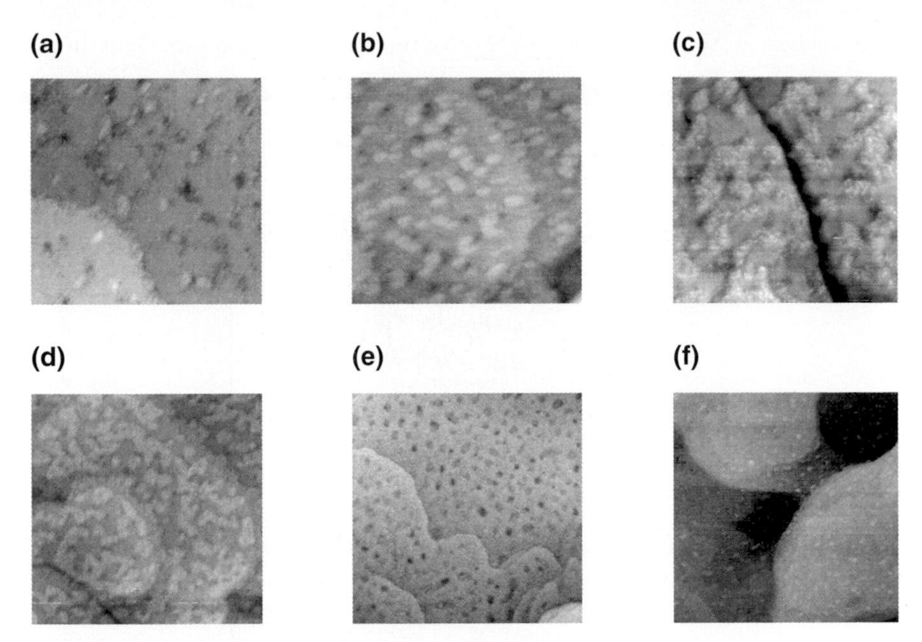

Fig. 7.8. STM images of octanethiol self-assembled monolayers (*SAMs*) after **(a)** 0.1 mL, **(b)** 0.2 mL, **(c)** 0.4 mL, **(d)** 0.5 mL, **(e)** 0.7 mL and **(f)** 1.1 mL Ag deposition. The area is $110 \times 110\,\text{nm}^2$ in **(a)** and **(c)** and $215 \times 215\text{nm}^2$ in **(d)** and **(f)** [90]

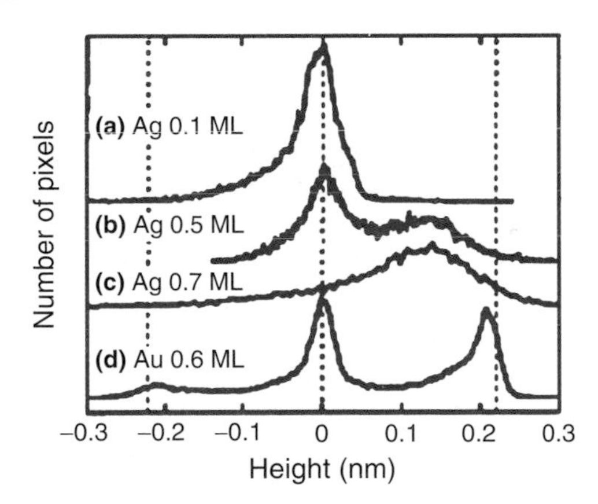

Fig. 7.9. Height distribution of SAMs/Au(111) surface after **(a)** 0.1 mL, **(b)** 0.5 mL and **(c)** 0.7 mL Ag deposition and **(d)** 0.6 mL Au deposition. Only one terrace is selected from Fig. 7.8 to remove the overlap of the height distribution between terraces. The distribution in trace **(d)** is from Fig.2c in [99] [90]

XPS analysis of this structure (SAMs/Ag monolayer/Au) shows that the Ag $3d_{5/2}$ binding energy is shifted 0.3 eV with respect to bulk Ag, the C 1s binding energy is 0.3 eV higher than that before Ag deposition, and the S $2p_{3/2}$ binding energy exhibits little shift before and after deposition (Fig. 7.10).

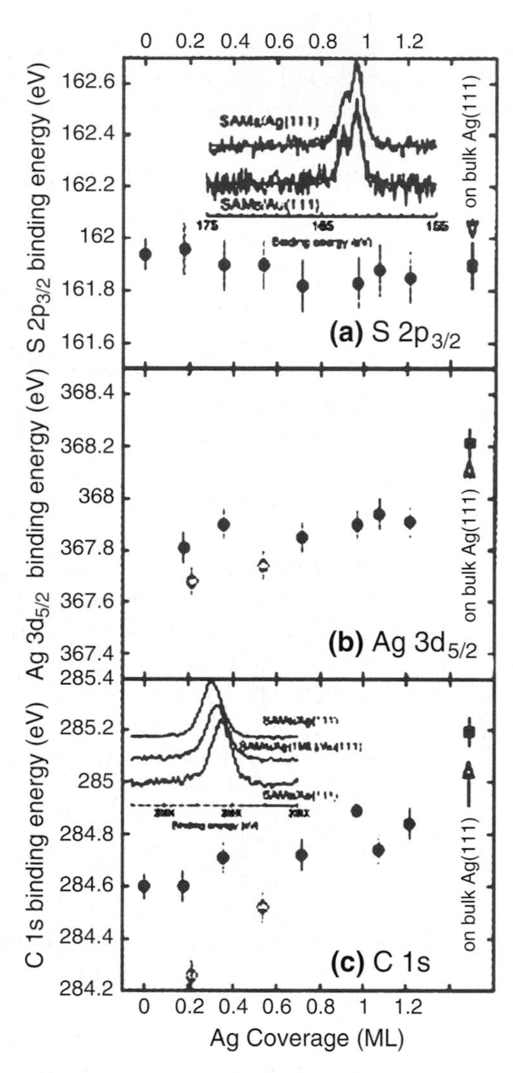

Fig. 7.10. The Ag coverage dependence of the core level shift of S $2p_{3/2}$ **(a)**, Ag $3d_{5/2}$ **(b)** and C 1s **(c)**. *Rectangles* denote the binding energies obtained from SAMs/Ag(111). *Circles* are from slightly oxidized samples. (0.2 and 0.5 mL Ag deposited samples were exposed to air for 7 and 4 days, respectively.) *Inset* in **(a)**: XPS spectra of S 2p from SAMs on Ag(111) and Au(111) substrate. *Inset* in **(c)**: XPS spectra of C 1s from SAMs/Ag(111), SAMs/Ag(1 ML)/Au(111) and SAMs/Au(111) [90]

The origin of the shift can be explained by the change of the dipole at the interface and the electrical isolation of alkyl chains from the surroundings.

7.5 Linear Chain Polymerization

Okawa and Aono [115] proposed a promising new approach to address the issue of the interconnection of devices. It is known that if a thin molecular film is stimulated by a biased scanning tunneling microscope tip, a local chemical reaction or reorientation is sometimes induced, as has been demonstrated in several cases [123–127]. Therefore, when the biased scanning tunneling microscope tip is scanned, continuous modification along the scan is possible. However, they have instead explored the possibility of inducing a nonlocal or extended chain chemical reaction initiated by a scanning tunneling microscope tip, because the chain reaction guarantees high structural perfection of the reaction product and because the entire process occurs in a short time span as a single event. Recently, Lopinski et al. [128] demonstrated a spontaneous growth of a molecular styrene line from a Si dangling bond created by a scanning tunneling microscope tip. Okawa et al. showed another type of chain reaction, chain polymerization, which can produce linear conjugated polymer nanowires. They have succeeded in realizing this idea using a monomolecular layer of 10,12-pentacosadiynoic acid, which is a diacetylene compound, adsorbed on a cleaved face of graphite. Diacetylene compounds are known to polymerize by appropriate stimulation in the solid state [129, 130]. The polymerization produces polydiacetylene compounds.

A conjugated polymer nanowire can be created at any designated position in a monomolecular layer by initiating chain polymerization using a scanning tunneling microscope tip with a spatial precision of the order of 1 nm (Fig. 7.11). The demonstration has been presented for a self-ordered monomolecular layer of 10,12-pentacosadiynoic acid (Figs. 7.12, 7.13), which is a diacetylene compound, adsorbed on a graphite surface. The polymer nanowires created have a length ranging from 5 to 300 nm, the length being controlled by domain boundaries (Fig. 7.14) or artificial defects in the molecular layer. The frequency of occurrence of chain polymerization is measured against the pulsed bias voltage (Fig. 7.15), which suggests that the excitation of the molecule is caused by the inelastically tunneling electrons.

7.6 Self-Assembly of Cyanine Fibers on Conducting Substrates

Cyanine dye molecules are able to self-assemble [150–152] into long ropes up to a few micrometers in length, a few tens of nanometers in width and a few nanometers in height. This self-assembly process occurs in solution [151,152], but it may be enhanced on a substrate surface [153]. The size of

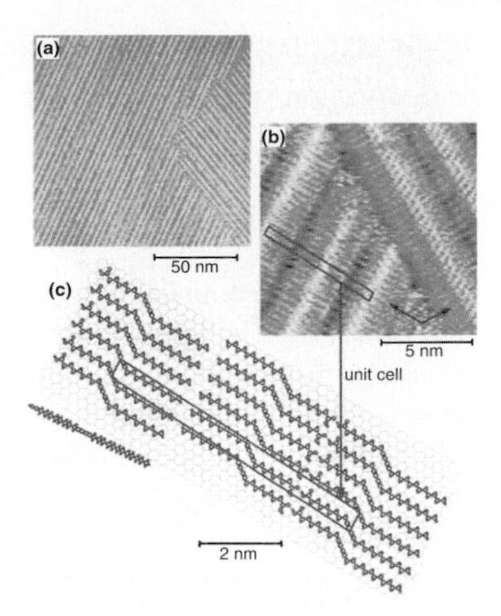

Fig. 7.11. **(a)** STM image of a 10,12-pentacosadiynoic acid layer on graphite (sample bias voltage $V_s = -1.0\,\mathrm{V}$ and tunneling current $I_t = 0.07\,\mathrm{nA}$). **(b)** Magnified STM image of the layer ($V_s = 0.5\,\mathrm{V}$, $I_t = 1.0\,\mathrm{nA}$). The *arrows* indicate the main crystal axes of graphite. **(c)** Top and side views of a model of the molecular arrangement. The parallelograms in **(b)** and **(c)** represent the unit cell [115]

the ropes strongly depends on the type of dye [153] and the surface morphology [153, 154] as well as the surface charge density of the substrate. Adding oppositely charged polymers to the dye solution enhances the rope formation [155]. A schematic view of the molecular ropes is shown in Fig. 7.16a. It is known that the dye molecules will form a brick-stone-like structure with a herringbone arrangement of the ring systems [155]. The assembly of cyanine dye molecules into such ropelike structures could provide a unique tool for bridging electrodes with molecular fibers which could have important applications in the development of nanoscopic electrical or electroluminescence devices. If the π-systems of the rope are stacked properly, the ropes could behave like a one-dimensional wire. Additionally, such a system suggests a study of the dependence of the optical contrast of the cyanine ropes when an electrical field is applied on the electrodes. Blumentritt et al. reported the successful assembly of individual ropes of cyanine dye molecules on chemically modified surfaces, HOPG, gold-coated mica and insulating substrates with a gold–palladium (AuPd) electrode pattern. The chemical structure of the cyanine dye used in the experiments is shown in Fig. 7.16b.

To assemble molecular ropes on HOPG, the surface of a freshly cleaved substrate was oxidized with CrO_3 (2mg/mL in 20% H_2SO_4) for 15 min at room temperature. This treatment is expected to create carboxylate (COO–) groups on the hydrophobic HOPG surface to which positively charged molecular ropes can attach. A 0.1 mM cyanine (1, l'-diethyl-2,2'-cyanine iodide) solution (produced by dissolving the dye in 1:1 water/ethanol) containing an

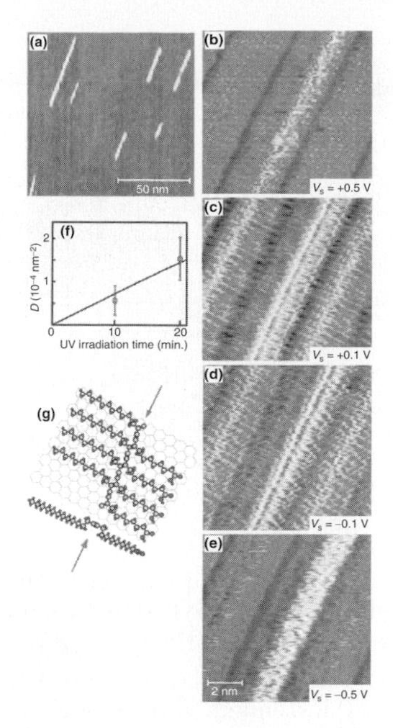

Fig. 7.12. **(a)** STM image obtained after UV irradiation ($V_s = -1.0\,\mathrm{V}$, $I_t = 0.07$ nA). **(b–e)** Magnified STM images of a polymer obtained at the same tunneling current ($I_t = 0.1$ nA) and various sample biases: **(b)** $V_s = 0.5\,\mathrm{V}$; **(c)** $V_s = 0.1\,\mathrm{V}$; **(d)** $V_s = -0.1\,\mathrm{V}$; **(e)** $V_s = -0.5\,\mathrm{V}$. **(f)** The number density of the photopolymerized polydiacetylenes D against the duration of UV irradiation. **(g)** Top and side views of the proposed structural model of a polymer. The backbone of the polymer is indicated by the *arrows* [115]

equimolar amount of sodium polyacrylate was spread on the surface. After evaporation of the solvents, the sample was transferred to a scanning tunneling microscope operated in ultrahigh vacuum (UHV) for further studies. A second type of sample was prepared for the assembly of the ropes on an atomically flat gold surface. For these samples, 100 nm of gold was evaporated onto a freshly cleaved mica substrate. After the evaporation, the substrate was annealed for 1.5 h at approximately 573 K and then immediately immersed in a 1 mM aqueous solution of 3-mercaptopropionic acid (3-MPA) for 3 h. This treatment attaches carboxylate groups to the gold surface which should bind the positively charged cyanine ropes. A 0.05 mM cyanine dye solution was prepared by dissolving the cyanine dye in 1:1 water/ethanol and adding an equimolar amount of sodium polyacrylate. The 3-MPA-treated substrate was immersed in this solution for 1 h, blown dry and put into the UHV scanning tunneling microscope. In a third experiment, cyanine fibers were attached to a substrate bearing an electrode pattern. The electrodes were prepared using electron beam lithography (EBL) and standard lift-off technique: 25-μm-long lines, approximately 100 nm in width

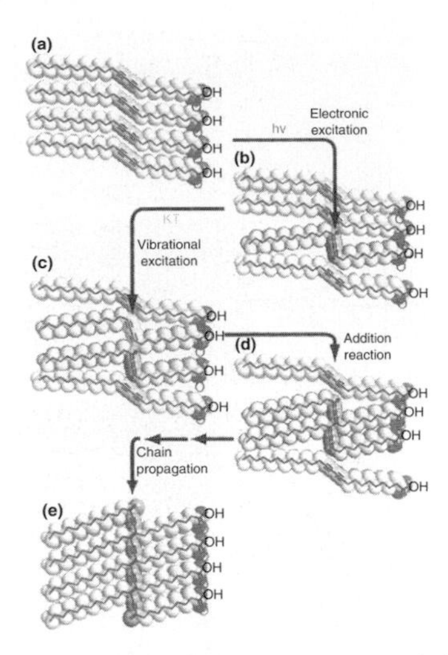

Fig. 7.13. The mechanism of photopolymerization. **(a)** Array of the monomer molecules. **(b)** Diradical formation by optical excitation. **(c)** Approach of a neighboring diacetylene moiety by vibrational excitation. **(d)** Radical dimer formation by an addition reaction. **(e)** Extended polymer formation by chain propagation reaction [115]

and 100 nm apart were fabricated on an insulating substrate (e.g., Si with a native oxide layer or GaAs). The electrodes were formed by evaporating approximately 8-nm AuPd (3/2). The assembly process of the molecular ropes was the same as described before for gold on mica. For the STM investigations, mechanically cut Pt–Ir tips were used. Atomic force microscope (AFM) measurements on the ropes attached to patterned substrates were performed with an AFM in tapping mode operating with commercially available Si_3N_4 tips. The assembly of the molecular ropes on HOPG was studied using a UHV scanning tunneling microscope. The STM images confirmed the presence of ropes which were 200–400 nm in length, approximately 25 nm in width and 1–3-nm high. However, the ropes were moved on the surface by the STM tip. For the tunneling currents (0.25 and 0.5 nA) and bias voltages (20–80mV) used, it was impossible to prevent the ropes from moving. They attributed this mobility to a low surface affinity of the charged ropes on the incompletely oxidized HOPG surface, which is manifested by the poor wettability of the surface. In order to perform spectroscopic measurements, the binding interaction between the cyanine dye molecule and the substrate surface has to be strong enough so that the molecules adopt stable positions on the substrate surface. To achieve this, the affinity of sulfur to gold was exploited: a thiol with negatively charged end groups was assembled on the gold surface to provide a high surface density of carboxylate groups for the binding of molecular ropes. In a first study, the assembly of 3-MPA on gold was investigated with STM. The influence of the 3-MPA treatment is mani-

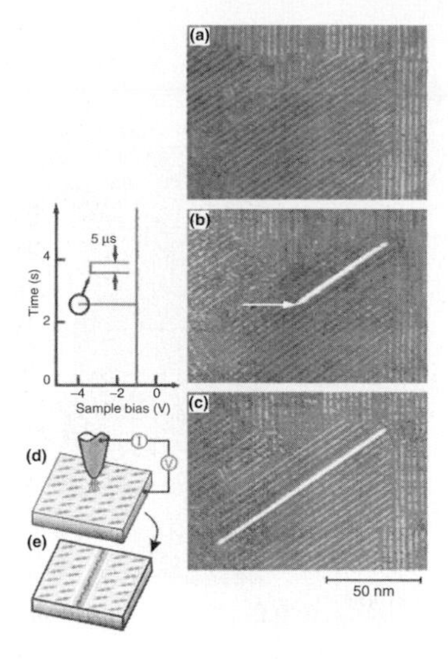

Fig. 7.14. STM image demonstrating that the length of the polymer can be controlled by an artificial defect ($V_s = -1.0\,V$, $I_t = 0.07\,nA$). First, an artificial defect in the form of a 6-nm-wide hole was created at the center of the image using a scanning electron microscope tip. Next, a pulsed sample bias voltage was applied at the point indicated by the *arrow*, which initiated a chain polymerization. The chain polymerization was terminated at the artificial defect [115]

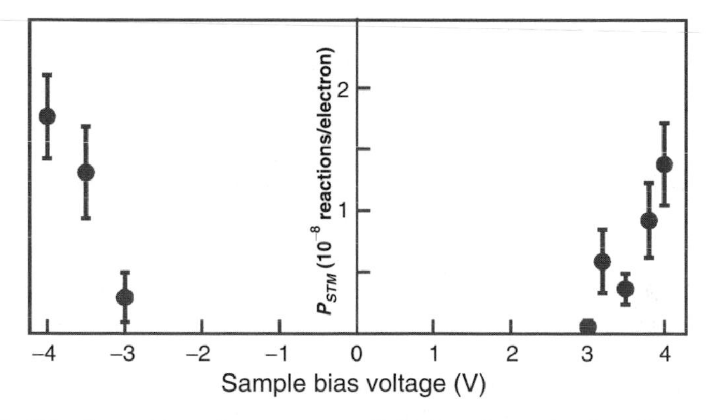

Fig. 7.15. The frequency of occurrence of chain polymerization P_{STM} in units of reactions per electron against the height of the pulsed sample bias V_s [115]

fested by the formation of small vacancies in the gold surface [156]. In a next step, a 3-MPA-modified substrate treated with the cyanine dye solution was investigated. In Fig. 7.16c, the STM image of a molecular rope observed on such a surface is shown. The image was obtained with a tunneling current of 0.16 nA and a bias voltage of 640 mV. The length of this molecular rope is about 170 nm, its width is 20 nm and its height is 3.2 nm. The sizes of other

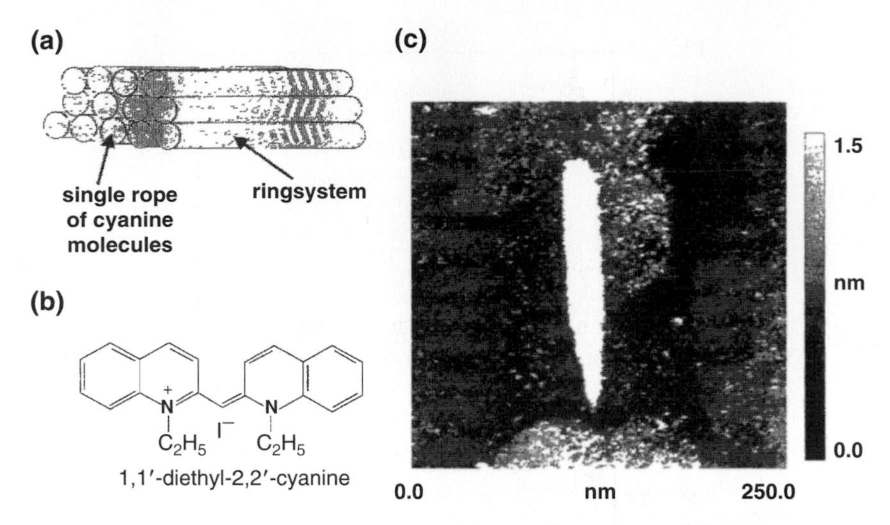

Fig. 7.16. (**a**) The herringbone arrangement of the cyano ring systems in a rope is displayed schematically with the circular tubes illustrating the long axis of the n-stack The polymer is omitted for clarity. (**b**) Chemical structure of the cyanine dye (1,1'-diethyl-2,2'-cyanine iodide); approximate size of the molecule, 2.0 nm in length, 1.0 nm in height and 0.34 nm in width. (**c**) STM image of a molecular rope on an atomically flat gold surface The rope is about 170-nm long, 20-nm wide and 1.3-nm high. The image was taken with a tunneling current of 0.16 nA and a bias voltage of −0.64 V [150]

ropes observed ranged from 65 to 195 nm in length, with widths of 20 nm and heights of 1.3–3.0 nm. From these values they estimated that a bundle of cyanine ropes contains 40–100 individual ropes. The similar width (20 nm) might indicate a convolution of the tip and the substrate. Rope motion was not observed on the 3-MPA–gold surface. The results of scanning tunneling spectroscopy (STS) performed after assembling the molecular ropes on gold are shown in Fig. 7.17. Both on the molecular rope (Fig. 7.17a) and away from it (Fig. 7.17b) the I–V curves exhibit either gaplike or metallic behavior (ohmic characteristics). It is important to note that on the pure and 3-MPA-modified gold surfaces exclusively metallic behavior was observed. The gap-like behavior and the current jumps seen for the molecular ropes (curve A) are tentatively assigned to resonant tunneling processes through molecular energy states. Such phenomena were, for example, discussed by Lindsay [157] and Mujica et al. [158]. In their theoretical studies, it is shown that the matching of the highest occupied or lowest unoccupied molecular orbital with the Fermi energy of the contacts leads to abrupt current increases. Experimentally, Dekker et al. [159] reported a current step in the STS measurements on a metallophthalocyanine and attributed this to a resonant process. The metallic behavior observed on some positions of the molecular ropes (curve B) might be due to local differences in the tilt of the ring systems or locally

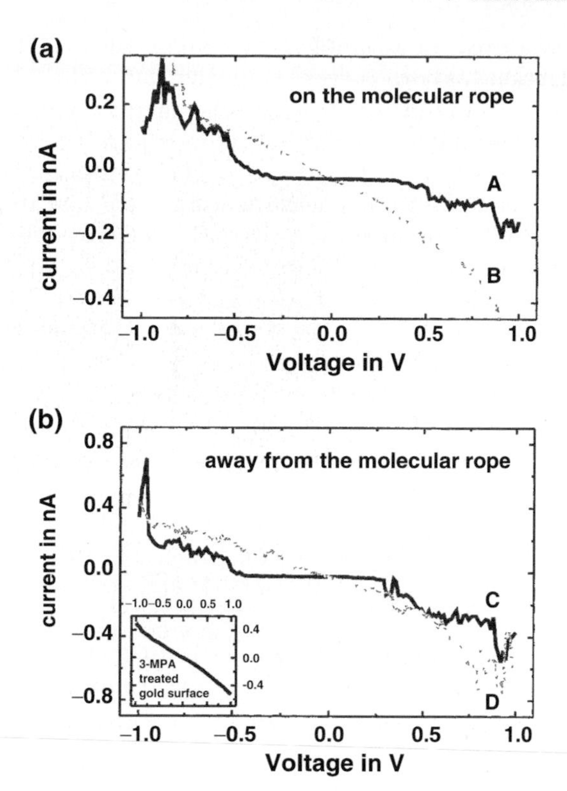

Fig. 7.17. Representative I–V characteristics **(a)** on the molecular rope, with A and B displaying I–V characteristics measured at two different positions on the rope, and **(b)** away from the molecular rope, where C and D are displaying I–V characteristics measured at two different positions away from the molecular rope. STM parameters used, bias voltage −0.64 V, tunneling current 0.16 nA, temperature at the sample 44 K. The curve in the *inset* was measured on a gold sample treated only with 3-mercaptopropionic acid solution for 3 h. Scanning tunneling spectroscopy was performed at a tunneling current of 0.2 nA and a bias voltage of 1 V at room temperature [150]

changed coupling between the metallic states of the tip (or substrate) and the molecular energy levels. If the coupling is strong, a broadening of the local state energy is expected: this leads to electron transmission at low bias [157]. The gaplike behavior observed away from the molecular rope (curve C) might be due to the presence of a (partial) monolayer of cyanine molecules which is not resolved in the STM image. Another possibility is that the tip picks up molecules [160] while scanning over the molecular rope The presence of molecules on either the tip or the substrate could also explain the more featureless image compared with the pure gold, where distinctive gold terraces and vacancies are visible. Curve D displays a metallic I–V characteristic, which is as one would expect on a gold surface. The inset of Fig. 7.17b shows an I–V curve measured on a sample treated only with 3-MPA solution for comparison. The slope of both curves is quite similar, although the tunneling current, bias voltage and temperature are different. The assumed presence of the cyanine molecules might be responsible for the fact that the curve obtained for the 3-MPA-treated surface is smoother than curve D. For the future goal of electrical transport measurements, molecular ropes have to be bridged over electrode structures on insulating substrates. Figure 7.18 shows a SEM image of a molecular rope assembled on lateral AuPd electrodes. The picture was taken under 45° to enhance the contrast. The rope is approximately 500 nm-long and 40-nm wide. From that it was estimated that the number of individual ropes in the bundle was about 200. Figure 7.18 shows a SEM image of a patterned substrate with attached molecular ropes. It seems that the rope bends over the 8-nm-high electrodes without breaking. The molecular ropes can also be imaged by tapping-mode AFM, which allows one to determine the arrangement of ropes on the electrodes prior to transport measurements. SEM investigations, on the other hand, may influence transport measurements because of carbon contamination. Ropes of cyanine dye molecules were assembled successfully on HOPG, gold-coated mica and lateral electrode patterns. In the case of a partially oxidized HOPG surface, the ropes moved while scanning with the STM tip. In contrast, STM images of ropes assembled on a thiol-modified gold surface confirm that they are not moved by the tip. STS performed on this system shows, both on the molecular rope and away from it, gaplike or metallic behavior. The gaplike behavior on and away from the molecular rope is attributed to the presence of cyanine molecules, whereas the metallic behavior on the rope is assigned to a locally different electronic coupling between metallic tip–rope or rope–metallic substrate. Rope assembly is also possible on substrates with electrode patterns,

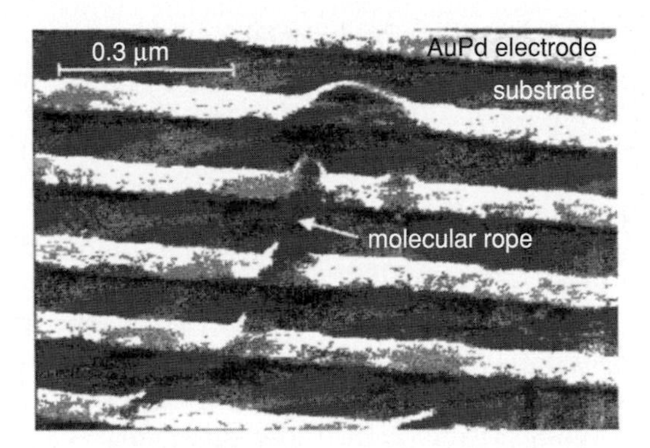

Fig. 7.18. Scanning electron microscope (*SEM*) image of a molecular rope assembled on lateral AuPd electrodes. The picture was taken under 45° to enhance the contrast. The rope is approximately 500-nm long and 40-nm wide. From that we estimated the number of individual ropes in the bundle to be about 200 [150]

offering the possibility of transport measurements on the molecular ropes in a lateral device configuration.

7.7 Inclusion Complex of Polyaniline Covered by Cyclodextrins for Molecular Devices

The main challenge in molecular electronics is to establish that single molecules or a finite number of self-assembled molecules can perform all the basic functions of conventional electronics components such as wires, diodes and transistors [161–163]. There have been significant advances in the fabrication and demonstration of molecular electronic wires [164, 165], molecular diodes and two-terminal electrical switches made from single molecules [166, 167]. In parallel with the progress of more effective fabrication technologies, theoretical study of promising molecular structures based on quantum mechanical calculations is also one of the key factors for designing new electronic devices with the desired physical characteristics. The theoretical studies have been mainly dedicated to the understanding of conduction mechanisms through molecular wires and how the electronic structure of the molecule and the geometry of the molecule–metal interface affect conduction characteristics [168, 170]. Density functional calculations have shown that the addition of sulfur and metal atoms to the terminal carbon atoms of a short phenylene-ethynylene oligomer does not modify the main characteristics of the isolated molecules [170]. However, in the case of long molecular wires, the interfacial interaction between molecules and metallic surfaces and possible charge transfer may affect the stability and conduction properties of the molecule. Therefore, it would be better if the long molecules were encapsulated by the insulated nanotube. One of the possible approaches for realizing isolated molecular wires is the formation of self-assembly supramolecular structures such as inclusion complexes [171]. It was reported that cyclic cyclodextrin (CD) molecules were threaded onto a polymer chain and formed a molecular necklace structure owing to very small cavities and close packing of the CD molecules [171–173]. In such molecular nanotubes, the conformation of the polymer chain remains rodlike (all-*trans* conformation). The other feature of this inclusion complex is that the polymer molecules can be isolated from each other by insulated CD molecules. Recently, the formation of a new inclusion complex (Fig. 7.19), namely, an insulated molecular wire in which a conjugated conducting polymer covered by CDs, was realized [174].

The structural configuration of polyaniline in β-CDs has been investigated by the two-layered own N-layered integrated molecular orbital and molecular mechanics (ONIOM) method, which includes both quantum mechanics and molecular mechanics calculations. It has been found that the configuration of polyaniline (Fig. 7.20) in β-CDs (Fig. 7.21) is similar to a planar structure. There is no charge transfer between β-CD and polyaniline; also, the electronic structure of polyaniline in β-CDs is almost the same as that

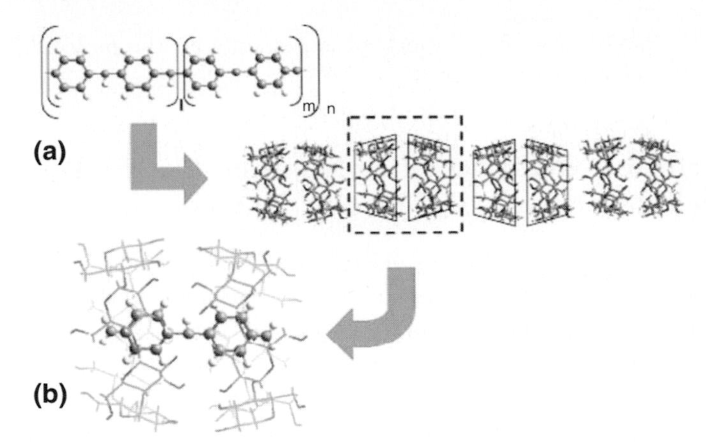

Fig. 7.19. (**a**) Inclusion complex formation of cyclodextrins (*CDs*) and conducting polymer. (**b**) Cluster model used for calculations [161]

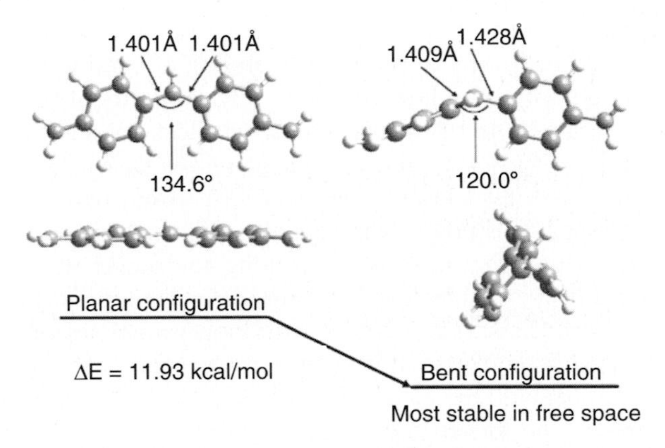

Fig. 7.20. Structural analysis of different conformations of a polyaniline fragment in free space [161]

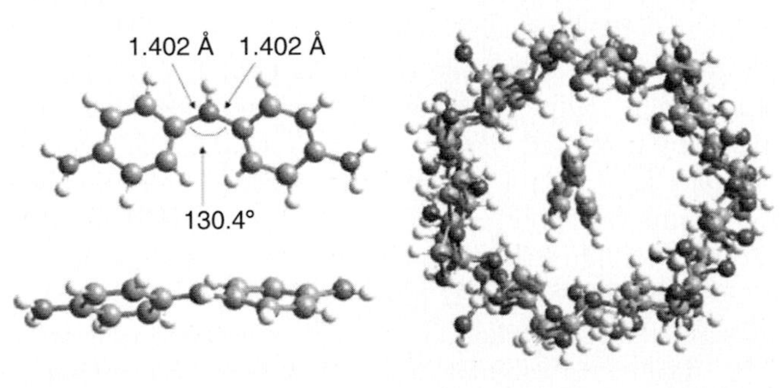

Fig. 7.21. Structural analysis of polyaniline in β-CDs [161]

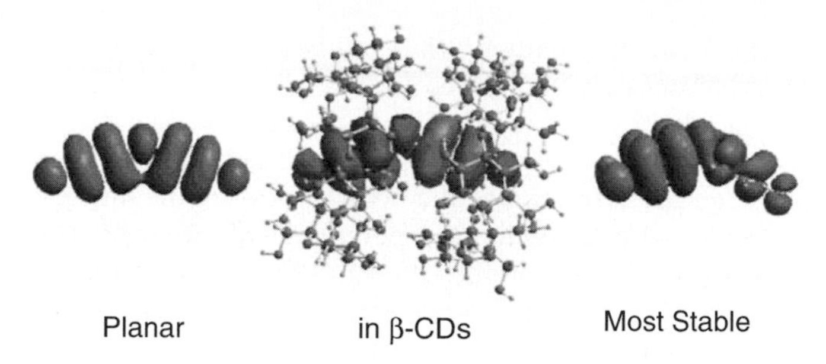

Fig. 7.22. Molecular orbital diagram of different conformations of a polyaniline fragment [161]

of the planar conformation. This indicates that β-CD molecules (Fig. 7.22) can be used as insulated molecular nanotubes for stabilization of the isolated near-planar configuration of polyaniline, which is in good agreement with experimental results. The stoichiometry of this inclusion compound is 1:1 (β-CD to polyaniline) and it can support the realization of an insulated molecular wire based on this complex for molecular electronics devices.

7.8 Manipulation of Insulated Molecular Wire with an Atomic Force Microscope

Molecular electronics, which is used to assemble organic molecules, has attracted the attention of many researchers in recent years [180]. When they fabricate the molecular devices, a molecular wire is important for connecting functional organic molecules such as molecular diodes and electrodes for input and output signals of the molecular devices. At present, there are several candidates for the molecular wire, including CNTs [181,182], DNA [183,184] and a conducting polymer [185, 186]. Of these, a conducting polymer has the advantage of combining organic molecules with the molecular wire by the technique of chemical synthesis. However, not even the conductance of a long single conducting polymer chain has been measured yet. This is mainly because it is difficult to isolate a single conducting polymer chain. Conducting polymers in general have low solubility owing to the large attractive interaction among polymer chains. Accordingly, they form fibrillar structures complicatedly intertwined with each other. Certainly, some conducting polymers with alkyl side chains are soluble in organic solvents and can be isolated in very dilute solution [187,188]; however, the resulting polymer chain then has a coiled conformation because of the large conformational entropy. The coiled conformation consists of a random distribution of *trans* and *gauche* configurations. The π-electron conjugation along the polymer chain is enhanced by the *trans* configuration and is broken down by the *gauche* configuration; therefore, high conductivity is expected not in the coiled conformation but in the all-*trans* configuration, namely, the rodlike conformation. Consequently, to use the conducting polymer for molecular wire, it is necessary to isolate a single polymer chain and extend it to a rodlike conformation. To resolve this difficulty, Yoshida et al. [189,190] have recently proposed a novel kind of molecular wire: an insulated molecular wire of polyaniline and β-CD. In the insulated molecular wire, a conducting polymer chain is covered by insulated cyclic molecules [191–193]; therefore, the intermolecular interaction among conducting polymers is reduced considerably and one can easily isolate a long single polymer chain. Moreover, a conducting polymer chain confined to the extremely narrow inside of CD has a rodlike conformation, i.e., the all-*trans* configuration. This suggests that the π-electron conjugation can spread over the entire chain. Very recently, they formed an insulated molecular wire of polyaniline and a molecular nanotube. The molecular nanotube was synthesized by cross-linking adjacent CD units [194,195] and can include various polymer chains in the cavity [196–199]. Since the molecular nanotube includes polyaniline more strongly than β-CD, the inclusion complex formation occurs at room temperature with high yield, while β-CD forms the inclusion complex at temperatures below approximately 275 K. In addition, this insulated molecular wire was connected in series to form a rodlike structure of over 1 μm in length.

They have reported the formation of an insulated molecular wire of a conducting polymer (Fig. 7.23), emeraldine base polyaniline, and a molecular nanotube synthesized from α-CD. In that study, they manipulated the insulated molecular wire with the cantilever tip of an AFM (Fig. 7.24). They found that the insulated molecular wire was moved or cut off by the manipulation process. The results of manipulation with varying AFM tip loading forces indicated that the insulated molecular wire is cut off at loading forces larger than approximately 30 nN (Fig. 7.25).

7.9 Production of Individual Suspended Single-Walled Carbon Nanotubes

Recently the electrical and (electro-)mechanical properties of one-dimensional nanostructures such as CNTs [203, 204], synthetic nanowires [205] and polymer nanotubes [206,207] have been extensively studied. The emphasis is now partially shifting from studies of the intrinsic properties of these materials to the exploration of the possibility of making electrical devices superior to silicon-based devices [208–210] or making nanoscale electromechanical systems [211,212]. There are several important issues that should be considered if these materials are to be practically useful as the building blocks of a new generation of devices: (1) how to control the position and the orientation of the one-dimensional nanostructure on a substrate; (2) how to remove any side effects from the electrical contact problem of the nanostructures

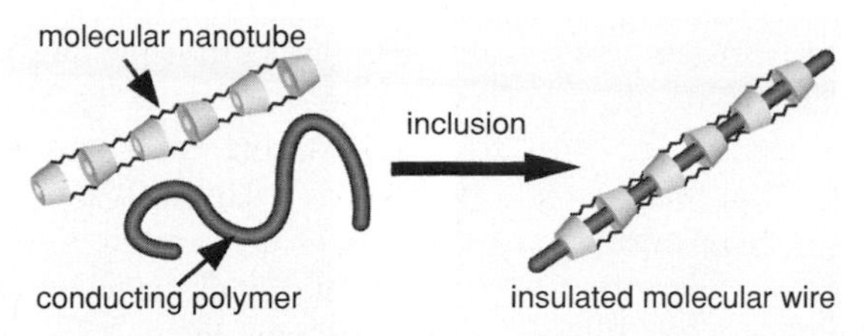

Fig. 7.23. The inclusion complex formation of a molecular nanotube and a conducting polymer [180]

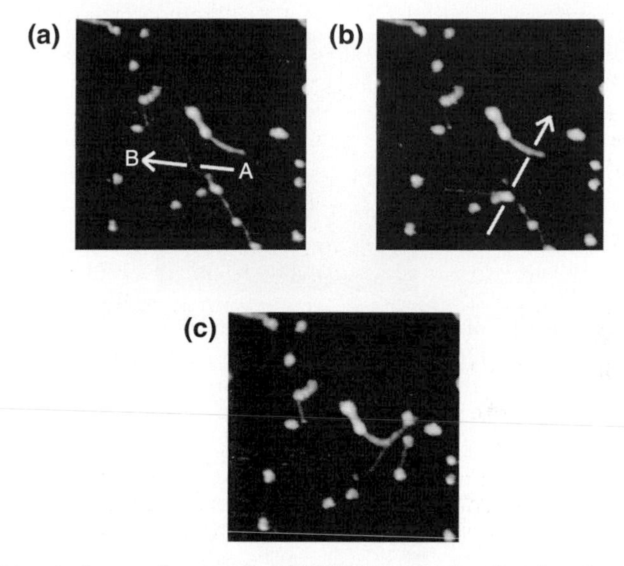

Fig. 7.24. Atomic force microscope (*AFM*) images of an insulated molecular wire on a mica substrate (scan size 620 nm × 620 nm). The sequence depicts the AFM manipulation process. The *white arrow* represents the movement of the cantilever tip in contact mode. The insulated molecular wire is cut off by the tip [180]

with electrodes and the interaction with the substrate; and (3) how to reproducibly suspend the nanostructures, for example, for nanoelectromechanical systems (NEMS). Several groups have reported the positioning or aligning of CNTs using various electrical or mechanical techniques, for example, by applying a direct current electric field during CVD growth [213], applying an alternating current (ac) electric field to the CNT suspension [214] or using the molecular combing technique [215]. Among these methods, the CNT alignment technique using an applied ac field has several advantages, such as a short time for the alignment and the possible application to different kinds of one-dimensional nanostructures, synthesized not only with the chemical vapor deposition method but also by other methods. Krupke et al. [216] have reported very recently that the majority of SWNTs deposited by an applied

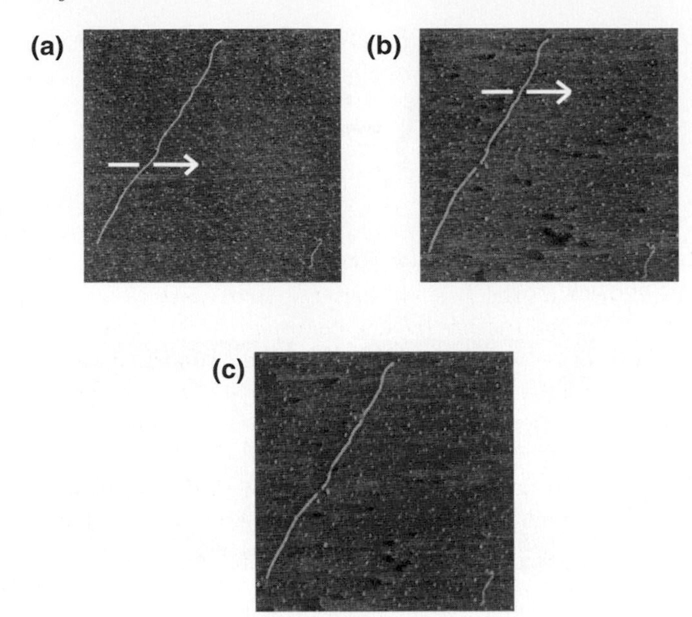

Fig. 7.25. AFM images of an insulated molecular wire on a mica substrate (scan size 2300 nm × 2300 nm). The sequence depicts the AFM manipulation process. The *white arrow* represents the movement of the cantilever tip in contact mode. The insulated molecular wire is moved by the tip [180]

Fig. 7.26. AFM image of aligned carbon nanotubes (*CNTs*) after applying an alternating current (*ac*) voltage between the two electrodes [203]

ac field at 10 MHz are metallic, so the ac dielectrophoresis method can be used to separate metallic from semiconducting SWNTs. The results were analyzed on the basis of the difference between the relative dielectric constants of the metallic and semiconducting SWNTs with respect to the solvent. They used a Raman spectroscopy study to prove the effectiveness of separation. Independently, Lee et al. studied the alignment of CNTs using a 13-MHz ac field and characterized the aligned CNTs by measuring the I–V characteristics. However, it is difficult to investigate the intrinsic properties of these materials by measuring the I–V characteristics because of contact problems, the influence of the interaction between sample and substrate [217], and the bending of the sample owing to the effect of the deposited electrodes [218]. A solution that could eliminate these problems is to lift the sample from the substrate and suspend it freely between the electrodes. Such suspended structures are useful for the production of many NEMS. So far, several trials for making suspended nanotubes have been reported. The most widely used method to make the suspended structure is to etch the substrate after connecting the CNT to the electrodes [219,220]; however, this method cannot be more generally applied to organic materials since the acid used for etching the substrate could destroy the organic material. A nonetching method has been developed [221], which prevents organic nanomaterials from being destroyed by acid. With this method, since CNTs (as well as other nanostructures) are dispersed with a random orientation, it is difficult to define the position and orientation of the nanostructures. To align the suspended CNT structures in a controllable manner, they have developed a new method by combining the ac electrophoresis technique with conventional EBL processes. This approach allows them to make a suspended, aligned structure without using acid. The method can be applied to make suspended structures of any materials that can be aligned in an ac field.

Lee et al. have developed a new method for aligning individual suspended SWNTs using a combination of the ac electrophoresis technique and EBL (Fig. 7.26). A poly(methyl methacrylate) (PMMA) underpinning was used in the region between the Au electrodes to prevent the CNT from falling down onto the substrate. O_2 plasma ashing was used to control the height of the underpinning PMMA layer so that it was same as that of the predefined electrodes. The biggest advantage of this method is that one can easily align the suspended nanostructure in a controllable manner (Fig. 7.27). This method can also be applied to making suspended structures of organic materials that are sensitive to acid treatment. They measured the temperature-dependent I–V characteristics of the suspended SWNTs and found that most of the aligned SWNTs were metallic.

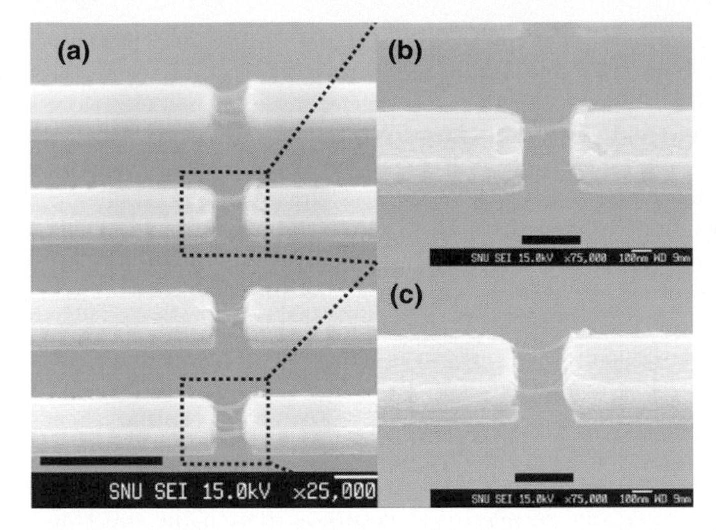

Fig. 7.27. SEM images of suspended CNTs. **(a)** CNTs are well aligned and attached between each electrode pair. The *scale bar* is 1 µm. **(b)**, **(c)** The number of CNTs controlled in the time for which the ac voltage was applied and by the density of the CNT sodium dodecyl sulfate water solution. The *scale bar* is 300 nm [203]

7.10 Low-Energy Electron Point Source Microscope: As a Tool for Transport Measurements of Free-Standing Nanometer-Scale Objects

Electrical measurements of wirelike nanometer-scale organic and inorganic objects are an important task in modern biology and applied physics [224]. A number of experimental methods for providing local electrical contacts to such objects have been developed recently and applied for transport measurements of DNA molecules [225, 226], nanotubes [227, 228], metal nanowires [229], etc. Dorozhkin et al. reported the development of an experimental technique for measuring the electrical properties of nanometer-scale objects. The method is based on the low-energy electron point source (LEEPS) microscope [226, 230] and it extends the conventional imaging capabilities of the microscope. They realized a procedure for making a well-controlled electrical contact between the field emission tip of the microscope and the nanometer-sized free-standing object for performing transport measurements. The method developed has a number of advantages compared with previously applied techniques (Fig.7.28).

The field emission tip of the microscope is used as a movable electrode to make a well-defined local electrical contact at a controlled place of a nanometer-sized object (Figs. 7.29, 7.30). This allows transport measurements

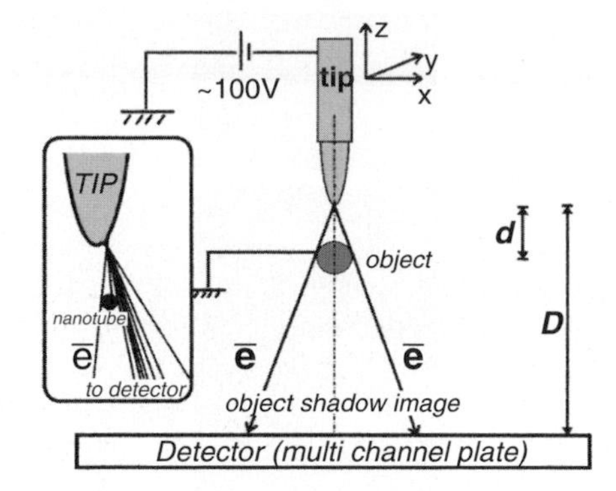

Fig. 7.28. Setup principle of the low-energy electron point source (*LEEPS*) microscope (not to scale). *Inset*: relative position of tip and sample at high microscope magnification [224]

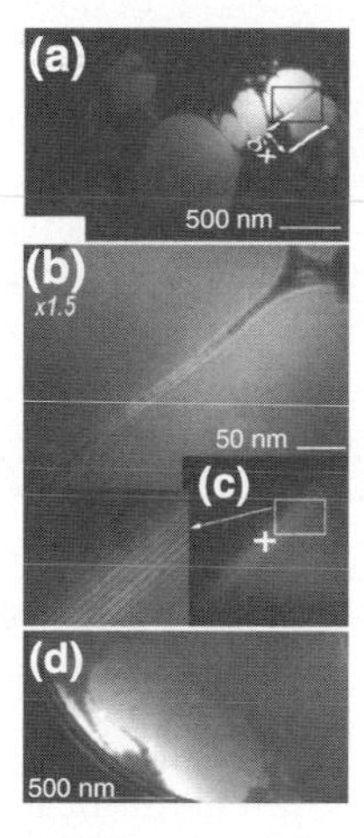

Fig. 7.29. **(a)**–**(c)** Sequence of LEEPS microscope images demonstrating the process of tip approach towards a single free-standing nanotube rope with both ends fixed. The *square* in **(a)** indicates the rope to be touched; **(b)** is enlarged 1.5 times; the *cross* in **(c)** marks the geometrical center of the detector screen, the area of the rope around the center is to be contacted. After tip positioning, the tip voltage was reduced to 50 mV and the tip was moved in the Z-direction towards the rope until the electrical contact was detected. **(d)** The rope was destroyed after a voltage of 5 V was applied to it during transport measurements; cf. **(a)** [224]

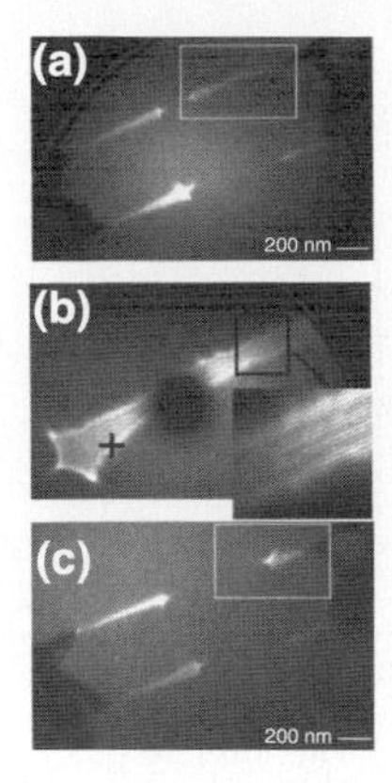

Fig. 7.30. (**a**), (**b**) Sequence of LEEPS microscope images demonstrating the process of the tip approach for contacting a single free-standing nanotube rope with one free end. The *cross* in (**b**) marks the geometrical center of the detector screen; the *black spot* in (**b**) is a defect of the detector. (**c**) The rope shortened after a voltage of 6 V was applied to it during transport measurements; cf. (**a**) [224]

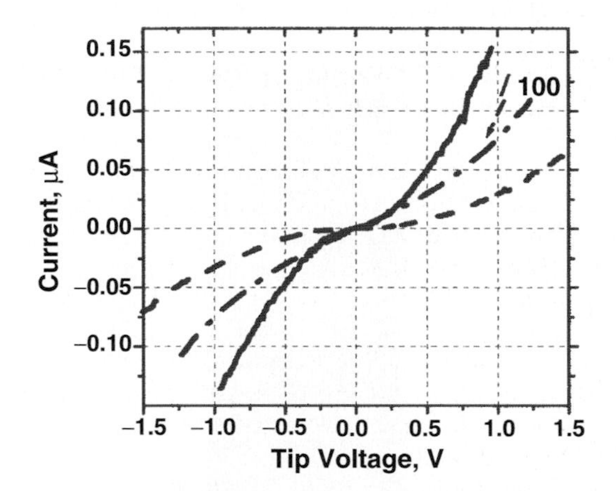

Fig. 7.31. *Solid line* I–V curve for the nanotube rope presented in Fig. 7.30; *dashed line* and *dash-dotted line* typical I–V curves for high- and low-resistance nanotube ropes, respectively; the *dash-dotted line* is divided by a factor of 100 [224]

of the object to be conducted (Fig.7.31). The technique was applied to CNT ropes and could be applied to any suspended nanowires.

References

1. V. Derycke, R. Martel, J. Appenzeller, P. Avouris: Nano Lett. **1**, 453 (2001)
2. A. Bachtod, P. Hadley, T. Nakanishi, C. Dekker: Science **294**, 1317 (2001)
3. D.H. Cobden: Nature **409**, 32 (2001)

4. D.I. Gittins, D. Bethell, D.J. Schiffrin et al.: Nature **408**, 67 (2000)
5. O. Marchenko, J. Cousty: Phys. Rev. Lett. **84**, 5363 (2000)
6. A. Marchenko, Z.X. Xie, J. Cousty et al.: Surf. Interface Anal. **30**, 167 (2000)
7. D.N. Futaba, S. Chiang: J. Vac. Sci. Technol. A **15**, 1295 (1997)
8. J. Kim, T.M. Swager: Nature **411**, 1030 (2001)
9. Z.J. Donhauser, B.A. Mantooth, K.F. Kelly et al.: Science **292**, 2303 (2001)
10. Y. Chen, D.A.A. Ohlberg, X. Li, D.R. Stewart, R.S. Williams, J.O. Jeppesen, K.A. Nielsen, J.F. Stoddart, D.L. Olynick, E. Anderson: Appl. Phys. Lett. **82**, 1610 (2003)
11. J.K. Gimzewski, S. Modesti, R.R. Schlittler: Phys. Rev. Lett. **72**, 1036 (1994)
12. T.A. Jung, R.R. Schlittler, J.K. Gimzewski et al.: Science **271**, 181 (1996)
13. M. Eremtchenko, J.A. Schaefer, F.S. Tautz: Nature **425**, 602 (2003)
14. B.C. Stipe, M.A. Rezaci, W. Ho: Phys. Rev. Lett. **81**, 1263 (1998)
15. B.C. Stipe, M.A. Rezaci, W. Ho: Science **280**, 1732 (1998)
16. B.C. Stipe, M.A. Rezaci, W. Ho: Science **279**, 1907 (1998)
17. T.R. Ohno, Y. Chen, S.E. Harvey et al.: Phys. Rev. B **44**, 13747 (1991)
18. E. Altman, R.J. Colton: Phys. Rev. B **48**, 18244 (1993)
19. T. Hashizume, K. Motai, X.D. Wang et al.: Phys. Rev. Lett. **71**, 2959 (1993)
20. S.J. Chase, W.S. Bacsa, M.G. Mitch et al.: Phys. Rev. B **46**, 7873 (1992)
21. R. Berndt, R. Gaisch, W.D. Schneider et al.: Surf. Science **307–309**, 1033 (1994)
22. R. Gaisch, R. Berndt, J.K. Gimzewski et al.: Appl. Phys. A **57**, 207 (1993)
23. R. Gaisch, R. Berndt, W.-D. Schneider et al.: J. Vac. Sci. Technol. B **12**, 2153 (1994)
24. J.G. Hou, Y. Jinlong, W. Haiqian et al.: Phys. Rev. Lett. **83**, 3001 (1999)
25. B. Wang, X. Xiao, X. Huang et al.: Appl. Phys. Lett. **77**, 1179 (2000)
26. C. Zeng, H. Wang, B. Wang et al.: Appl. Phys. Lett. **77**, 3595 (2000)
27. H. Wang, J.G. Hou: Phys. Rev. B **61**, 2199 (2000)
28. B. Wang, H. Wang, H. Li et al.: Phys. Rev. B **63**, 035403 (2000)
29. H. Wang, C. Zeng, Q. Li et al.: Surf. Science **442**, L1024 (1999)
30. J.K. Gimzewski, C. Joachim: Science **283**, 1683 (1999)
31. M. Matus, H. Kuzmany, E. Sohmen: Phys. Rev. Lett. **68**, 2822 (1992)
32. K. Harigaya: Chem. Phys. Lett. **189**, 79 (1992)
33. K. Harigaya: Phys. Rev. B **48**, 2765 (1993)
34. K. Harigaya: Chem. Phys. Lett. **253**, 420 (1996)
35. S. Kato, T. Kodama, M. Oyama et al.: Chem. Phys. Lett. **186**, 35 (1991)
36. H. Prinzbach, A. Weiler, P. Landenberger et al.: Nature **407**, 60 (2000)
37. R. Yamachika, M. Grobis, A. Wachowiak et al.: Science **304**, 281 (2004)
38. H. Park, J. Park, A.K.L. Lim et al.: Nature **407**, 57 (2000)
39. J.W.G. Wildöer, L.C. Verema, A.G. Rinzler et al.: Nature **391**, 59 (1998)
40. L.C. Venema, J.W.G. Widöer, H.L.J. Teminck et al.: Appl. Phys. Lett. **71**, 2629 (1997)
41. L. Grigorian, K.A. Williams, S. Fang et al.: Phys. Rev. Lett. **80**, 5560 (1998)
42. G.S. Dwesberg, W. Blau, H.J. Byrne et al.: Synth. Met. **103**, 248 (1999)
43. G.S. Dwesberg, J. Muster, H.J. Byrne et al.: Appl. Phys. A **69**, 269 (1999)
44. M. Burghhard, G. Dwesberg, G. Philipp et al.: Adv. Mater. **10**, 584 (1998)
45. J. Muster, G.S. Dwesberg, S. Roth et al.: Appl. Phys. A **69**, 261 (1999)
46. J. Muster, M. Burghard, S. Roth et al.: J. Vac. Sci. Technol. B **16**, 2796 (1998)
47. A.Y. Kasumov, R. Deblock, M. Kociak et al.: Science **284**, 1508 (1999)

48. B. Lounis, W.E. Moerner: Nature **407**, 491 (2000)
49. H.E. Katz, A.J. Lovinger, J. Johnson et al.: Nature **404**, 478 (2000)
50. J.H. Schon, C. Kloc, E. Bucher et al.: Nature **403**, 408 (2000)
51. P.K.H. Ho, Ji-S. Kim, J. H. Burroughes et al.: Nature **404**, 481 (2000)
52. H. Tang, E. Li, J. Schinar: Appl. Phys. Lett. **71**, 2560 (1997)
53. H. Yanagi, S. Okamoto: Appl. Phys. Lett. **71**, 2563 (1997)
54. C. Zenz, W. Graupner, S. Tasch et al.: Appl. Phys. Lett. **71**, 2566 (1997)
55. P.J.M. van Bentum, R.T.M. Smkers, H. van Kempen: Phys. Rev. Lett. **60**, 2543 (1988)
56. R.P. Andres, T. Bein, M. Dorogi et al.: Science **272**, 1323 (1996)
57. S.W. Wu, N. Ogawa, W. Ho: Science **312**, 1362 (2006)
58. D. Drews, W. Ehrfeld, M. Lacher et al.: Nanotechnology **10**, 61 (1999)
59. W. Noell, M. Abraham, K. Mayer et al.: Appl. Phys. Lett. **70**, 1236 (1997)
60. W.E. Moerner, M. Orrit: Science **283**, 1670 (1999)
61. V.T. Binh, N. Garcia, S.T. Purcell: In: *Advances in Imaging Electron Physics*, vol. 95 (Academic, New York 1996)
62. S. Datta: *Electronic Transport in Mesoscopic Systems* (Cambridge University Press, Cambridge 1995)
63. A. Mayer, J.-P. Vigneron: J. Phys. Condens. Matter **10**, 869 (1998)
64. A. Mayer, J.-P. Vigneron: Phys. Rev. B **56**, 12599 (1997)
65. H.Y. Sheng, Z.C. Dong, D. Fujita, T. Ohgi, H. Nejoh: Ultramicroscopy **73**, 195 (1998)
66. D.P.E. Smith, J.K.H. Horber, G. Binnig, H. Nejoh: Nature **344**, 641 (1990)
67. H. Nejoh: Appl. Phys. Lett. **57**, 2907 (1990)
68. G.C. McGonigal, R.H. Bernhardt, D.J. Thomson: Appl. Phys. Lett. **57**, 28 (1990)
69. J.P. Rabe, S. Bochholz: Science **253**, 424 (1991)
70. J.P. Rabe, S. Bohholz, L. Askadskaya: Synth. Met. **54**, 339 (1993)
71. A.P. Gunning, A.R. Kirby, X. Mallard, V.J. Morris: J. Chem. Soc. Faraday Trans. **90**, 2552 (1994)
72. B. Venkataraman, G.W. Flynn, J.L. Wilbur, J.P. Folkers, G.M. Whitesides: J. Phys. Chem. **99**, 8684 (1995)
73. H. Wolf, H. Ringsdorf, E. Delamareche, T. Takami, H. Kang, B. Michel, Ch. Gerber, M. Jaschke, H.-J. Butt, E. Bamberg: J. Phys. Chem. **99**, 7102 (1995)
74. A.J. Groszek: Proc. R. Soc. Lond. Ser. A **314**, 473 (1970)
75. H.Y. Sheng, D. Fujita, T. Ohgi, Z.C. Dong, Q.D. Jiang, H. Nejoh: Appl. Surf. Sci. **121/122**, 129 (1997)
76. A. Aviram (ed.): *Molecular Electronics: Science and Technology*, Conference Proceedings no. 262 (American Institute of Physics, New York 1992)
77. L.A. Bumm, J.J. Arnold, M.T. Cygan, T.D. Dunbar, T.P. Burgin, L. Jones II, D.L. Allara, J.M. Tour, P.S. Weiss: Science **271**, 1705 (1996)
78. U. During, P. Ziiger, B. Michel, L. Hiussling, H. Ringsdorf: Phys. Rev. B **48**, 1711 (1993)
79. G.E. Poirier, E.D. Pylant: Science **272**, 1145 (1996)
80. K. Mullen, K. Carton: Anal. Chem. **66**, 478 (1994)
81. A.P. Gunning, A.R. Kirby, X. Maltard, V.J. Morris: J. Chem. Soc. Faraday Trans. **90**, 2551 (1994)
82. B. Venkataraman, G. Flynn, J. Wilbur, J. Folkers, G. Whitesides: J. Phys. Chem. **99**, 8684 (1995)

83. J. Rabe, S. Buchholz: Science **253**, 424 (1991)
84. S. Taki, K. Ishida, H. Okabe, K. Matsushige: J. Cryst. Growth **131**, 13 (1993)
85. K. Ishida, S. Taki, H. Okabe, K. Matsushige: Jpn. J. Appl. Phys. **34**, 3846 (1995)
86. H. Wolf, H. Ringsdorf, E. Delamarche, T. Takami, H. Kang, B. Michel, Ch. Gerber, M. Jaschke, H.-J. Butt, E. Bamberg: J. Phys. Chem. **99**, 7102 (1995)
87. W.B. Caldwell, D.J. Campbell, K. Chen, B.R. Herr, C.A. Mirkin, A. Malik, M.K. Durbin, P. Dutta, K.G. Huang: J. Am. Chem. Soc. **117**, 6071 (1995)
88. C.D. Wagner: J. Electron Spectrosc. Relat. Phenom. **32**, 99 (1983)
89. S.J. Stranick, A.N. Parikh, Y.-T. Tao, D.L. Allara, P.S. Weiss: J. Phys. Chem. **98**, 7636 (1994)
90. T. Ohgi, D. Fujita, W. Deng, Z.-C. Dong, H. Nejoh: Surf. Sci. **493**, 453 (2001)
91. M.J. Tarlov: Langmuir **8**, 80 (1992)
92. E.L. Smith, C.A. Alves, J.W. Anderegg, M.D. Porter, L.M. Siperko: Langmuir **8**, 2707 (1992)
93. G.C. Herdt, A.W. Czanderna: J. Vac. Sci. Technol. A **17**, 3415 (1999)
94. D.R. Jung, A.W. Czanderna, G.C. Herdt: J. Vac. Sci. Technol. A **14**, 1779 (1996)
95. D. Anselmetti, T. Richmond, A. Barato., G. Borer, M. Dreier, M. Bernasconi, H.-J. Guntherodt: Europhys. Lett. **25**, 297 (1994)
96. M. Dorogi, J. Gomez, R. Osifchin, R.P. Andres, R. Reifenberger: Phys. Rev. B **52**, 9071 (1995)
97. C. Baulas, J.V. Davidovits, F. Rondelez, D. Vuillaume: Phys. Rev. Lett. **17**, 4797 (1996)
98. C. Zhou, M.R. Deshpande, M.A. Reed, J. Jones II, J.M. Tour: Appl. Phys. Lett. **71**, 611 (1997)
99. T. Ohgi, H.-Y. Sheng, Z.-C. Dong, H. Nejoh: Surf. Sci. **442**, 277 (1999)
100. G.K. Jennings, P.E. Laibinis: Langmuir **12**, 6173 (1996)
101. G.K. Jennings, P.E. Laibinis: J. Am. Chem. Soc. **119**, 5208 (1997)
102. E. Delamarche, B. Michel, C. Gerber, D. Anselmetti, H.-J. Guntherodt, H. Wolf, H. Ringsdorf: Langmuir **10**, 2869 (1994)
103. G.E. Poirier, M.J. Tarlov: Langmuir **10**, 2853 (1994)
104. C. Schonenberger, J.A.M. Sontag-Huethorst, J. Jorritsma, L.G.J. Fokkink: Langmuir **10**, 611 (1994)
105. P.M.A. Sherwood, In: *Practical Surface Analysis*, 2nd ed., vol. 1, ed. D. Briggs, M.P. Seah (Wiley, Chichester 1990)
106. M.M. Dovek, C.A. Lang, J. Nogami, C.F. Quote: Phys. Rev. B **40**, 11973 (1989)
107. A. Dhirani, M.A. Hines, A.J. Fisher, O. Ismail, P. Guyot-Sionnest: Langmuir **11**, 2609 (1995)
108. M.-H. Hsieh, C.-H. Chen: Langmuir **16**, 1729 (2000)
109. M.P. Seah, W. Dench: Surf. Interface Anal. **1**, 2 (1979)
110. J. Thome, M. Himmelhause, M. Zharnikov, M. Grunze: Langmuir **14**, 7435 (1998)
111. T. Ishida, M. Hara, I. Kojima, S. Tsuneda, N. Nishida, H. Sasabe, W. Knoll: Langmuir **14**, 2092 (1998)
112. H.-J. Himmel, C. Woll, R. Gerlach, G. Polanski, H.-G. Rubahn: Langmuir **13**, 602 (1997)
113. J.F. Moulder, W.F. Stickle, P.E. Sobol, K.D. Bomben: *Handbook of X-ray Photoelectron Spectroscopy*, (Physical Electronics Institute, Minnesota 1995)

114. S. Narioka, H. Ishii, D. Yoshimura, M. Sei, Y. Ouchi, K. Seki, S. Hasegawa, T. Miyazaki, Y. Harima, K. Yamashita: Appl.Phys. Lett. **67**, 1899 (1995)
115. Y. Okawa, M. Aono: J. Chem. Phys. **115**, 2317 (2001)
116. A. Aviram, M. A. Ratner, Chem. Phys. Lett. **29**, 277 (1974)
117. D.J. Goldhaber-Gordon, M.S. Montemerlo, J.C. Love, G.J. Opiteck, J.C. Ellenbogen: Proc. IEEE **85**, 521 (1997)
118. C.P. Collier, E.W. Wong, M. Belohradsky, F.M. Raymo, J.F. Stoddart, P.J. Kuekes, R.S. Williams, J.R. Heath: Science **285**, 391 (1999)
119. J. Chen, M.A. Reed, A.M. Rawlett, J.M. Tour: Science **286**, 1550 (1999)
120. V.J. Langlais, R.R. Schlittler, H. Tang, A. Gourdon, C. Joachim, J.K. Gimzewski: Phys. Rev. Lett. **83**, 2809 (1999)
121. Y. Wada, M. Tsukada, M. Fujihira, K. Matsushige, T. Ogawa, M. Haga, S. Tanaka: Jpn. J. Appl. Phys. **39**, 3835 (2000)
122. C. Joachim, J.K. Gimzewski, A. Aviram: Nature **408**, 541 (2000)
123. G. Dujardin, R.E. Walkup, P. Avouris: Science **255**, 1232 (1992)
124. B.C. Stipe, M.A. Rezaei, W. Ho, S. Gao, M. Persson, B.I. Lundqvist: Phys. Rev. Lett. **78**, 4410 (1997)
125. L.P. Ma, W.J. Yang, S.S. Xie, S.J. Pang: Appl. Phys. Lett. **73**, 3303 (1998)
126. S.N. Patitsas, G.P. Lopinski, O. Hulko, D.J. Moffatt, R.A. Wolkow: Surf. Sci. **457**, L425 (2000)
127. H.J. Gao, K. Sohlberg, Z.Q. Xue, H.Y. Chen, S.M. Hou, L.P. Ma, X.W. Fang, S.J. Pang, S.J. Pennycook: Phys. Rev. Lett. **84**, 1780 (2000)
128. G.P. Lopinski, D.D.M. Wayner, R.A. Walkow: Nature **406**, 48 (2000)
129. G. Wegner: Makromol. Chem. **154**, 35 (1972)
130. B. Tieke, G. Lieser, G. Wegner: J. Polym. Sci. Polym. Chem. Ed. **17**, 1631 (1979)
131. J.P. Rabe, S. Buchholz, L. Askadskaya: Synth. Met. **54**, 339 (1993)
132. P.C.M. Grim, S. De Feyter, A. Gesquie're, P. Vanoppen, M. Rucker, S. Valiyaveettil, G. Moessner, K. Mullen, F.C. De Schryver: Angew. Chem. Int. Ed. Engl. **36**, 2601 (1997)
133. T. Takami, H. Ozaki, M. Kasuga, T. Tsuchiya, Y. Mazaki, D. Fukushi, A. Ogawa, M. Uda, M. Aono: Angew. Chem. Int. Ed. Engl. **36**, 2755 (1997)
134. M. Hibino, A. Sumi, I. Hatta: Jpn. J. Appl. Phys. **34**, 610 (1995)
135. A.J. Groszek: Proc. R. Soc. Lond. Ser. A **314**, 473 (1970)
136. G.C. McGonigal, R.H. Bernhardt, D.J. Thomson, Appl. Phys. Lett. **57**, 28 (1990)
137. J.P. Rabe, S. Buchholz: Science **253**, 424 (1991)
138. B. Venkataraman, J.J. Breen, G.W. Flynn: J. Phys. Chem. **99**, 6608 (1995)
139. W. Neumann, H. Sixl: Chem. Phys. **58**, 303 (1981)
140. T. Takabe, M. Tanaka, J. Tanaka: Bull. Chem. Soc. Jpn. **47**, 1912 (1974)
141. M. Bertault, J.L. Fave, M. Schott: Chem. Phys. Lett. **62**, 161 (1979)
142. R.R. Chance, G.N. Patel: J. Polym. Sci. Polym. Phys. Ed. **16**, 859 (1978)
143. Y. Okawa M. Aono: Nature **409**, 683 (2001)
144. T.R. Albrecht, M.M. Dovek, M.D. Kirk, C.A. Lang, C.F. Quate, D.P.E. Smith: Appl. Phys. Lett. **55**, 1727 (1989)
145. W.R. Salaneck, M. Fahlman, C. Lapersonne-Meyer, J.-L. Fave, M. Schott, M. Logdlund, J.L. Bredas: Synth. Met. **67**, 309 (1994)
146. J.L. Bredas, R.R. Chance, R. Silbey, G. Nicolas, P. Dur: J. Chem. Phys. **75**, 255 (1981)

147. A. Karpfen: J. Phys. C **13**, 5673 (1980)
148. K.J. Donovan, R.V. Sudiwala, E.G. Wilson: Thin Solid Films **211**, 271 (1992)
149. D.R. Day, J.B. Lo: J. Appl. Polym. Sci. **26**, 1605 (1981)
150. S. Blumentritt, M. Burghard, S. Roth et al.: Surf. Sci. **397**, L280 (1998)
151. G. Schelbe: Angew. Chem. **49**, 563 (1936)
152. E.E. Jelly: Nature **138**, 1009 (1936)
153. H. Saljo, M. Shlojin, J. Irnagmg: Sci. Technol. **40**, 111 (1996)
154. M. Kawasakl, H. Ishn: Chem. Lett. **40**, 1079 (1994)
155. D.A. Hlggins, J. Kerlmo, D.A. Vanden Bout, P.F Barbara: J. Am. Chem. Soc **118**, 4049 (1996)
156. J.G.E. Polraer: Langmuir **13**, 2019 (1997)
157. S.M. Lindsay: *Scanning Tunneling Microscopy and Spectroscopy, Theory, Techniques and Applications* (VHC, New York, 1993) p. 335
158. V. Mujlca, M. Kemp, M.A. Ratner: J. Chem. Phys. **101**, 6856 (1994)
159. C. Dekker, S.J. Tans, B. Oberndorff. R. Meyer, L.C.Venema: Synth. Met. **84**, 853 (1997)
160. J. Resh, D. Sarkar, J. Kuhk, J. Brueck, A. Ignatxev, N.J. Halas: Surf. Sci. **316**, L1061 (1994)
161. R.V. Belosludov, H. Mizuseki, K. Ichinoseki, Y. Kawazoe: Jpn. J. Appl. Phys. **41**, 2739 (2002)
162. C. Joachim, J.K. Gimzewski, A. Aviram: Nature **408**, 541 (2000)
163. M.A. Reed: Sci. Am. June, 69 (2000)
164. M.A. Reed, C. Zhou, C.J. Muller, T.P. Burgin, J.M. Tour: Science **278**, 252 (1997)
165. L.A. Bumm, J.J. Arnold, M.T. Cygan, T.D. Dunbar, T.P. Burgin, L. Jones II, D.L. Allara, J.M. Tour, P.S. Weiss: Science **217**, 1705 (1996)
166. J. Chen, M.A. Reed, A.M. Rawlett, J.M. Tour: Science **286**, 1550 (1999)
167. R.M. Mertzger: Acc. Chem. Res. **32**, 950 (1999)
168. S.N. Yaliraki, M. Kemp, M.A. Ratner: J. Am. Chem. Soc. **121**, 3428 (1999)
169. M. Magoga, C. Joachim: Phys. Rev. B **56**, 4722 (1997)
170. J.M. Seminario, A.G. Zacarias, P.A. Derosa: J. Phys. Chem. A **105**, 792 (2001)
171. G. Wenz: Angew Chem. Int. Ed. Engl. **33**, 803 (1994)
172. A. Harada, J. Li, M. Kamachi: Nature **356**, 325 (1994)
173. Y. Okumura, K. Ito, R. Hayakawa: Phys. Rev. Lett. **80**, 5003 (1998)
174. K. Yoshida, T. Shimomura, K. Ito, R. Hayakawa: Langmuir **15**, 910 (1999)
175. M.L. Bender, M. Komiyama: *Cyclodextrin Chemistry* (Springer, Berlin Heidelberg New York 1978)
176. S. Humbel, S. Sieber, K. Morokuma: J. Chem. Phys. **105**, 1959 (1996)
177. M.J. Frisch, G.W. Trucks, H.B. Schlegel, G.E. Scuseria, M.A. Robb, J.R. Cheeseman, V.G. Zakrzewski, J.A. Montgomery Jr., R.E. Stratmann, J.C. Burant, S. Dapprich, J.M. Millam, A.D. Daniels, K.N. Kudin, M.C. Strain, O. Farkas, J. Tomasi, V. Barone, M. Cossi, R. Cammi, B. Mennucci, C. Pomelli, C. Adamo, S. Clifford, J. Ochterski, G.A. Petersson, P.Y. Ayala, Q. Cui, K. Morokuma, P. Salvador, J.J. Dannenberg, D.K. Malick, A.D. Rabuck, K. Raghavachari, J.B. Foresman, J. Cioslowski, J.V. Ortiz, A.G. Baboul, B.B. Stefanov, G. Liu, A. Liashenko, P. Piskorz, I. Komaromi, R. Gomperts, R.L. Martin, D.J. Fox, T. Keith, M.A. Al-Laham, C.Y. Peng, A. Nanayakkara, M. Challacombe, P.M.W. Gill, B. Johnson, W. Chen, M.W. Wong, J.L. Andres, C. Gonzalez, M. Head-Gordon, E.S. Replogle, J.A. Pople: Gaussian 98, revision A.11.1. Gaussian, Pittsburgh, 2001

178. P.W. Atkins: *Quanta: A Hand book of Concepts* 2nd ed. (Oxford University Press, Oxford 1992)
179. Y. Wada, M. Tsukada, M. Fujihira, K. Matsushige, T. Ogawa, M. Haga, S. Tanaka: Jpn. J. Appl. Phys. **39**, 3835 (2000)
180. T. Akai, T. Abe, T. Shimomura, K. Ito: Jpn. J. Appl. Phys. **40**, L1327 (2001)
181. Z. Yao, H.W.C. Postma, L. Balents, C. Dekker: Nature **402**, 273 (1999)
182. P. Avouris, T. Hertel, R. Martel, T. Schmidt, H.R. Shea, R.E. Walkup: Appl. Surf. Sci. **141**, 201 (1999)
183. H.W. Fink, C. Schonenberger: Nature **398**, 407 (1999)
184. L.T. Cai, H. Tabata, T. Kawai: Appl. Phys. Lett. **77**, 3105 (2000)
185. T. Ito, H. Shirakawa, S. Ikeda: J. Polym. Sci. Polym. Chem. Ed. **12**, 11 (1974)
186. C.K. Chang, C.R. Fincher Jr., Y.W. Park, A.J. Heeger, H. Shirakawa, E.J. Louis, S.C. Gau, A.G. MacDiamid: Phys. Rev. Lett. **39**, 1098 (1977)
187. R.E. Gregory: In: *Handbook of Conducting Polymers*, eds. T.A. Skotheim, R.L. Elsenbaumer, J.R. Reynolds (Dekker, New York 1998) Chap. 18
188. S. Hotta, K. Ito: In: *Handbook of Polythiophenes and Oligothiophenes*, ed. D. Fichou (Wiley, New York 1998) Chap. 2
189. K. Yoshida, T. Shimomura, K. Ito, R. Hayakawa: Langmuir **15**, 910 (1999)
190. T. Shimomura, K. Yoshida, K. Ito, R. Hayakawa: Polym. Adv. Technol. **11**, 837 (2000)
191. A. Harada, M. Kamachi: Macromolecules **23**, 2821 (1990)
192. A. Harada, J. Li, M. Kamachi: Nature **356**, 325 (1992)
193. G. Wenz: Angew. Chem. Int. Ed. Engl. **33**, 803 (1994)
194. A. Harada, J. Li, M. Kamachi: Nature **364**, 516 (1993)
195. M. Ceccato, P. L. Nostro, C. Rossi, C. Bonechi, A. Donati, P. Baglioni: J. Phys. Chem. B **101**, 5094 (1997)
196. E. Ikeda, Y. Okumura, T. Shimomura, K. Ito, R. Hayakawa: J. Chem. Phys. **112**, 4321 (2000)
197. T. Ikeda, T. Ooya, N. Yui: Macromol. Rapid Commun. **21**, 1257 (2000)
198. M. Saito, T. Shimonura, Y. Okumura, K. Ito: J. Chem. Phys. **114**, 1 (2001)
199. T. Ikeda, E. Hirota, T. Ooya, N. Yui: Langmuir **17**, 234 (2001)
200. T. Hertel, R. Martel, P. Avouris: J. Phys. Chem. B **102**, 910 (1998)
201. L.T. Hansen, A. Kuhle, A.H. Sorensen, J. Bohr, P.E. Lindelof: Nanotechnology **9**, 337 (1998)
202. Y. Okumura, K. Ito, R. Hayakawa, T. Nishi: Langmuir **16**, 10278 (2000)
203. S.W. Lee, D.S. Lee, H.Y. Yu, E.E.B. Campbell, Y.W. Park: Appl. Phys. A **78**, 283 (2004)
204. S. Iijima: Nature **354**, 56 (1991)
205. J. Muster, G.T. Kim, V. Krstic, J.G. Park, Y.W. Park, S. Roth, M. Burghard: Adv. Mater. **12**, 420 (2000)
206. C.R. Martin: Acc. Chem. Res. **28**, 61 (1995)
207. J.G. Park, S.H. Lee, B. Kim, Y.W. Park: Appl. Phys. Lett. **81**, 4625 (2002)
208. V. Derycke, R. Martel, J. Appenzeller, P. Avouris: Nano Lett. **1**, 453 (2001)
209. J.G. Park, G.T. Kim, V. Krstic, S.H. Lee, B. Kim, S. Roth, M. Burghard, Y.W. Park: Synth. Met. **119**, 469 (2001)
210. G.T. Kim, J. Muster, V. Krstic, J.G. Park, Y.W. Park, S. Roth, M. Burghard: Appl. Phys. Lett. **76**, 1875 (2000)
211. J.M. Kinaret, T. Nord, S. Viefers: Appl. Phys. Lett. **82**, 1287 (2003)
212. P. Kim, C.M. Lieber: Science **286**, 2148 (1999)

213. A. Ural, Y. Li, H. Dai: Appl. Phys. Lett. **81**, 3464 (2002)
214. K. Yamamoto, S. Akita, Y. Nakayama: J. Phys. D Appl. Phys. **31**, L34 (1998)
215. S. Gerdes, T. Ondaruhu, S. Cholet, C. Joachim: Europhys. Lett. **48**, 292 (1999)
216. R. Krupke, F. Hennrich, H. Lohneysen, M. Kappes: Science **301**, 344 (2003)
217. R. Martel, T. Schmidt, H.R. Shea, T. Hertel, P. Avouris: Appl. Phys. Lett. **73**, 2447 (1998)
218. A. Rochefort, P. Avouris, F. Lesage, D.R. Salahub: Phys. Rev. B. **60**, 13824 (1999)
219. T.W. Tombler, C. Zhou, L. Alexseyev, J. Kong, H. Dai, L. Liu, C.S. Jayanthi, M. Tang, S. Wu: Nature **405**, 769 (2000)
220. J. Nygård, D.H. Cobden: Appl. Phys. Lett. **79**, 4216 (2001)
221. G.T. Kim, G. Gu, U. Waizmann, S. Roth: Appl. Phys. Lett. **80**, 1815 (2002)
222. G.S. Duesberg, J. Muster, V. Krstic, M. Burghard, S. Roth: Appl. Phys. A **67**, 117 (1998)
223. H.Y. Yu: Ph.D. Thesis, Seoul National University (2003)
224. P. Dorozhkin, H. Nejoh, D. Fujita: J. Vac. Sci. Technol. B **20**, 1044 (2002)
225. D. Porath, A. Bezryadin, S. Vries, C. Dekker: Nature **403**, 635 (2000)
226. H.-W. Fink C. Schonenberger: Nature **398**, 407 (1999)
227. S.J. Tans, M.H. Devoret, H.J. Dai, A. Thess, R.E. Smalley, L.J. Geerligs, C. Dekker: Nature **386**, 474 (1997)
228. M. Bockrath, D.H. Cobden, P.L. McEuen: Science **275**, 1922 (1997)
229. H. Ohnishi, Y. Kondo, K. Takayanagi: Nature **395**, 780 (1998)
230. Vu Thien Binh, V. Semet: Ultramicroscopy **73**, 107 (1998)
231. C. Adessi, M. Devel, Vu Thien Binh, P. Lambin, V. Meunier: Phys. Rev. B **61**, 13385 (2000)
232. J. Spence, W. Qian, X. Zhang: Ultramicroscopy **55**, 19 (1994)
233. A. Thess, R. Lee, P. Nikolaev, H. Dai, P. Petit, J. Robert, C. Xu, Y.H. Lee, S.G. Kim, A.G. Rinzler, D.T. Colbert, G.E. Scuseria, D. Tomanek, J.E. Fischer, R.E. Smalley: Science **273**, 483 (1996)
234. M. Seah, W. Dench: Surf. Interface Anal. **1**, 2 (1979)
235. G.M. Shedd: J. Vac. Sci. Technol. A **12**, 2595 (1994)
236. W. Lai, A. Degiovanni, R. Morin: Appl. Phys. Lett. **74**, 618 (1999)
237. Vu Thien Binh, S. Purcell, N. Garcia, J. Doglioni: Phys. Rev. Lett. **69**, 2527 (1992)
238. H.-W. Fink, W. Stocker, H. Schmid: J. Vac. Sci. Technol. B **8**, 1323 (1990)

Index